XX International School on Nuclear Physics, Neutron Physics and Applications

(Varna 2013)

Journal of Physics: Conference Series Volume 533

Varna, Bulgaria
16 – 22 September 2013

ISBN: 978-1-63439-343-0
ISSN: 1742-6588

Printed from e-media with permission by:

Curran Associates, Inc.
57 Morehouse Lane
Red Hook, NY 12571

Some format issues inherent in the e-media version may also appear in this print version.

Printed by Curran Associates, Inc. (2014)

Institute of Physics
Dirac House, Temple Back
Bristol BS1 6BE UK

Phone: 44 1 17 929 7481
Fax: 44 1 17 920 0979

techtracking@iop.org

Additional copies of this publication are available from:

Curran Associates, Inc.
57 Morehouse Lane
Red Hook, NY 12571 USA
Phone: 845-758-0400
Fax: 845-758-2634
Email: curran@proceedings.com
Web: www.proceedings.com

TABLE OF CONTENTS

Journal of Physics: Conference Series **533** (2014) 011001 doi:10.1088/1742-6596/533/1/011001

Preface

The present volume contains the lectures and short talks given at the XX International School on Nuclear Physics, Neutron Physics and Applications. The School was held from 16–22 September 2013 in 'Club Hotel Bolero' located in 'Golden Sands' (Zlatni Pyasaci) Resort Complex on the Black Sea coast, near Varna, Bulgaria. The School was organized by the Institute for Nuclear Research and Nuclear Energy of Bulgarian Academy of Sciences. Co-organizer of the School was the Bulgarian Nuclear Regulatory Agency and the Bogoliubov Laboratory of Theoretical Physics of Joint Institute for Nuclear Research – Dubna. Financial support was also provided by the Bulgarian Ministry of Education and Science.

According to the long-standing tradition the School has been held every second year since 1973. The School's program has been restructured according to our enlarged new international links and today it is more similar to an international conference than to a classical nuclear physics school. This new image attracts many young scientists and students from around the world. This year, 2013, we had the pleasure to welcome more than sixty distinguished scientists as lecturers. Additionally, twenty young colleagues received the opportunity to present a short contribution. Ninety-four participants altogether enjoyed the scientific presentations and discussions as well as the relaxing atmosphere at the beach and during the pleasant evenings.

The program of the School ranged from latest results in fundamental areas such as nuclear structure and reactions to the hot issues of application of nuclear methods, reactor physics and nuclear safety.

The main topics have been the following:

- Nuclear excitations at various energies.
- Nuclei at high angular moments and temperature.
- Structure and reactions far from stability
- Symmetries and collective phenomena
- Methods for lifetime measurements
- Astrophysical aspects of nuclear structure
- Neutron nuclear physics
- Nuclear data
- Advanced methods in nuclear waste treatment
- Nuclear methods for applications

A special session in honor of the late Mario Stoitsov, was also part of the program. Many colleagues of Mario from all over the world came to Varna to pay tribute to this prominent scientist and loyal friend.

Several colleagues contributed to the organization of the School. We would like to thank them and especially the Scientific Secretary of the School Dr Elena Stefanova and the members of the Organizing Committee Dr Dimitar Tarpanov and Peter Zivkov for their cordiality and high level assistance. We are also grateful to Dr Jacek Dobaczewski, who reached out to the collaborators of Mario Stoitsov on behalf of the conference.

Sofia, 20 March 2014

Co-chair persons of the Organizing Committee
Prof Dr Sc Ch Stoyanov
Prof Dr Sc S Dimitrova

Journal of Physics: Conference Series **533** (2014) 011001 doi:10.1088/1742-6596/533/1/011001

Organization and Committees

Organizers

- Institute for Nuclear Research and Nuclear Energy
- Bulgarian Academy of Sciences

Co-organizers

- Bulgarian Nuclear Regulatory Agency
- Joint Institute for Nuclear Research, Dubna, Russia
- Ministry of Education, Youth and Science

International Advisory Committee

F Azaiez (IPN, Orsay)
R Casten (Yale, USA)
J Dobaczewski (U Warsaw)
W Furman (FLNP, Dubna)
M Harakeh (KVI Groningen)
A Jokinen (U Jyväskylä)
L Kostov (NRA, Bulgaria)
Y Penionzhkevich (FLNR, Dubna)
A Richter (TU, Darmstadt)
D Tonev (INRNE, Sofia)
V Zelevinsky (MSU, USA)
V Voronov (BLTP, Dubna)
A Bracco (U Milan)
G De Angelis (LNL, Italy)
J Draayer (SURA)
S Gales (CNRS)
M Itkis (JINR, Dubna)
J Jolie (U Koeln)
N Pietralla (TU, Darmstadt)
M Ploszajczak (GANIL)
H Sakai (RIKEN)
H Stoecker (GSI)
N V Zamfir (NIPNE-Bucharest)
S Zhou (Chinese Academy of Sc)

Local Organizing Committee

Ch Stoyanov (chairperson)
S Dimitrova (co-chairperson)
E Stefanova (scientific secretary)
D Tarpanov
P Zhivkov

Journal of Physics: Conference Series **533** (2014) 011002 doi:10.1088/1742-6596/533/1/011002

List of participants

1. Alan Dzhioev
   Bogoliubov Laboratory of Theoretical Physics, JINR-Dubna, Russia
   dzhioev@theor.jinr.ru
2. Alberto Ventura
   INFN, Sezione di Bologna, Italy
   alberto.ventura@enea.it
3. Alfred Dewald
   Cologne University, Germany
   dewald@ikp.uni-koeln.de
4. Anabel Morales
   INFN-Sezione di Milano, Italy
   AnaIsabel.Morales@mi.infn.it
5. Andrea Jungclaus
   Instituto de Estructura de la Materia, CSIC, Spain
   andrea.jungclaus@csic.es
6. Andrew Stuchbery
   The Australian National University, Department of Nuclear Physics, Australia
   Andrew.Stuchbery@anu.edu.au
7. Andrey Blazhev
   Institute for nuclear physics, University of Cologne, Germany
   a.blazhev@ikp.uni-koeln.de
8. Angela Bracco
   Dipartimento di Fisica, Università di Milano, Italy
   Angela.Bracco@mi.infn.it
9. Anthony Cowley
   Department of Physics, Stellenbosch University, South Africa
   aac@sun.ac.za
10. Anton N Antonov
    Institute for Nuclear Research and Nuclear Energy, Bulgaria
    aantonov@inrne.bas.bg
11. Anu Kankainen
    University of Edinburgh, UK
    anu.k.kankainen@jyu.fi

Journal of Physics: Conference Series **533** (2014) 011002 doi:10.1088/1742-6596/533/1/011002

12. Aysegul Ertoprak
Science Faculty, Department of Physics, Istanbul University, Turkey
aysegulertoprak@gmail.com
13. Baki Akkus
Science Faculty, Department of Physics, Istanbul University Turkey
akkus@istanbul.edu.tr
14. Bozjidar Dimitrov
Institute for Nuclear Research and Nuclear Energy, Bulgaria
bzhdrdimitrov@gmail.com
15. Bozkurt Kutsal
Yildiz Technical University, Turkey
b_kutsal@yahoo.com
16. Chavdar Stoyanov
Institute for Nuclear Research and Nuclear Energy, Bulgaria
stoyanov@inrne.bas.bg
17. Çigdem GÖk
Yildiz Technical University, Physics Department, Turkey
cigdemgk87@gmail.com
18. Davide Bianco
U. Napoli and INFN, Italy
davidebianco1@gmail.com
19. Danilo Gambakurta
GANIL, France
gambacurta@ganil.fr
20. Dimitar Tarpanov
Institute for Nuclear Research and Nuclear Energy, Bulgaria and
University of Warsaw, Poland
dimitert@yahoo.co.uk
21. Dimitar Tonev
Institute for Nuclear Research and Nuclear Energy, Bulgaria
mitko@inrne.bas.bg
22. Elena Zemlyanaya
Joint Institute for Nuclear Research, Russia
elena@jinr.ru
23. Elena Stefanova
Institute for Nuclear Research and Nuclear Energy, Bulgaria
elenas@inrne.bas.bg
24. Enrico Fioretto
INFN – Laboratori Nazionali di Legnaro, Italy
enrico.fioretto@lnl.infn.it
25. Ernest Grodner
Faculty of Physics, University of Warsaw, Poland
ernest.grodner@fuw.edu.pl
26. Esra Yüksel
Physics Department, Yildiz Technical University, Turkey
esrayu@hotmail.com

27. Evgeny Balbutsev
BLTP, JINR – Dubna, Russia
balbuts@theor.jinr.ru

28. Filip Kondev
Nuclear Engineering Division Argonne National Laboratory, USA
kondev@anl.gov

29. Galina Asova
Institute for Nuclear Research and Nuclear Energy, Bulgaria
gasova@inrne.bas.bg

30. Georgi Gavrilov
Institute for Nuclear Research and Nuclear Energy, Bulgaria
georgi.gavrilov@gmail.com

31. Georgi Georgiev
CSNSM, IN2P3/CNRS, France
Georgi.Georgiev@csnsm.in2p3.fr

32. Georgi Rainovski
Faculty of physics, University of Sofia, Bulgaria
rig@phys.uni-sofia.bg

33. Gerfried Kumbartzki
Department of Physics and Astronomy Rutgers,
The State University of New Jersey, USA
kum@physics.rutgers.edu

34. Guastalla Giulia
TU Darmstadt and GSI, Germany
G.Guastalla@gsi.de

35. Haozhao Liang
Theoretical Nuclear Physics Laboratory, RIKEN Nishina Center, Japan
haozhao.liang@riken.jp

36. Huben Ganev
Joint Institute for Nuclear Research, Russia
huben@theor.jinr.ru

37. Ivan Sivácek
Institue of Physics, Slovak Academy of Sciences, Slovakia
and Flerov Laboratory of Nuclear Reactions, JINR-Dubna, Russia
ivan.sivacek@savba.sk

38. Jacek Dobaczewski
University of Warsaw, Poland
dobaczew@fuw.edu.pl

39. Jan Jolie
University of Cologne, Germany
jolie@ikp.uni-koeln.de

40. János Timár
Institute for Nuclear Research , Hungarian Academy of Sciences, Hungary
timar@atomki.mta.hu

41. Jan Taprogge
IEM CSIC, Spain
jan.taprogge@uam.es
42. Jitka Vrzalová
Nuclear Physics Institute of ASCR, Czech Republic
vrzalova@ujf.cas.cz
43. Jorge Dukelsky
Instituto de Estructura de la Materia, CSIC, Spain
dukelsky@iem.cfmac.csic.es
44. Jun Terasaki
University of Tsukuba, Japan
terasaki@ccs.tsukuba.ac.jp
45. Junchen Pei
Peking University, China
jcpei.pku@gmail.com
46. Korneliya Spasova
Institute for Nuclear Research and Nuclear Energy
and University of Shumen, Bulgaria
nelispas@gmail.com
47. Kumar Raju
Andhra University, Nuclear Physics Department, India
kumar8284@yahoo.com
48. Marcella Grasso
IPN Orsay, France
grasso@ipno.in2p3.fr
49. Marco Brenna
Università degli Studi di Milano and INFN Milano, Italy
marco.brenna@mi.infn.it
50. Marcus Scheck
IKP, TU Darmstadt, Germany
mscheck@ikp.tu-darmstadt.de
51. Marek Ploszajczak
GANIL, France
ploszajczak@ganil.fr
52. Mariana Petris
Horia Hulubei National Institute of Physics and Nuclear Engineering, Romania
mpetris@nipne.ro
53. Marilena Avrigeanu
Horia Hulubei National Institute of Physics and Nuclear Engineering, Romania
Marilena.Avrigeanu@nipne.ro
54. Markus Kortelainen
Department of Physics, University of Jyväskylä, Finland
e.markus.kortelainen@gmail.com
55. Maria Yavahchova
Institute for Nuclear Research and Nuclear Energy, Bulgaria
yavahchova@inrne.bas.bg

56. Michal Warda
Maria Curie-Sklodowska University, Poland
warda@kft.umcs.lublin.pl

57. Michelangelo Sambataro
Istituto Nazionale di Fisica Nucleare, Sezione di Catania, Italy
michelangelo.sambataro@ct.infn.it

58. Mikael Reponen
University of Manchester, UK
and University of Jyväskylä, Finland
mikael.h.t.reponen@jyu.fi

59. Milena Mineva
Institute for Nuclear Research and Nuclear Energy, Bulgaria
dr.mineva@gmail.bg

60. Mitko Gaidarov
Institute for Nuclear Research and Nuclear Energy, Bulgaria
gaidarov@inrne.bas.bg

61. Michael Pelliciali
IPHC, CNRS, France
michel.pelliciolli@iphc.cnrs.fr

62. Naftali Auerbach
Tel Aviv University, School of Physics and Astronomy, Israel
auerbach@post.tau.ac.il

63. Negrea Daniel Ciprian
Horia Hulubei National Institute of Physics and Nuclear Engineering, Romania
negrea.daniel@theory.nipne.ro

64. Nguyen Van Giai
Institut de Physique Nucleaire, Universite Paris-Sud, France
nguyen@ipno.in2p3.fr

65. Nicolae Sandulescu
National Institute of Physics and Nuclear Engineering, Romania
sandulescu@theory.nipne.ro

66. Nicolas Michel
Michigan State University, USA
nicolas.lj.michel@gmail.com

67. Nobuaki Yoshida
Faculty of Informatics, Kansai University, Japan
yoshida@res.kutc.kansai-u.ac.jp

68. Norbert Pietralla
TU Darmstadt, Germany
pietralla@ikp.tu-darmstadt.de

69. Ovidiu Tesileanu
National Institute of Physics and Nuclear Engineering, Romania
ovidiu.tesileanu@eli-np.ro

70. Paolo Gomes
Instituto de Fisica Universidade Federal Fluminense, Brazil
paulogom@if.uff.br

Journal of Physics: Conference Series **533** (2014) 011002 doi:10.1088/1742-6596/533/1/011002

71. Pavel Petkov
Institute for Nuclear Research and Nuclear Energy, Bulgaria
petkov@inrne.bas.bg

72. Pedro Sarriguren
Instituto de Estructura de la Materia , CSIC, Spain
pedro.sarriguren@csic.es

73. Peter Ring
Physics Department, Technical University Munich, Germany
peter.ring@tum.de

74. Petr Vesely
Universita di Napoli Federico II, Dipartimento di Fisica, Italy
vesely@na.infn.it

75. Petra Kudejova
Technische Universität München, Germany
petra.kudejova@frm2.tum.de

76. Petar Zhivkov
Institute for Nuclear Research and Nuclear Energy, Bulgaria
peter_jivkov@inrne.bas.bg

77. Petko Krustev
Institute for Nuclear Research and Nuclear Energy, Bulgaria
pkrastev2002@abv.bg

78. R.Burcu Cakirli
Istanbul University, Science Faculty, Department of Physics, Turkey
rburcu@istanbul.edu.tr

79. R.V.Jolos
Laboratory of Theoretical Physics, JINR-Dubna, Russia
jolos@theor.jinr.ru

80. Richard Casten
Yale University, USA
richard.casten@yale.edu

81. Serguei Zemlyanoy
Flerov laboratory of nuclear reactions, JINR – Dubna, Russia
zemlya@jinr.ru

82. Sevdalina Dimitrova
Institute for Nuclear Research and Nuclear Energy, Bulgaria
sevdim@inrne.bas.bg

83. Shan-Gui Zhou
Institute of Theoretical Physics, Chinese Academy of Sciences, China
sgzhou@itp.ac.cn

84. Stefan Wild
Mathematics and Computer Science Division, Argonne National Laboratory, USA
wild@mcs.anl.gov

85. Stijn de Baerdemacker
Ghent University, Center for Molecular Modelling, Belgium
stijn.debaerdemacker@ugent.be

Journal of Physics: Conference Series **533** (2014) 011002 doi:10.1088/1742-6596/533/1/011002

86. Stoyan Mishev
Bogoliubov Laboratory of Theoretical Physics, JINR-Dubna, Russia
mishev@theor.jinr.ru

87. Stuart Pittel
Department of Physics and Astronomy, University of Delaware, USA
pittel@bartol.udel.edu

88. Takashi Nakatsukasa
RIKEN Nishina Center, Japan
nakatsukasa@riken.jp

89. Thorsten Kröll
TU Darmstadt, Institut für Kernphysik, Germany
tkroell@ikp.tu-darmstadt.de

90. Valter Furman
JINR-Dubna, Russia
furman@dubna.ru

91. Vladimir Ponomarev
Institut für Kernphysik, TU Darmstadt, Germany
ponomare@crunch.ikp.physik.tu-darmstadt.de

92. Vladimir Wagner
Nuclear Physics Institute of ASCR, Czech Republic
wagner@ujf.cas.cz

93. Xian-Rong Zhou
Department of Physics, Xiamen University, China
xrzhou@xmu.edu.cn

94. Yuri Penionzhkevich
Joint Institute for Nuclear Research-Dubna, Russia
pyuer@jinr.ru

Peer review statement

All papers published in this volume of *Journal of Physics: Conference Series* have been peer reviewed through processes administered by the proceedings Editors. Reviews were conducted by expert referees to the professional and scientific standards expected of a proceedings journal published by IOP Publishing.

Journal of Physics: Conference Series **533** (2014) 012001 doi:10.1088/1742-6596/533/1/012001

Isospin corrections to super-allowed beta decays in nuclei

N. Auerbach
School of Physics and Astronomy, Tel Aviv University
Tel Aviv 69978, Israel

Abstract: Isospin corrections to the super-allowed beta decay matrix elements are evaluated in perturbation theory using the notion of the giant isovector monopole state. The calculation avoids the separation into different contributions and therefore presents a consistent, systematic and more transparent approach. Explicit expressions for δ_c as a function of the mass number A are derived. These corrections affect the values of the V_{ud} matrix element in the Cabbibo-Maskawa-Kobayashi matrix. Also, it is pointed out that in some nuclei with a low number of excess neutrons (protons) Coulomb mixing with the anti-analog state can introduce significant isospin impurities in the isobaric analog state.

1. Introduction

Recently there is much activity to determine the corrections one has to introduce when one studies the beta-decay matrix elements for super-allowed decays in $T=1,\ T_z=+1$ (or $T_z=-1$) nuclei [1,2] .This is important because using the measured ft values one can relate these to the u-quark to d-quark transition matrix element (m.el.) V_{ud} in the Cabibbo-Kobayashi-Maskawa (CKM) matrix. In the Standard Model (SM) this matrix satisfies the unitarity condition:

$$V_{ud}^2+V_{us}^2+V_{ub}^2=1 \tag{1}$$

In order to use the experimental ft values to determine V_{ud} one has to introduce corrections [1, 2]. (There are radiative corrections which we will not treat here, discussions of these can be found abundantly in the literature [1].) The second type of correction, which is usually termed as the isospin symmetry breaking term, denoted as δ_c :

$$|M_F|^2=|M_F^0|^2(1-\delta_c) \tag{2}$$

where M_F is the physical Fermi matrix element:

$$M_F=\langle\Psi_1|T_+|\Psi_2\rangle \tag{3}$$

$|\Psi_1\rangle$ and $|\Psi_2\rangle$ are the parent and daughter physical states, M_F^0 stands for the Fermi matrix element obtained in the limit when in the Hamiltonian all the charge-dependent parts are put to zero, and the wave functions are eigenstates of the charge-independent Hamiltonian. In the

Journal of Physics: Conference Series **533** (2014) 012001 doi:10.1088/1742-6596/533/1/012001

present approach we start from a charge-independent Hamiltonian so that the matrix element in eq. (1) is exactly $\sqrt{2T}$ and we then treat the Coulomb force in perturbation theory. In the way we approach the problem there is no need to break up the contribution of the Coulomb interaction into various separate components. All the effects of Coulomb mixing (such as isospin mixing, the change in the radial part of the wave functions, etc.) are taken into account in a single term.

2. Coulomb mixing

We start by introducing a nuclear charge independent Hamiltonian - H_0. The eigenstates of this Hamiltonian with isospin T and T_z will be denoted as $|T,T_z\rangle$ and:

$$H_0|T,T_z\rangle = E_T|T,T_z\rangle \tag{4}$$

The action of the isospin lowering and raising operators, T_-, T_+, gives:

$$T_-|T,T\rangle = \sqrt{2T}|T,T-1\rangle \ ; \ T_+|T,T-1\rangle = \sqrt{2T}|T,T\rangle \tag{5}$$

We now add to the charge independent Hamiltonian a charge dependent part V_{CD}.

The dominant part in the charge dependent interaction is the charge asymmetric Coulomb force V_C. Because of the long range nature of the Coulomb force, the prevailing part will be in such cases the one-body part. Of interest to us here is the isovector part of the potential. For a uniform charge distribution of radius R any off-diagonal matrix element between two states of the isovector part is:

$$\langle 0|V_C|n\rangle = \frac{Ze^2}{2R^3}\langle 0|\sum_i r_i^2 t_z(i)|n\rangle \equiv \frac{Ze^2}{2R^3}\langle 0|M_0^{(1)}|n\rangle \tag{6}$$

where $M_0^{(1)}$ denotes the z-component of the isovector monopole operator. If state $|n\rangle$ is the giant isovector monopole [3] , that is the state obtained by acting with the $M_0^{(1)}$ operator on the ground state $|0\rangle$ and normalizing, then the above matrix element will be proportional to $\langle 0|M_0^{(1)}M_0^{(1)}|0\rangle$ and thus will exhaust the isovector part ($r^2 t_z$) of the Coulomb sum rule [3].

We now find in perturbation theory the effect of the charge-dependent part on the wave functions of the two members of the isomultiplet, $|T,T\rangle$ and $|T,T-1\rangle$.

$$\Psi_1 = \left(|T,T\rangle + \varepsilon_T|M_{T,T}\rangle + \varepsilon_{T+1}|M_{T+1,T}\rangle\right)N_1^{-1} \tag{7a}$$

$$\Psi_2 = \left(|T,T-1\rangle + \eta_{T-1}|M_{T-1,T-1}\rangle + \eta_T|M_{T,T-1}\rangle + \eta_{T+1}|M_{T+1,T-1}\rangle\right)N_2^{-1} \tag{7b}$$

where $|M_{T',T_z'}\rangle$, are the T',T_z' components of the isovector monopole, and where

$$N_1 = \sqrt{1+\varepsilon_T^2+\varepsilon_{T+1}^2} \text{ and } N_2 = \sqrt{1+\eta_{T-1}^2+\eta_T^2+\eta_{T+1}^2} \tag{8}$$

The admixtures are given in perturbation theory by the equations:

Journal of Physics: Conference Series **533** (2014) 012001 doi:10.1088/1742-6596/533/1/012001

$$\varepsilon_i = \frac{\langle T,T | V_C^{(1)} | M_{T+i,T} \rangle}{E_{M_{T+i,T}} - E_0}, \; i = 0,1 \tag{9}$$

where E_0 is g.s. energy

$$\eta_i = \frac{\langle T,T-1 | V_C^{(1)} | M_{T+i,T-1} \rangle}{E_{M_{T+i,T-1}} - E_1}, \; i = -1,0,1 \tag{10}$$

Here E_1 is the energy of the analog state.

One derives finally for T=1 isotriplets (the detailed derivation can be found in ref. [2]):

$$\delta_c = 8 \frac{V_1}{41\xi A^{2/3}} \varepsilon_1^2 \tag{11}$$

where $V_s = \frac{V_1}{A}(\vec{t} \bullet \vec{T})$ is the symmetry potential and ξ is a numerical factor which depends on the model used to describe the isovector monopole. The range of values for this parameter is between 3 and 4 [3].

3. Results

The isospin impurities in the ground state of a nucleus ε_1^2 where computed in the past [3] using collective models for the isovector monopole state. Using these impurities one finds the results shown in the table:

Table1. Values of δ_c in % for several mass numbers A for the four models discussed in ref. [2, 3].

δ_c \ A	10	40	80
Hydrodynamical	0.006%	0.1%	0.4%
NEWSR	0.001%	0.04%	0.20%
EWSR	0.005%	0.09%	0.4%
Microscopic	0.009%	0.06%	0.25%

These results are a factor 3-8 smaller than in ref. [1].

4. Isospin mixing and the role of the anti- analog state

In some nuclei with a low number of excess neutrons (protons) Coulomb mixing with the anti-analog state can introduce significant isospin impurities in the isobaric analog state. The corrections are particularly sizeable in those medium heavy nuclei that are off the stability line having a small number of excess neutrons. The enhancement of the isospin admixture in the analog can reach a few percent.

Journal of Physics: Conference Series **533** (2014) 012001 doi:10.1088/1742-6596/533/1/012001

Consider a simple parent state in which n_1 excess neutrons occupy orbit j_1 and n_2 neutrons orbit j_2.
The parent state is:

$$|\pi\rangle = |j_1^{n_1}(n) j_2^{n_2}(n)\rangle \tag{12}$$

The analog is:

$$|A\rangle = \frac{1}{\sqrt{2T}}\left[\sqrt{n_1}\,|j_1^{n_1-1}(n) j_1(p) j_2^{n_2}(n)\rangle + \sqrt{n_2}\,|j_1^{n_1}(n) j_2^{n_2-1}(n) j_2(p)\rangle\right] \tag{13}$$

The anti-analog $|\overline{A}\rangle$ is then:

$$|\overline{A}\rangle = \frac{1}{\sqrt{2T}}\left[\sqrt{n_2}\,|j_1^{n_1-1}(n) j_1(p) j_2^{n_2}(n)\rangle - \sqrt{n_1}\,|j_1^{n_1}(n) j_2^{n_2-1}(n) j_2(p)\rangle\right] \tag{14}$$

The letters n and p denote neutrons and protons.
We will consider here parent nuclei with simple configurations: for even-even nuclei the n_1 and n_2 are even and in each orbit the excess nucleons are coupled to $J = 0^+$ and in odd-even nuclei n_1 is odd and n_2 is even.
The one-body Coulomb matrix element between the analog and anti-analog is then:

$$\langle\overline{A}|V_c|A\rangle = \frac{\sqrt{n_1 n_2}}{2T}\left[\langle j_1|V_c|j_1\rangle - \langle j_2|V_c|j_2\rangle\right] \tag{15}$$

If the excess neutrons occupy orbits belonging to different major shells, this matrix element is sizable. But even if this is not the case, because of binding energy effects in a finite potential well and angular momentum, this matrix element is of the order of several hundred keV [3,4].
The energy splitting between the analog and anti-analog is given by the symmetry potential. It is easy to show that in this case:
$\delta_c = \varepsilon^2$, the admixture of the anti-analog into the analog state.
For low T and large Z nuclei this contribution might be significant. For example: for T=3/2 nuclei with the excess of three nucleons occupying two orbits from different major shells we select ${}^{17}_{7}N_{10}$ and find from $\delta_c = 0.04\%$, for ${}^{41}_{22}Ti_{19}$, $\delta_c = 0.7\%$ and for ${}^{79}_{38}Sr_{41}$, $\delta_c = 3.3\%$.
For T=2 nuclei one can look at the example of ${}^{80}_{38}Sr_{42}$, here $\delta_c = 2.1\%$. In the examples chosen we selected nuclei in which the excess nucleons occupy two orbits belonging to different harmonic oscillator shells.
Other examples for T=2 states, for mass A=40, δ_c=0.4%, for A=60, δ_c=1.1% and for A=100, δ_c=4.3%.

References

[1] Towner I S and Hardy J C, 2008 *Phys. Rev.* C **77** 025501
[2] Auerbach N, 2009 *Phys. Rev.* C **79** 035502
[3] Auerbach N, 1983 *Phys. Rep.* **98** 273
[4] Bertsch G F and Mekjian A, 1972 *Annu. Rev. Nucl. Sci.* **22** 25

Journal of Physics: Conference Series **533** (2014) 012002 doi:10.1088/1742-6596/533/1/012002

Superheavy magic nuclei in Relativistic Hartree-Fock-Bogoliubov theory

Jiajie Li[1,2], Wenhui Long[1], Jérôme Margueron[3], Nguyen Van Giai[2,*]

[1]School of Nuclear Science and Technology, Lanzhou University, Lanzhou 730000, China
[2]Institut de Physique Nucléaire, IN2P3-CNRS, Université Paris-Sud, F-91406 Orsay, France
[3] Université Lyon-1, F-69622 Villeurbanne Cedex and IN2P3-CNRS, UMR 5822, IPN Lyon, France

E-mail: nguyen@ipno.in2p3.fr

Abstract. The occurrence of the spherical shell closures in superheavy nuclear systems is explored within the relativistic Hartree-Fock-Bogoliubov (RHFB) theory with density-dependent meson-nucleon couplings. We use the two-nucleon gaps $\delta_{2n(p)}$ and pairing gaps $\Delta_{\nu(\pi)}$ to characterize the shell effects. The results depend slightly on the forces used, but the general set of magic numbers beyond ^{208}Pb are $Z = 120$, 138 and $N = 172$, 184, 228 and 258. Our calculations are in favor of the nuclide $^{304}120$ as the next spherical doubly magic nucleus. Combined with the bulk properties of symmetric matter, we find that the shell effects are sensitive to the values of both scalar mass and effective mass. These two masses essentially determine the spin-orbit splittings and single-particle level density, respectively. Furthermore, the breaking of relativistic pseudo-spin symmetry influences the level structure and the occurrence of closed shells.

1. Introduction

In this contribution we report on our recent investigations of the possible occurrence of new, not yet observed superheavy nuclei beyond $Z \geq 120$. The motivations for this endeavour are based on the new possibilities offered by the Relativistic Hartree-Fock-Bogoliubov (RHFB) theory which has made a major step forward in the recent years [1].

The advantages of a covariant mean field approach for such a study are well established. In the non-relativistic self-consistent mean field method, the central and spin-orbit average potentials depend on different parts of the energy density functional (EDF) and therefore, on independently adjusted parameters of the EDF[2]. In the relativistic self-consistent mean field, on the other hand, all parameters of the EDF contribute to the Lorentz scalar and vector mean fields - U_s and U_v - and consequently the central potential $U_0 = U_s + U_v$, and the spin-orbit potential $U_{so} = U_s - U_v$ of the nucleonic Dirac equation are not independent. This aspect is quite crucial obviously when one tries to predict magic nuclei in unknown regions of the periodic table.

In the literature, there are quite many versions of effective Lagrangians to be used in the relativistic Hartree-Bogoliubov context, we will call them the Relativistic Hartree-Bogoliubov (RHB) models [3]. Some of the results that we discuss in this contribution are obtained by using representative RHB models, namely the PKDD[4] and DD-ME2[5] Lagrangians. On the other hand, there are also several RHFB Lagrangians which are doing well in describing the general

Journal of Physics: Conference Series **533** (2014) 012002 doi:10.1088/1742-6596/533/1/012002

properties (binding energies, radii, shell closures) of a vast number of nuclei throughout the periodic table, and we wish to include them in this survey of superheavy magic nuclei. Thus, the RHFB Lagrangians PKO1, PKO2, PKO3[6] and PKA1[7] are also used in this study. All these covariant Lagrangians do not contain originally a term describing the pairing effects. Here, we have used as pairing interaction the original D1S Gogny force[8] modified by an adjusted strength factor f=0.9, so that the odd-even mass differences of Pb isotopes are reproduced in our RHFB calculations.

2. Shell closures in the regions $N \geq 160$, $Z \geq 110$

To identify the magic shells it is convenient to look at the two-nucleon gaps - δ_{2p} for protons and δ_{2n} for neutrons - because they provide a quantitative measure of the shell effects. These gaps are the differences of two-proton (two-neutron) separation energies between isotones (isotopes) differing by two units:

$$\begin{aligned} \delta_{2p}(N,Z) &= S_{2p}(N,Z) - S_{2p}(N,Z+2) \\ \delta_{2n}(N,Z) &= S_{2n}(N,Z) - S_{2n}(N+2,Z) \end{aligned} \tag{1}$$

Larger values of δ_{2p} or δ_{2n} occur whenever S_{2p} or S_{2n} undergo a sudden jump, which is the signal of a magic shell occurrence.

We have thus explored, using the five Lagrangians mentioned in Sec.1, the evolution of δ_{2p} and δ_{2n} for the nuclei covering the ranges $110 \leq Z \leq 140$ and $140 \leq N \leq 280$. The calculated values of δ_{2p} and δ_{2n} vary from 1 to 5 MeV. As a general result, PKA1-RHFB is the model which predicts the larger two-nucleon gaps (3 to 5 MeV) for $Z = 120, 126, 138$ and $N = 184, 258$. These are the magic numbers corresponding to PKA1. The other Lagrangians also indicate a well marked proton closed shell at $Z = 120$ and $Z = 138$.

Concerning the neutron shells the predictions vary with the particular model Lagrangian employed. At N=172 and 228, neutron shell closures are found with the various Lagrangians considered here, although the shell effects appear to be rather weak. Other predicted neutron magic numbers are N=184 and 258 (except with PKO2). By examining the predictions of the various models discussed here, one may tentatively conclude that the nuclide $^{304}120_{184}$ could be a doubly magic system, and $^{292}120_{172}$ might be another, less stable candidate.

3. The role of pseudo-spin symmetry and its violation

It is interesting to examine the significance of the magic numbers Z=120, N=184 in the light of the pseudo-spin symmetry (PSS) [9] and its breaking [7]. The PSS is a property of the solutions of the Dirac equation that one can schematically describe as follows: if the Lorentz scalar and vector potentials, U_s and U_v, of the system are such that:

$$U_s(\mathbf{r}) + U_v(\mathbf{r}) = \text{constant} \tag{2}$$

at all values of $\mathbf{r}$, then the pair of solutions (nlj) and $(n' = n+1, l' = l-2, j' = j-1)$ are degenerate. In finite nuclei, the constant value on the right hand side of Eq.(2) is necessarily zero since both $U_s(\mathbf{r})$ and $U_v(\mathbf{r})$ tend to zero asymptotically. Obviously, the condition (2) is generally not realized in a nuclear system having bound states, because $U_s(\mathbf{r}) + U_v(\mathbf{r})$ must be negative inside the nucleus with typical values in the range of -50 MeV near the center, while it tends to zero outside the nuclear region. Thus, the amount of PSS violation will affect the relative positions of the partners (nlj) and $(n'l'j')$, and therefore the N and Z numbers corresponding to filled subshells.

This is illustrated in Fig.1 by the single-particle spectra calculated in the nuclide $^{304}120_{184}$ with the various Lagrangians used in this work. One can see that the PSS is sometimes fairly well

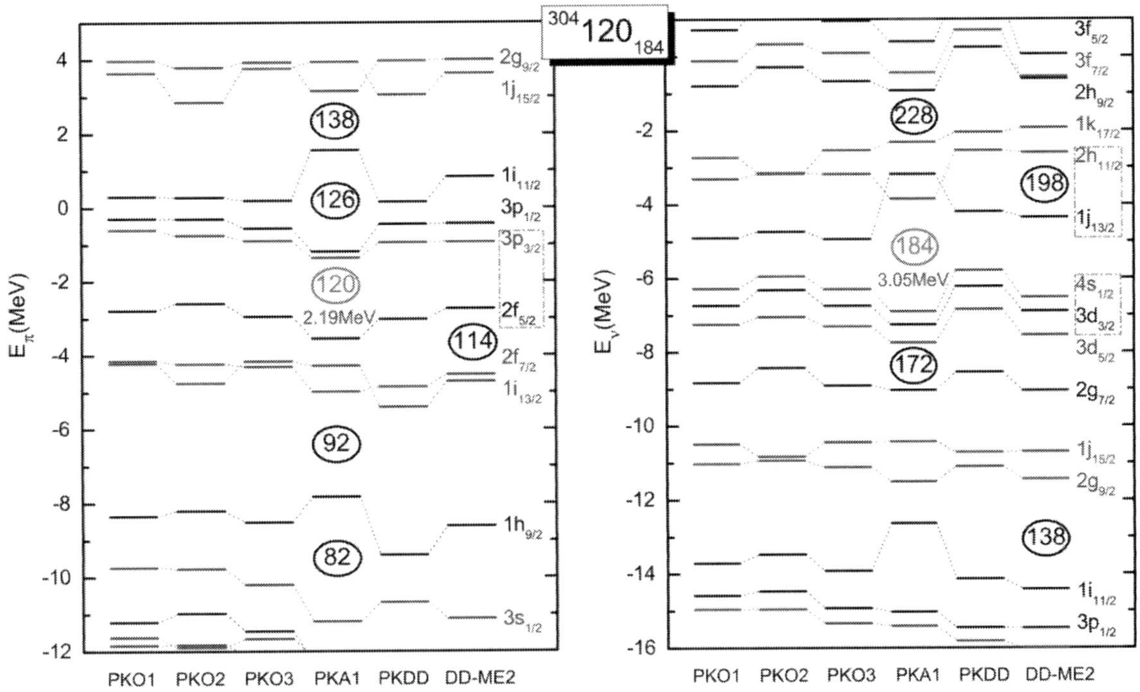

Figure 1. (Color online) Proton (left panel) and neutron (right panel) single-particle levels in the superheavy nucleus $^{304}120_{184}$. They correspond to RHFB calculations performed with the Lagrangians PKOn and PKA1, and RHB calculations with PKDD and DD-ME2. The pairing interaction used is the D1S force with a reduction factor f=0.9.

obeyed (neutron states 4s1/2 - 3d3/2) and sometimes strongly broken (neutron states 2h11/2 - 1j13/2, proton states 3p3/2 - 2f5/2). In the case of the proton states, the magic number $Z = 120$ just corresponds to the lower partner 2f5/2 completely filled and the upper partner 3p3/2 empty. Indeed, for all the models shown in Fig.1 there is a large splitting - of the order of 2 MeV - for the proton pseudo-spin partners 2f5/2 - 3p3/2, while the spin-orbit splitting of the 3p3/2 - 3p1/2 proton levels is fairly small.

On the neutron side, we observe that the prediction of PKA1 for the pseudo-spin partners 1j13/2 - 2h11/2 is very different from that of the 5 other models, with a smaller splitting and a reversed order. We recall that the PKA1 Lagrangian contains an extra degree of freedom as compared to the other models, namely a Lorentz tensor ρ-N coupling. This is probably the reason for the different 1j13/2 - 2h11/2 neutron splitting.

4. Conclusion

We have explored the regions of the nuclear chart around (Z=120, N=184) and beyond, in search for doubly closed-shell systems. We have taken advantage of a newly developed tool, the RHFB code in spherical symmetry, and of the most recent meson-nucleon effective Lagrangians adjusted to the properties of known nuclei.

We find that several effective Lagrangians do agree in predicting some new magic numbers, even though the detailed single-particle spectra show some model dependence. In particular, the results from the four Hartree-Fock type of models (PKO1, PKO2, PKO3, PKA1) and from the two Hartree type of models (PKDD, DD-ME2) indicate a well-marked shell closure at Z=120,

N=184.

Acknowledgements

This work is partly supported by the National Natural Science Foundation of China under Grant No. 11075066, the Fundamental Research Funds for the Central Universities under Contracts No. lzujbky-2012-k07, and the Program for New Century Excellent Talents in University under Grant No. NCET-10-0466.

References

[1] Long WH, Ring P, Van Giai N, and Meng J 2010 *Phys. Rev. C* **81** 024308
[2] Bender M, Heenen P-H and Reinhard P-G 2003 *Revs. Mod. Phys.***75** 121
[3] Ring P 1996 *Prog. Part. Nucl. Phys.* **37** 193
[4] Long WH, Meng J, Van Giai N and Zhou SG 2004 *Phys. Rev. C* **69** 034319
[5] Lalazissis GA, Nikšić T, Vretenar D and Ring P 2005 *Phys. Rev. C* **71** 24312
[6] Long WH, Sagawa H, Meng J and Van Giai N 2008 *Europhys. Lett. B* **82** 12001
[7] Long WH, Sagawa H, Van Giai N and Meng J 2007 *Phys. Rev. C* **76** 034314
[8] Berger JF, Girod M and Gogny D 1984 *Nucl. Phys. A* **428** 23
[9] Ginocchio JN 2005 *Phys. Reports* **414** 165

Probing the pairing interaction through two-neutron transfer reactions

M Grasso[1], D Beaumel[1], E Khan[1], D Lacroix[1], J Margueron[2], A Vitturi[3]

[1]Institut de Physique Nucléaire, IN2P3-CNRS, Université Paris-Sud, F-91406 Orsay Cedex, France
[2]Institut de Physique Nucléaire, IN2P3-CNRS, Université Lyon 1, 69622 Villeurbanne, France
[3]Dipartimento di Fisica, Università di Padova, IT-35131 Padova, Italy; Istituto Nazionale di Fisica Nucleare, Sezione di Padova, IT-35131 Padova, Italy

Abstract. We analyze two–neutron transfers in nuclei within a mean–field approach. We first show that the response associated with pairing excitations can be affected by the surface/volume localization of the employed effective pairing interaction. Then, ground–state–to–ground–state transitions are studied and some improvements on the currently used theoretical formulas are discussed. Finally, Cr isotopes at the drip line are analyzed. The analysis of pair–transfer reactions in these nuclei may improve our understanding of two aspects: the spatial distibution of pairing correlations in nuclei and the debated question of the persistence of pairing at the drip lines.

1. Introduction

The spatial localization properties of the pairing interaction are analyzed within the mean–field framework. A Skyrme interaction is used together with a zero–range density–dependent interaction for the pairing channel in the Hartree–Fock–Bogoliubov (HFB) model. Two–neutron transfers are studied within the quasiparticle random–phase approximation (QRPA) model from the ground state of a nucleus A to excited states of the nucleus $A \pm 2$: (p,t) reactions for very neutron–rich Sn isotopes are suggested as good cases where the spatial localization (surface/volume) of the pairing interaction may be explored. Ground-state to ground-state transitions are also analyzed by using the HFB wave functions and different formulas are compared. Finally, the interplay of the spatial localization of the interaction and the surface properties of low–l states is investigated in weakly bound Cr isotopes located close to the drip line. The manuscript is organized as follows: In Sec. 2 the transitions from the ground state of the nucleus A to excited states of the nucleus $A \pm 2$ are studied. In Sec. 3 the ground-state to ground-state transitions are investigated. The interplay of surface effects at the drip line is illustrated in Sec. 4. Conclusions are drawn in Sec. 5.

2. From the ground state of the nucleus A to excited states of the nucleus $A \pm 2$

A detailed analysis may be found in Ref. [1]. Similar studies have been also published in Refs. [2,3].

The pairing interaction employed in the HFB + QRPA calculations is the following:

$$V(\vec{r}_1, \vec{r}_2) = V_0 \left[1 - x \left(\frac{\rho(r)}{\rho_0}\right)^{\gamma}\right] \delta(\vec{r}_1 - \vec{r}_2), \quad (1)$$

where V_0 is the strength of the interaction to be tuned and ρ_0 is the saturation density. The value of γ is chosen equal to 1. A cutoff of 60 MeV is employed for the quasiparticle states

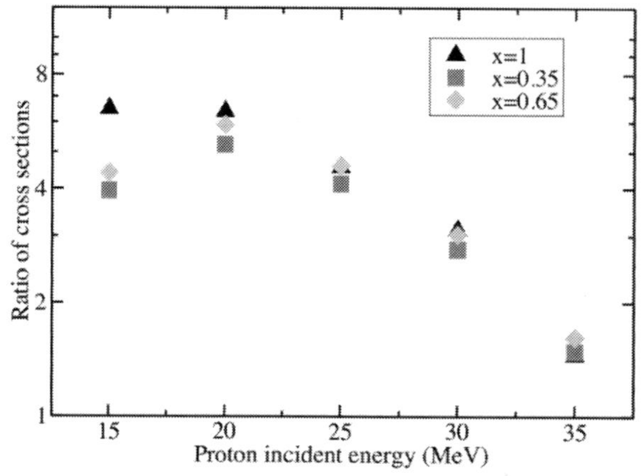

Figure 1. Ratio of the cross sections associated with the transitions to the first two excited states, obtained by using different pairing interactions (different values of x) and for different energies of the incident proton

and the parameter x, that controls the surface/volume mixing of the interaction, is chosen equal to 0.35, 0.65 (mixed interactions), and 1 (surface–peaked interaction). V_0 is adjusted to reproduce the two–neutron separation energy of Sn isotopes. In the HFB + QRPA calculations a box discretization is employed to treat the continuum states. The excited states associated with addition/removal two–neutron transfer reactions are analyzed by using different pairing interaction in the HFB-QRPA model. In the QRPA model, the particle–particle and the hole–hole sectors are employed to study the addition and the removal modes, respectively. The QRPA transition densities are used as form factors in zero–range DWBA reaction calculations to evaluate (p,t) cross sections. The limitation of such reaction calculations is that the absolute cross sections cannot be calculated and that inelastic excitations and two–step processes (corresponding to sequential particle transfers) are not included. The reaction $^{136}\mathrm{Sn}(p,t)^{134}\mathrm{Sn}$ is investigated. Since absolute cross sections cannot be evaluated, we have analyzed the ratios of cross sections. We show in Fig. 1 the ratio of the cross sections associated with the transitions to the first two excited states, obtained by using different pairing interactions (different values of x) and for different energies of the incident proton.

We observe that, at low incident proton energy (15 MeV), some differences between the results obtained by using a surface–peaked pairing interaction and a mixed pairing interaction are visible. These results are an indication that, for very neutron–rich Sn isotopes, one may expect measurable effects in the cross sections related to the nature of the pairing interaction.

3. From the ground state of the nucleus A to the ground state of the nucleus $A \pm 2$

The analysis of ground–state to ground–state transition probabilities from the nucleus A to the nucleus $A\pm2$ has also been discussed in Ref. [4]. In Ref. [5], we have improved the expressions to be used for the calculation of the transition probabilities. Within the approximation employed in Ref. [4], the removal and addition transfer probabilities are calculated as:

$$P_{GS}^{Rem}(A) = P_{GS}^{Add}(A) = |\int dr \sum_{nlj}(2j+1)u_{nlj}^{A}(r)v_{nlj}^{A}(r)|^2, \quad (2)$$

where the HFB wave functions of the nucleus A are used. This approximation is clearly not valid for closed–shell nuclei, where the transition probabilities calculated with Eq. (2) are equal to zero. By using a slightly improved approximation, we have introduced different expressions to calculate the probabilities, namely:

$$P_{GS}^{Rem}(A) = |\int dr \sum_{nlj}(2j+1)u_{nlj}^{A-2}(r)v_{nlj}^{A}(r)|^2, \quad (3)$$

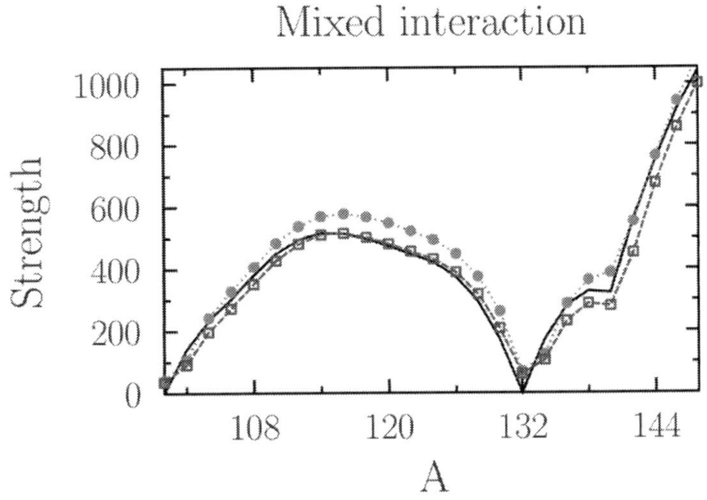

Figure 2. Removal transfer probabilities for Sn isotopes obtained with a mixed surface/volume pairing interaction by using different levels of approximation. The black solid line is obtained with the formula using only the wave functions of the nucleus A. The red circles are obtained with the improved formula. The blue squares represent the results obtained with the particle number projection.

$$P_{GS}^{Add}(A) = |\int dr \sum_{nlj}(2j+1)u_{nlj}^{A}(r)v_{nlj}^{A+2}(r)|^2. \tag{4}$$

It can be observed that now the wave functions of all implied nuclei (A and $A \pm 2$) are used. In Ref. [5] effects associated with particle number projection are also studied. Some results are summarized in Fig. 2.

The black line corresponds to the results obtained with Eq. (2), the red circles to the results obtained with Eq. (3), and the blue squares are the results obtained by performing the particle number projection. One observes that the improved formula (red circles) provides enhanced probabilities with respect to the black line and, in particular, probabilities that are not equal to zero at shell closures. The particle number projection has the effect to reduce the probabilities at mid-shell, still keeping finite values at shell closures.

4. Interplay between surface effects

A large odd–even staggering has been found in reaction cross sections of Ne and Cr isotopes in recent studies [6,7]. This staggering is found for weakly bound nuclei located at the drip lines and is related to the presence of a low-l (s state). In Ref. [8] this interplay between surface effects related to the presence of a weakly bound s state and different spatial localizations of the pairing interaction is investigated for Cr isotopes in the transfer probabilities. With the employed Skyrme interaction SkM*, the drip line Cr isotope is ^{82}Cr, where the last bound neutron state is the $3s_{1/2}$ state. In Fig. 3 we show the ground–state to ground–state transition probabilities for Cr isotopes by using Eq. (2) (dotted lines) and Eq. (4) (solid lines); η corresponds to the parameter x introduced above. Let us consider only the results obtained with the improved formulas. In the case of a surface–peaked pairing interaction (c) a change of slope is found when approaching the drip lines in the transition probabilities. This change of slope may be related to the enhancement of surface effects that takes place in the calculation of the probability for the nucleus ^{80}Cr due to the wave function of the weakly bound neutron s state that enters into play.

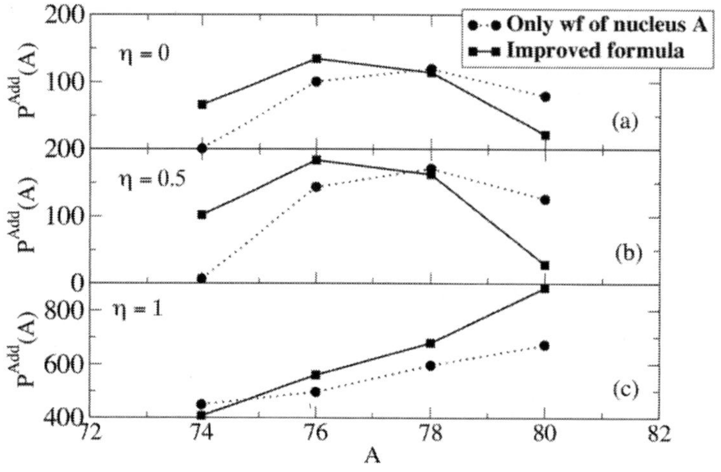

Figure 3. Addition transfer probabilities for Cr isotopes calculated with a volume (a), a mixed (b), and a surface–peaked (c) pairing interaction.

5. Conclusions

The spatial localization properties of the pairing interaction have been analyzed within the mean–field framework. Two–neutron transfers have been studied within the QRPA model from the ground state of a nucleus A to excited states of the nucleus $A \pm 2$: (p,t) reactions for very neutron–rich Sn isotopes have been suggested as good cases where the spatial localization (surface/volume) of the pairing interaction may be explored. Ground-state to ground-state transitions have been also analyzed by using the HFB wave functions and different formulas have been compared. Finally, the interplay of the spatial localization of the interaction and the surface properties of weakly bound s states has been investigated for Cr weakly bound nuclei located at the drip line, where the last neutron bound state is the $3s_{1/2}$ state.

References

[1] E. Pllumbi, M. Grasso, D. Beaumel, E. Khan, J. Margueron, and J. Van de Wiele, Phys. Rev. C 83, 034613 (2011).
[2] M. Matsuo and Y. Serizawa, Phys. Rev. C 82, 024318 (2010).
[3] B. Avez, C. Simenel, and P. Chomaz, Phys. Rev. C 78, 044318 (2008).
[4] H. Shimoyama and M. Matsuo, Phys. Rev. C 84, 044317 (2011).
[5] M. Grasso, D. Lacroix, and A. Vitturi, Phys. Rev. C 85, 034317 (2012).
[6] K. Hagino and H. Sagawa, Phys. Rev. C 84, 011303 (R) (2011).
[7] K. Hagino and H. Sagawa, Phys. Rev. C 85, 014303 (2012).
[8] M. Grasso, Phys. Rev. C 87, 064308 (2013).

Journal of Physics: Conference Series **533** (2014) 012004 doi:10.1088/1742-6596/533/1/012004

Direct processes effects on deuteron activation cross sections

M. Avrigeanu and V. Avrigeanu

"Horia Hulubei" National Institute for Physics and Nuclear Engineering, P.O. Box MG-6, 76900 Bucharest, Romania

Abstract. An extended analysis of reaction mechanisms involved in deuterons interaction with target nuclei from ^{27}Al till ^{231}Pa, at incident energies up to 60 MeV, is presented. Increased attention is devoted to direct processes, concerning the breakup, stripping, and pick–up contributions to the deuteron activation cross sections. Finally, the pre-equilibrium and evaporation cross sections, corrected for the initial flux leakage towards direct processes, have completed the deuteron interaction analysis. The overall agreement of the measured data and model calculations proves the correctness of nuclear mechanism description.

1. Introduction

The request of the accurate activation cross sections by several on-going strategic research programmes for international large-scale facilities as ITER [1], IFMIF [2], and SPIRAL-2 [3] focus the attention on the deuteron data, less extensive and mature than the neutron ones.

Description of the deuteron interaction process is still a challenge for basic research, the difficulties to describe the complex deuteron–induced reaction data in terms of the usual reaction mechanism models being the result of the high enhancement of a variety of reactions, even at low bombarding energies, induced by the deuteron–breakup nucleons. An increased attention is devoted in the present work to direct interactions, namely the breakup, stripping, and pick–up mechanisms (e.g., Refs. [4, 5, 6, 7, 8, 9, 10]), usually neglected or very poorly taken into account in spite of their importance. For energies below and around the Coulomb barrier, the interaction of deuterons with target nuclei proceeds largely through direct reactions (DR) mechanism, while pre-equilibrium (PE) and evaporation (CN) processes become more important with the incident energy increase. Moreover, the deuteron breakup (BU) mechanism is quite important along the whole incident-energy range [5, 6, 7, 9, 10].

2. Deuteron breakup

The physical picture of the deuteron breakup, in the Coulomb and nuclear fields of the target nucleus, considers two distinct chains. These are the elastic–breakup (EB) in which the target nucleus remains in its ground state and none of the deuteron nucleons interacts with it, and the inelastic–breakup or breakup fusion (BF), where one of the deuteron nucleons interacts with the target nucleus while the remaining one is detected.

An empirical parametrization of the total proton-emission breakup fraction of the deuteron total reaction cross section σ_R, $f^{(p)}_{BU} = \sigma^p_{BU}/\sigma_R$, and elastic breakup fraction, $f_{EB} = \sigma_{EB}/\sigma_R$, have been obtained [4] by analysis of the experimental systematic [11] of the proton-emission

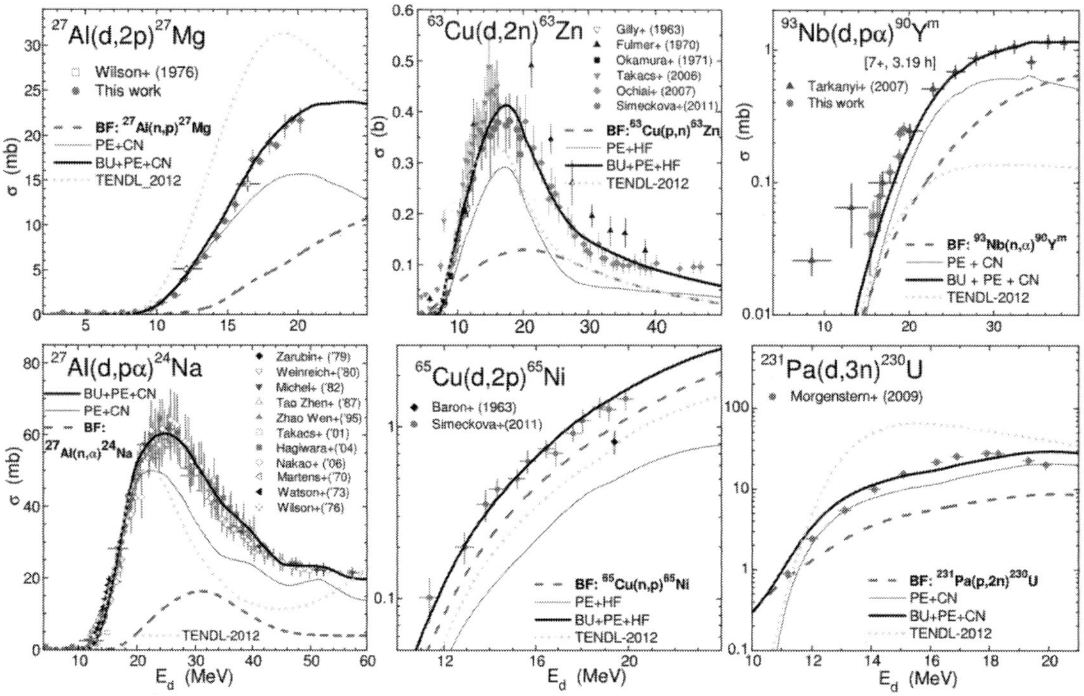

Figure 1. Comparison of measured deuteron activation cross sections ([5, 7, 10, 9] and references therein), complete analysis results (thick solid curves) taking into account the BF enhancement (dashed curves) and PE+CN contributions (thin solid curves), and the TENDL-2012 evaluations (dotted curves) (see text).

spectra and angular distributions of deuteron–induced reactions on target nuclei from Al to Pb, at incident energies from 15 to 80 MeV. Their dependence with respect to the charge Z and atomic number A of the target nucleus, as well as deuteron incident energy E is [4]:

$$f^{(p)}_{BU} = 0.087 - 0.0066Z + 0.00163ZA^{1/3} + 0.0017A^{1/3}E - 0.000002ZE^2, \tag{1}$$

$$f_{EB} = 0.031 - 0.0028Z + 0.00051ZA^{1/3} + 0.0005A^{1/3}E - 0.000001ZE^2. \tag{2}$$

Consequently, it results the inelastic–breakup fraction $f^p_{BF} = f^p_{BU}$ - f_{EB}.

Overall, there are actually two opposite effects of the deuteron breakup on the deuteron activation cross sections that should be considered. Firstly, the reaction cross section σ_R, that is shared among different outgoing channels, is reduced by the value of the total breakup cross section σ_{BU}. On the other hand, the BF component, where one of deuteron constituents interacts with the target nucleus leading to a secondary composite nucleus, brings contributions to different reaction channels [4, 5, 6, 7, 9]. Thus, the absorbed proton or neutron, following the deuteron breakup, contribute to the enhancement of the corresponding (d, xn), respectively (d, xp) reaction cross sections.

In order to calculate the BF enhancement of, e.g., the (d, xn) reaction cross sections, the proton BF cross section should be multiplied by the ratios $\sigma_{(p,x)}/\sigma^p_R$ corresponding to this reaction, convoluted with the Gaussian line shape distribution of the breakup protons energy for a given deuteron incident energy E_d. Finally the integration over the breakup proton energy,

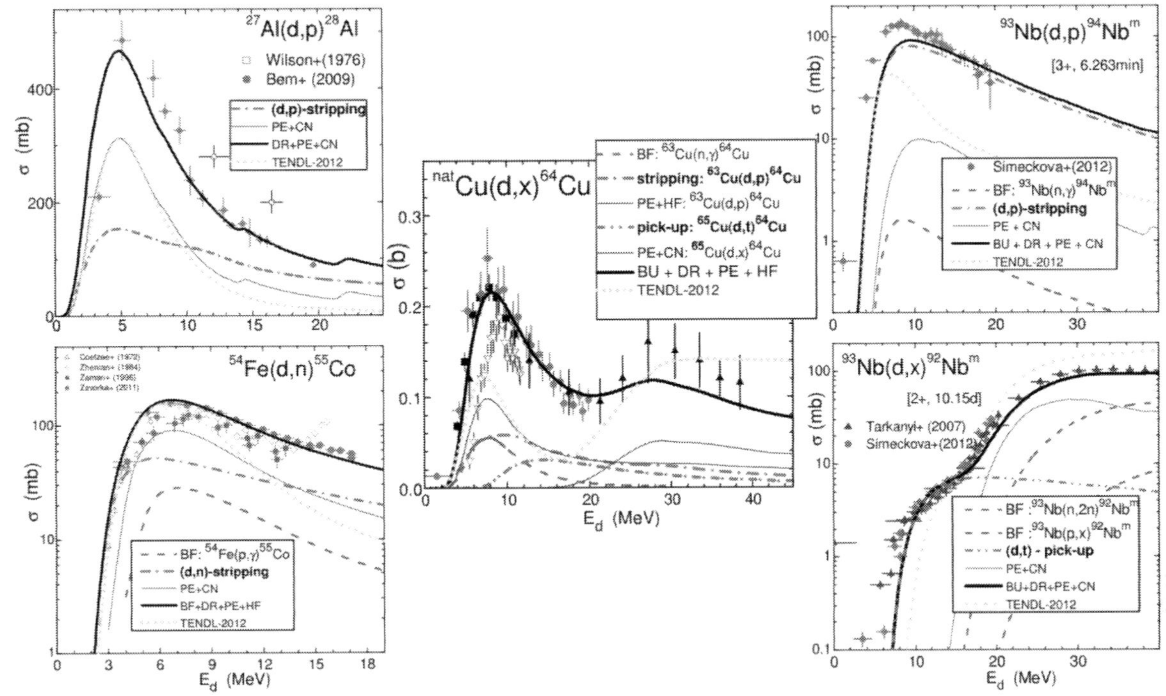

Figure 2. Comparison of measured deuteron activation cross sections ([5, 7, 10] and references therein), complete analysis results (thick solid curves) taking into account the BF enhancement (dashed curves), DR (dot-dashed and dot-dot-dashed curves) and PE+CN contributions (thin solid curves), and the TENDL-2012 evaluations (dotted curves) (see text).

E_p, led the corresponding BF enhancement cross section [7, 6, 9]:

$$\sigma^{p,x}_{BF}(E_d) = \sigma^{p}_{BF}(E_d) \int dE_p \; \frac{\sigma_{(p,x)}(E_p)}{\sigma^p_R} \frac{1}{(2\pi)^{\frac{1}{2}} w} exp[-\frac{(E_p - E^0_p(E_d))^2}{2w^2}] \; , \qquad (3)$$

where σ^p_R is the proton total reaction cross section, x stands for various γ, n, d, or α outgoing channels, while the Gaussian distribution parameters, w and E^0_p, are given by Kalbach [12].

The BF enhancements brought by the breakup neutron and proton interactions with ^{27}Al, ^{54}Fe, ^{63,65}Cu, ^{93}Nb and ^{231}Pa target nuclei are shown by dashed curves in Figs. 1 and 2.

3. Transfer reactions

Apart from the breakup contributions to deuteron interaction, an increased attention has to be devoted to the DR very poorly accounted so far in deuteron activation analysis. The calculations of the DR mechanisms contributions, like stripping and pick-up, that are important at the low energy side of the (d,p), (d,n), and (d,t) excitation functions [4, 5, 6, 7, 8, 10], have been performed in the frame of the CRC formalism by using the code FRESCO [13]. The n-p interaction in deuteron [14] as well as d-n interaction in triton [15] are assumed to have a Gaussian shape, while the transferred nucleon bound states were generated in a Woods–Saxon real potential [4, 7, 10]. A particular note should concern the pick-up essential contribution to the total (d,t) activation cross section, at the energies between its threshold and those for the (d,nd) and $(d,2np)$ reactions that lead to the same residual nucleus. Thus, the pick-up component of the (d,t) excitation function is critical for the data description at deuteron incident energies lower than 10 MeV, where PE and CN contributions are almost negligible. The importance of the stripping and pick–up reactions for the deuteron interaction process is evidenced in Fig. 2.

4. Statistical particle emission

Following the decay path, the PE and CN reaction cross sections have been calculated by means of the code STAPRE-H [17], taking into account the breakup and DR results discussed above. A consistent local parameter set [5, 7, 10] has also been used. The mark BU, rather than BF, for the sum of various contributions to an activation cross section in Figs. 1 and 2 underlines the consideration of both breakup effects, i.e., the overall decrease of σ_R, as well as the BF enhancement. On the other hand, the apparent discrepancies between the experimental data and corresponding TENDL-2012 [16] evaluation, shown in Figs. 1 and 2 too, stress out the effects of disregarding the direct processes within TENDL.

5. Conclusions

The overall agreement between the measured data and model calculations supports the description of nuclear mechanisms taken into account for the deuteron-nucleus interaction. However, while the associated theoretical frames are already settled for DR, PE and CN mechanisms, an increased attention should be given to microscopical description of the BF component. The improvement of deuteron breakup description requires complementary experimental studies involving deuterons, protons and neutron induced reactions too.

Acknowledgments

This work was partly supported by a grant of the Romanian National Authority for Scientific Research, CNCS - UEFISCDI, project No. PN-II-ID-PCE-2011-3-0450, and by Fusion for Energy through the Specific Grant Agreement GRT-168.01.

References

[1] http://www.iter.org/proj.
[2] http://www.ifmif.org/b/.
[3] http://pro.ganil-spiral2.eu/spiral2/instrumentation/nfs.
[4] M. Avrigeanu *et al.*, Fusion Eng. Design **84** (2009) 418.
[5] P. Bém *et al.*, Phys. Rev. C **79** (2009) 044610.
[6] M. Avrigeanu and V. Avrigeanu, EPJ Web of Conf. **2** (2010) 01004; EPJ Web of Conf. **21** (2012) 07003; J. Phys: Conf. Ser. **205** (2010) 012014; E. Šimečková *et al.*, EPJ Web of Conferences **8** (2010) 07002.
[7] E. Šimečková *et al.*, Phys. Rev. C **84** (2011) 014605.
[8] M. Avrigeanu and V. Avrigeanu, J. Korean Phys. Soc. **59** (2011) 903.J. Korean Phys.Soc. **59** (2011) 1928.
[9] M. Avrigeanu, V. Avrigeanu, and A. J. Koning, Phys. Rev. C **85** (2012) 034603.
[10] M. Avrigeanu *et al.*, Phys. Rev. C **88** (2013) 014612.
[11] J. Kleinfeller *et al.*, Nucl. Phys. **A370** (1981) 205; J. Pampus *et al.*, Nucl. Phys. **A311** (1978) 141; J. R. Wu *et al.*, Phys. Rev. C **19** (1979) 370; N. Matsuoka *et al.*, Nucl. Phys. **A345** (1980) 1; M. G. Mustafa *et al.*, Phys. Rev. C **35** (1987) 2077.
[12] C. Kalbach Walker, TUNL Progress Report **XLII** (2002-2003) p. 82-83, Triangle University Nuclear Laboratory; www.tunl.duke.edu/publications/tunlprogress/2003/.
[13] I. J. Thompson, Comput. Phys. Rep. **7**, 167 (1988); v. FRES 2.3 (2007).
[14] M. Kawai *et al.*, Prog. Theor. Phys. Suppl. **184**, 118 (1969).
[15] P. Guazzoni *et al.*, Phys. Rev. C **83**, 044614 (2011).
[16] A.J Koning *et al.*, *TENDL-2012: TALYS-based Evaluated Nuclear Data Library*, 11 January, 2013, http://www.talys.eu/tendl-2012/.
[17] M. Avrigeanu, V. Avrigeanu, IPNE Report NP-86-1995, Bucharest, 1995, Refs. therein; News NEA Data Bank **17**, 22 (1995).

Journal of Physics: Conference Series **533** (2014) 012005 doi:10.1088/1742-6596/533/1/012005

Current understanding of the reaction mechanism in two-nucleon transfer reactions

A A Cowley[1,2]

[1] Department of Physics, Stellenbosch University, Private Bag X1, Matieland 7602, South Africa

[2] iThemba Laboratory for Accelerator Based Sciences, P O Box 722, Somerset West 7129, South Africa

E-mail: aac@sun.ac.za

Abstract. The current understanding of the reaction mechanism of proton-induced two-nucleon transfer reactions is reviewed. The ideal is to employ a theoretical formulation which accurately reflects the important physics, but which is nevertheless still fairly simple. The influence of sequential transfer of the two nucleons involved in a pickup reaction is the most obvious complication that has to be considered. This is discussed in detail.

1. Introduction

Two-nucleon transfer reactions offer a powerful method to study pairing correlations in atomic nuclei. In principle reactions such as $(p,^3\mathrm{He})$ and (p,t) should proceed in a very simple way, but there may be many complications. One issue which is especially contentious, is to what extent a sequential process competes with a direct, simultaneous two-nucleon reaction. In the former process, applicable to for example a (p,t) reaction, the reaction is envisaged as a single-nucleon (p,d) reaction as a first step, followed next by a (d,t) reaction. Of course, this sequential mechanism can involve a number of intermediate states all ending in the same final state. Whereas some authors simply ignore the possibility of sequential transfer (See for example Ref. [1]), others (See for example Ref. [2]) believe that it is crucial to include it properly. Yet a third group (See for example Ref. [3]) readily admits the controversy, and then exploits the situation to justify their implementation of a simplistic theoretical analysis.

In this paper I review the historical development of our understanding of the two-nucleon pickup reaction. In addition, I will try to reconcile results of different studies which appear to be of a conflicting nature. Clearly a comprehensive review is not possible in a short contribution such as this, but I will attempt to include most of the salient insights which have contributed to our knowledge of the topic over the years.

2. Insight into proton-induced two-nucleon pickup

As would be expected, first attempts to interpret two-nucleon transfer made use of the same techniques previously employed successfully to analyze single-nucleon transfer, namely plane-wave Born approximation (PWBA), later distorted-wave Born approximation (DWBA). The nucleons in the target were consequently described as a single particle (di-nucleon) system bound to the core of the target system with quantum numbers based on conservation of oscillator

Journal of Physics: Conference Series **533** (2014) 012005 doi:10.1088/1742-6596/533/1/012005

quanta. However, from the start it was already clear that structure amplitudes [4] which may be extracted from wave functions generated in a microscopic model, need to be employed for a realistic assessment of results from two-nucleon pickup. In spite of various refinements to the theory, such as for example inclusion of finite range effects, it soon became clear that the theoretical analyses appear to underpredict absolute cross sections to a disconcerting extent. At roughly the same time as this was identified as a problem, the reaction ^{208}Pb$(p,t)^{206}$Pb to the first 3^+ state, which has unnatural parity and is thus forbidden in zero-range approximation (although not in finite range) was studied. The theoretical implications of sequential transfer was fully investigated and found to resolve the cross section problem [5]. A dissenting opinion emerged from the work of Nagarajan *et al.* [6].

Pinkston and Satchler summarize the issues that need to be addressed correctly in two-nucleon transfer as follows [7] :

- Exact treatment of finite range of interaction (not zero range)
- Include simultaneous plus sequential transfer
- Account for a non-orthogonality correction (which tends to cancel the one-step contribution)
- Allow intermediate system to be in a continuum state (breakup effect)
- Use correct potentials to generate distorted waves
- Use correct nuclear wave functions to construct two-nucleon overlaps

They point out that unless all these ingredients are treated correctly, results from such studies should be viewed with some caution. However, many recent papers (See for example Refs. [1, 3, 8]) very successfully analyze two-nucleon pickup in terms of a simplistic, single-step semi-microscopic cluster DWBA formulation. An example is shown in Fig. 1. It needs to be understood whether such a treatment is adequate, and if so, what its range of validity is in terms of target mass and incident energy.

As was mentioned before, the possible influence of a sequential transfer reaction mechanism is of profound importance. In order to investigate the relationship between cross section angular distributions which either include or exclude the sequential component, it is useful to consider the theoretical expressions. For this purpose we reproduce a brief summary of the theory, based on the formulation of Charlton [9] who writes the two-nucleon transfer T-matrix element, in a notation appropriate for a (p,t) reaction, as

$$T_{tp} = \langle \chi_t^{(-)} \left| \widehat{V}_t + \widehat{V}_t G^{(+)} \widehat{V}_p \right| \chi_p^{(+)} \rangle, \tag{1}$$

where the wave functions χ, potentials $\widehat{V}$, and Green's function have the meaning as described in Ref. [9] . This expression is then shown to be approximately equal to the sum of two terms:

$$T_{tp} \approx T_{tp}^{sim} + T_{tp}^{seq}, \tag{2}$$

where the simultaneous part is given by

$$T_{tp}^{sim} = \langle \chi_t^{(-)} \left| \widehat{V}_t \right| \chi_p^{(+)} \rangle, \tag{3}$$

and the sequential component by

$$T_{tp}^{seq} = \langle \chi_t^{(-)} \left| \widehat{V}_t G_d^{(+)} \widehat{V}_p \right| \chi_p^{(+)} \rangle. \tag{4}$$

The Green's function $G_d^{(+)}$ is related to $G^{(+)}$ in a simple way, as defined by Charlton [9].

As Satchler cautions [10], splitting the simultaneous and sequential parts of the mechanism in this way, is somewhat misleading. Nevertheless the expressions give some insight into the relationship between the simultaneous and sequential mechanisms involved in the reaction. Of

Journal of Physics: Conference Series **533** (2014) 012005 doi:10.1088/1742-6596/533/1/012005

course, Eq. 3 is the normal DWBA expression for simultaneous transfer. Also, keep in mind that Eq. 4 applies to each intermediate state ϕ_i participating in the sequential process. The full contribution for all intermediate states can therefore be written as

$$T_{tp}^{seq} = \sum_i \langle \chi_t^{(-)} \left| \widehat{V}_t \left| \phi_i \right\rangle G_d^{(+)i} \left\langle \phi_i \right| \widehat{V}_p \right| \chi_p^{(+)} \rangle. \tag{5}$$

As Satchler points out [10], these participating states are normally expected to be close in excitation energy, so that $G_d^{(+)i}$ in Eq. 5 can be replaced by an average Green's function $\overline{G}^{(+)}$ applicable to all states, and together with the closure condition

$$\sum_i \left| \phi_i \right\rangle \left\langle \phi_i \right| = 1, \tag{6}$$

it means that there is a very simple relationship between the simultaneous and sequential expressions, namely that

$$T_{tp}^{seq} \propto \overline{G}^{(+)} T_{tp}^{sim}. \tag{7}$$

Eq. 7 implies that both mechanisms in which a two-nucleon transfer reaction may proceed have the same form. Consequently their cross section angular distributions are sensitive to pair interactions in exactly the same way. The main manifestation of the importance of a sequential transfer mechanism should be reflected in the magnitude of the observed absolute cross section. However, Charlton [9] finds no convincing difference between calculations which include or exclude sequential transfer in the ^{208}Pb(p,t)^{206}Pb reaction for the natural parity transition to the 0^+ ground state. In fact, comparing two different sets of optical potentials used to generate distorted waves, he concludes that inclusion of a sequential transfer mechanism serves only as a parameterization to improve agreement between experimental data and calculation, and that good agreement can be obtained anyway by means of a careful selection of the optical potentials. On the other hand, Igarashi *et al.* [5] find that although sequential transfer is not as dominant in the case of the natural parity as in the unnatural parity case, it is still substantial. The energy-dependant ratios of two-step to one-step processes are shown in Fig. 2 for both unnatural and natural parity transitions. As was pointed out earlier, this way of presenting the relative importance of the two different mechanisms, based on a splitting of the mechanisms instead of a proper coherent addition, may not be reliable. Nevertheless, it still suggests a strong incident energy dependence, the effect of which has to be taken into account when estimating the importance of sequential pickup.

Analyzing power is an observable which is predicted to be very sensitive to the reaction mechanism of two-nucleon transfer [5], and it does indeed display a sign difference for the unnatural parity transition in ^{208}Pb(p,t)^{206}Pb depending on whether two-step pickup is taken into account or not. Furthermore, the progression of the analyzing power with incident energy is correctly reproduced [5]. On the other hand, the analyzing power angular distribution difference is not as clear for the natural parity transition.

3. Summary and conclusion

The extent to which the reaction mechanism for proton-induced two-particle transfer is understood was reviewed. Clearly important issues still need to be investigated, especially the importance of sequential transfer of the nucleon pair. This may have an important consequence for future applications of two-nucleon transfer reactions to reveal properties of nuclei far from stability.

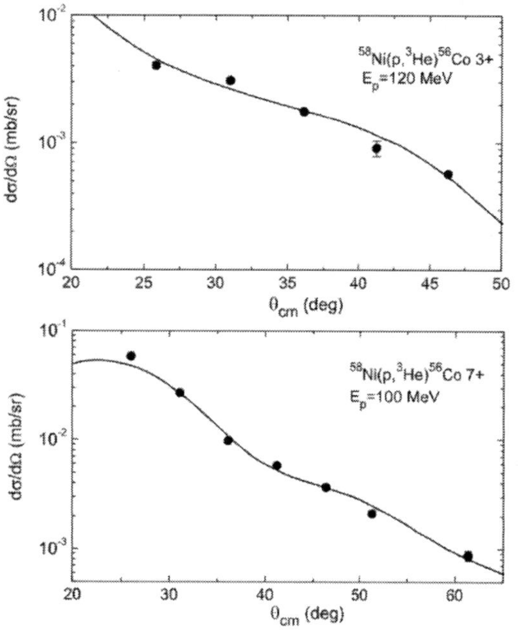

Figure 1. Two examples from Ref. [8] of cross section angular distribution for the reaction ^{58}Ni$(p,{}^{3}$He$)^{57}$Co at incident energies as indicated for excitations to a 3^{+} (top panel) and a 7^{+} state (bottom panel). The curves are zero-range semi-microscopic di-nucleon cluster DWBA calculations as described by van Zyl [8].

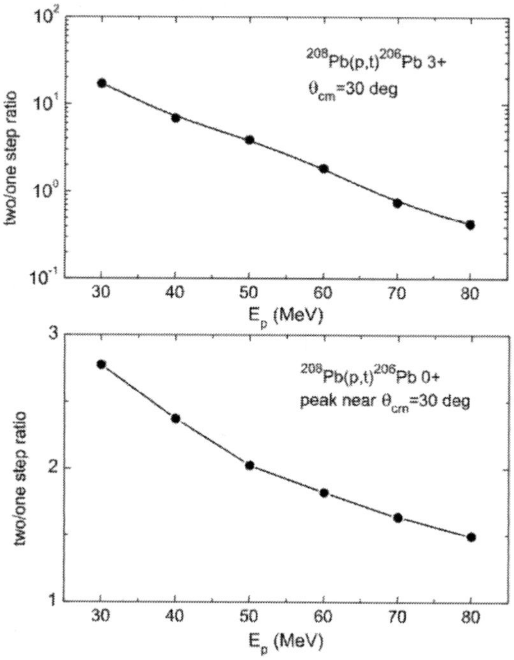

Figure 2. Ratio of two-step to one-step cross sections extracted from the work of Igarashi *et al.* [5] as a function of incident energy for the unnatural parity excitation (top panel) and natural parity excitation (bottom panel) for the reaction ^{208}Pb$(p,t)^{206}$Pb. Note the logarithmic Y-axis scale of the top panel.

Acknowledgement
This work was performed with funding from the South African National Research Foundation (NRF). The financial support is gratefully acknowledged.

References
[1] Chipps K A *et al.* 2012 *Phys. Rev.* C **86** 014329.
[2] Potel G, Adini A, Barranco F, Vigezzi E and Broglia R A. Proceedings of the 13th International Conference on Nuclear Reaction Mechanisms (Varenna, Italy) 11-15 June 2012 (Editors: F. Cerutti, M. Chadwick, A. Ferrari, T. Kawano, S. Bottoni, L. Pellegri, Publisher: CERN, Geneva, 2012) CERN Proceedings 2012-002 pp. 85–90.
[3] Guazzoni P, Zetta L, Covello A, Gargano A, Bayman B F, Faestermann T, Graw G, Hertenberger R, Wirth H-F and Jaskola M 2011 *Phys. Rev.* C **83** 044614.
[4] Glendenning N 1965 *Phys. Rev.* B **86** 102–113.
[5] Igarashi M, Kubo K-I and Yagi K 1991 *Physics Reports* **199** 1–72.
[6] Nagarajan M A, Strayer M R, and Werby M F. 1977 *Phys. Lett. B* **68** 421.
[7] Pinkston W T and Satchler 1982 *Nucl. Phys.* A **383** 61–76.
[8] Van Zyl J J 2012 PhD thesis, Stellenbosch University (unpublished).
[9] Charlton L A 1976 *Phys. Rev.* C **14** 506–513.
[10] Satchler G R in *Heavy Ion Collisions*, Lecture Notes in Physics **168** (Springer, Berlin, 1982) pp. 44–53.

Journal of Physics: Conference Series **533** (2014) 012006 doi:10.1088/1742-6596/533/1/012006

Sub-barrier transfer reactions studied with the magnetic spectrometer PRISMA

E Fioretto[1], L Corradi[1], S Szilner[2], D Montanari[3], C Michelagnoli[3], T Mijatović[2], G Montagnoli[3], F Scarlassara[3], A M Stefanini[1], C A Ur[3], G Pollarolo[4], N Soić[2]

[1] INFN - Laboratori Nazionali di Legnaro, Viale dell'Università 2, Legnaro (PD), I-35020, Italy
[2] Ruđer Bošković Institute, Bijenička 54, Zagreb, HR-10001, Croatia
[3] Dipartimento di Fisica dell'Università di Padova and INFN, Via Marzolo 8, Padova, I-35131, Italy
[4] Dipartimento di Fisica Teorica dell'Università di Torino and INFN, Via P. Giuria 1, Torino, I-10125, Italy

E-mail: fioretto@lnl.infn.it

Abstract. Transfer reactions in the sub-barrier region have been studied at Laboratori Nazionali di Legnaro (LNL) by using the large solid angle magnetic spectrometer PRISMA. In particular, excitation functions of multinucleon transfer channels have been measured for the systems ^{96}Zr+^{40}Ca (closed shell nuclei), ^{116}Sn+^{60}Ni (super-fluid nuclei) from the Coulomb barrier to 25% below. Results of these measurements are presented in this contribution.
An ancillary detection system has been designed and built in order to perform kinematic coincidence measurements in conjunction with PRISMA. It will allow to study the effects of secondary processes, such as evaporation and fission, which strongly affect the final yield of the reaction products. The status of this development is also reported.

1. Introduction

Transfer reactions play an important role in both nuclear structure and reaction dynamics studies. One- and two-nucleon transfer processes can be used to get useful information on the single particle structure and the pairing correlations in nuclei [1]. In the sub-barrier regime only few transfer channels are open and the Q-value distributions become narrower. These peculiar conditions should allow to extract more quantitative information on the mechanism of multiple transfer processes, in particular on the relative contribution of single particle and more complex degrees of freedom which include nucleon-nucleon correlations. However, available data for heavy ion transfer reactions in the sub-barrier region are extremely scarce or almost not existing due to their significant experimental difficulties. The use of large acceptance magnetic spectrometers such as PRISMA [2, 3, 4] and inverse kinematics reactions allows to perform this kind of experiments in an efficient way.
The study of the heavy neutron-rich nuclei is presently receiving a distinctive attention from the nuclear physics and nuclear astrophysics communities. One of the promising tool for the production of neutron-rich heavy nuclei is the use of multinucleon transfer reactions at energies close to the Coulomb barrier. Kinematic coincidence measurements are crucial to evaluate the production rate of both binary partners, and especially the relevance of secondary processes in population of exotic heavy neutron-rich nuclei.

Journal of Physics: Conference Series **533** (2014) 012006 doi:10.1088/1742-6596/533/1/012006

2. Sub-barrier transfer measurements

Transfer processes have been investigated in a wide range of energies for the $^{96}Zr+^{40}Ca$ and $^{116}Sn+^{60}Ni$ inverse kinematics reactions. The ^{96}Zr and ^{116}Sn beams were delivered by the Tandem/PIAVE-ALPI accelerator complex of the Laboratori Nazionali di Legnaro at energies ranging from the Coulomb barrier down to 25% below (from 330 MeV to 275 MeV for ^{96}Zr - from 500 MeV to 395 MeV for ^{116}Sn) with an average current of about 2 pnA. Light target-like ions ^{40}Ca and ^{60}Ni were detected at a very forward angle (θ_{lab}=20°) with the large acceptance magnetic spectrometer PRISMA. The use of inverse kinematics allowed to have at the same time target-like ions with enough kinetic energy and forward focused angular distributions resulting in good mass A and atomic number Z resolutions and high detection efficiency, respectively. Excitation functions have been measured for both systems and, making use of semi-classical conditions, transfer probabilities P_{tr} have been extracted for the most intense transfer channels (with P_{tr} defined as the ratio of transfer cross section to the elastic one). Experimental P_{tr} are plotted as a function of the distance of closest approach D in figure 1 for $^{96}Zr+^{40}Ca$ and in figure 2 for $^{116}Sn+^{60}Ni$. The experimental slopes agree well with those predicted by the binding energies of transferred nucleons. In both systems one finds two-neutron transfer probabilities higher than the simple product of probabilities for single neutron transfer, phenomenologically indicating degrees of freedom beyond independent particle transfer.

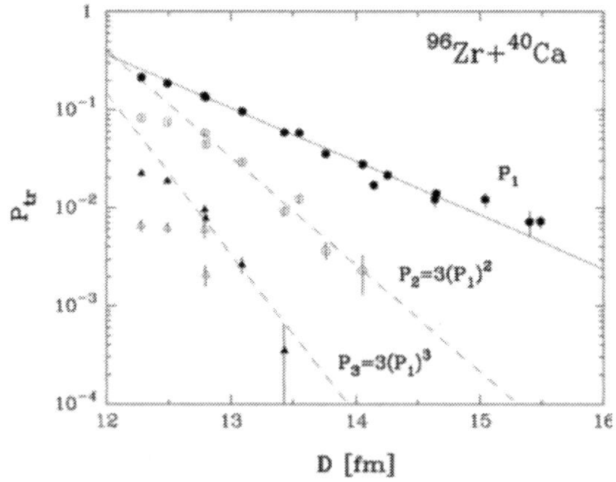

Figure 1. Transfer probabilities of $+1n$ (full circles), $+2n$ (empty circles), $+3n$ (full triangles) and $+4n$ (empty triangles) transfer channels, respectively. The full line is a fit of the data for the $+1n$ pick-up channel according to the expression P_{1n} = exp(13.95 − 1.25D). The dashed lines correspond to the shown equations.

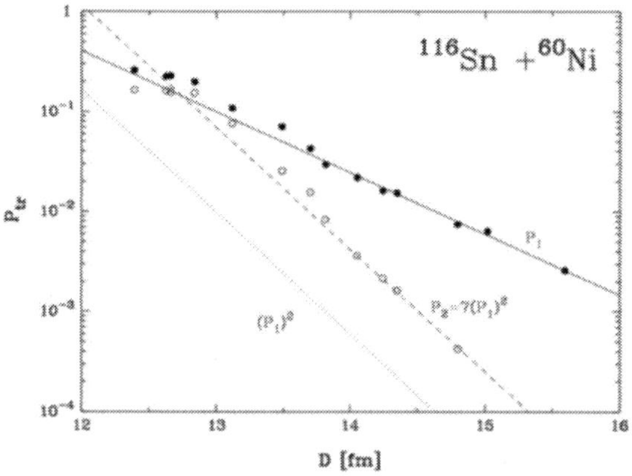

Figure 2. Transfer probabilities of $+1n$ (full circles), $+2n$ (empty circles) transfer channels, respectively. The full line is a fit of the data for the $+1n$ pick-up channel according to the expression P_{1n} = exp(13.86 − 1.22D). For the sake of comparison $(P_1)^2$ (dotted line) is also indicated in the plot.

Closed or near-closed shell nuclei represent a good reference for a quantitative comparison with theoretical calculations. Thus, the experimental data for $^{96}Zr+^{40}Ca$ have been compared with microscopic calculations performed within a semi-classical theory [5], where for the two particle transfer the 0^+ states of both projectile and target have been included. Figure 3 shows the results of these calculations where the full line represents the inclusive transfer probability for $+1n$ transfer channel, the dotted line the ground state to ground state transition for the $+2n$ transfer and the dashed line the transition to the 0^+ excited state at 5.76 MeV in ^{42}Ca. From the comparison with the experimental data it turns out that the contribution to the cross section of this excited 0^+ state in ^{42}Ca is much larger than the contribution of the ground state. Still, the experimental transfer probability is underestimated by a factor ~3. We presently ascribe this enhancement to the fact that the two-nucleon

transfer reaction does not populate only 0^+ states but it is much richer, so that more complicated two-particle correlations have to be taken into account.
At variance with the ^{96}Zr+^{40}Ca system, in the ^{116}Sn+^{60}Ni system (super-fluid nuclei) [6] the ground to ground state Q-values for neutron transfers are close to zero matching their optimum Q-value. Figure 4 shows TKEL spectra for the elastic, the one ($+1n$) and the two ($+2n$) neutron pick-up channels at different bombarding energies. The vertical dashed lines indicate the position of the Q-value for the elastic which is taken as a reference for all the other isotopes. TKEL distributions for neutron transfers are peaked around $Q = 0$, as it is expected. At the same time one sees a tail toward larger TKEL, more marked when more nucleons are transferred. These energy loss components tend to disappear far below the barrier. Microscopic calculations for this system are in progress and it will be interesting to see how calculations including only transfer of $J = 0^+$ pairs to 0^+_{gs} compare with the experimental data, improving our understanding of the origin of the enhancement factors.

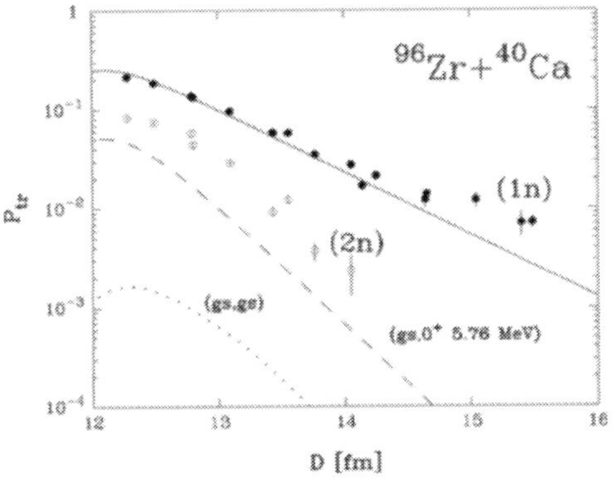

Figure 3. Theoretical transfer probabilities for one- and two-particle transfer (lines) in comparison with the experimental data (points).

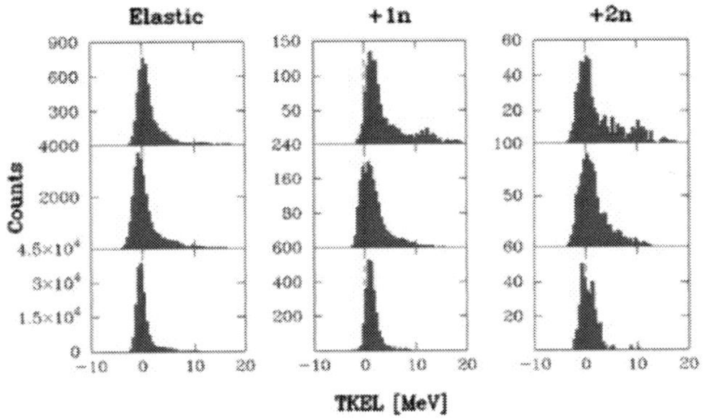

Figure 4. TKEL spectra for the elastic and $+1n$, $+2n$ transfer channels measured for the ^{116}Sn+^{60}Ni reaction at E_{lab}=500 MeV (upper row), E_{lab}=460 MeV (middle row) and E_{lab}=430 MeV (lower row).

3. The second arm of PRISMA

Multinucleon transfer reactions attracted much interest in the last decade due to the fact that they can be used as a powerful tool to populate heavy neutron-rich nuclei in mass regions where production by other methods is very difficult. To predict the optimal conditions for the production of specific nuclei, one has to get quantitative information on the lowering of the final yields due to the onset of secondary processes (such as nucleon evaporation and fission) of which the heavy partner is generally much more affected. An important step forward in this kind of experiments is the detection of both partners participating in the reaction. In order to perform kinematic coincidence measurements in conjunction with the magnetic spectrometer PRISMA a new detection system has been built at LNL.
It is mainly composed of a position sensitive Parallel Plate Avalanche Counter (PPAC) followed by an axial field ionization chamber also known as Bragg chamber (BC). The PPAC consists of two anode planes (10 μm diameter and 1 mm spacing) and a central cathode made of an aluminized mylar foil (1.5 μm thick and 20 μg/cm^2 Al on both sides). The distance cathode-anodes is 2 mm. The detector has an active area of 10x10 cm^2 and provides a timing signal for Time of Flight measurements and X and Y position information (delay-line readout).
The BC has a 32 cm active depth with a Frish grid (FG) to anode distance of 2 cm. The voltages for the 114 guard rings between the cathode and the FG are reduced by means of voltage dividers made of 10 MΩ resistors. As well known, axial field ionization chambers allow to measure the charge distribution produced along the ionization track (Bragg curve) and to perform a Bragg Curve Spectroscopy (BCS) first proposed and tested by Gruhn *et al* [7].

Preliminary bench-tests of the detectors have been performed in laboratory with α particles from a mixed ^{239}Pu-^{241}Am-^{244}Cm source. In these tests the classical method of analog readout of the Bragg chamber was used: a charge sensitive preamplifier followed by two spectroscopy amplifiers in parallel, each with different shaping times τ (4 and 0.5 μs, respectively). Typical two-dimensional scatter-plot Bragg peak (BP) vs E is shown in figure 5. An energy resolution of about 2% and position resolutions of 1 mm in both X (figure 6) and Y directions have been obtained, respectively. In-beam tests of this second arm will be carried out at LNL in the forthcoming months with different heavy ion beams. The use of digital techniques via a Flash ADC and the development of specific algorithms should improve the identification of the heavy binary partners providing the possibility of extracting additional parameters (such as the range R and the energy loss ΔE) from the Bragg curve.

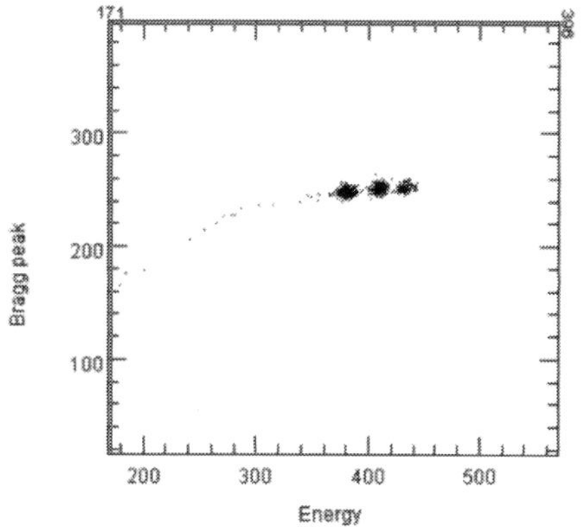

Figure 5. Two-dimensional scatter-plot BP (arb. units) vs E (arb. units) obtained with a mixed α source.

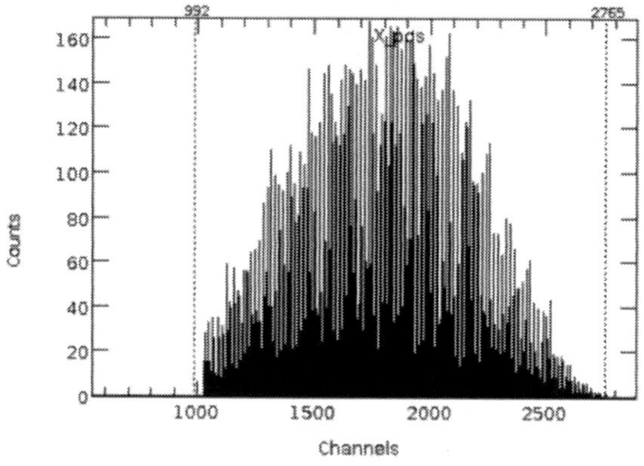

Figure 6. X position spectrum measured with the PPAC mounted in front of the axial ionization chamber.

4. Summary

We measured excitation functions for the ^{96}Zr+^{40}Ca and ^{116}Sn+^{60}Ni inverse kinematics reactions. Light target-like ions have been detected with the magnetic spectrometer PRISMA with a good mass and atomic number resolution in the whole energy range. Enhancement of the two-neutron transfer has been evidenced for both systems. Microscopic calculations for the ^{116}Sn+^{60}Ni system are in progress. The comparison between data and theory for both cases, namely near closed shell and super-fluid nuclei, will significantly improve our understanding of nucleon-nucleon correlations in the transfer process. A second arm of the PRISMA spectrometer has been built and tested in laboratory with α particles obtaining good energy and position resolutions. In-beam tests will allow to evaluate its Z resolving power and to improve the identification of the heavy binary partners.

Acknowledgements

The research leading to these results has received funding from the European Union Seventh Framework Programme FP7/2007- 2013 under Grant Agreement No. 262010 - ENSAR.

References

[1] Corradi L *et al* 2009 *J. of Phys.* G **36** 113101
[2] Stefanini A M *et al* 2002 *Nucl. Phys.* A **701** 217c
[3] Montagnoli G *et al* 2005 *Nucl. Instr. and Meth.* A **547** 455
[4] Beghini S *et al* 2005 *Nucl. Instr. and Meth.* A **551** 364
[5] Corradi L *et al* 2011 *Phys. Rev.* C **84** 034603
[6] Montanari D *et al* 2013 *INPC2013* (Firenze, Italy, 2-7 June 2013), ed. Signorini C et al in press
[7] Gruhn C R *et al* 1982 *Nucl. Instr. and Meth.* **196** 33

XX International School on Nuclear Physics, Neutron Physics and Applications (Varna2013) IOP Publishing
Journal of Physics: Conference Series **533** (2014) 012007 doi:10.1088/1742-6596/533/1/012007

Determination of the $B(E3, 0^+ \rightarrow 3^-)$-excitation strength in octupole-correlated nuclei near $A \approx 224$ by the means of Coulomb excitation at REX-ISOLDE

M. Scheck[1,2,3], L. P. Gaffney[3,4], P. A. Butler[3], A. B. Hayes[5], F. Wenander[6], M. Albers[7], B. Bastin[8], C. Bauer[9], A. Blazhev[7], S. Bönig[9], N. Bree[4], J. Cederkäll[10], T. Chupp[11], D. Cline[5], T. E. Cocolios[6], T. Davinson[12], H. De Witte[4], J. Diriken[4], T. Grahn[13,14], E. T. Gregor[1,2], A. Herzan[13], M. Huyse[4], D. G. Jenkins[15], D. T. Joss[3], N. Kesteloot[4], J. Konki[13], M. Kowalczyk[16], Th. Kröll[9], E. Kwan[17], R. Lutter[18,19], K. Moschner[7], P. Napiorkowski[16], J. Pakarinen[6,13,14], M. Pfeiffer[7], D. Radeck[7], P. Reiter[7], K. Reynders[4], S. V. Rigby[3], L. M. Robledo[20], M. Rudigier[7], S. Sambi[4], M. Seidlitz[7], B. Siebeck[7], T. Stora[6], P. Thoele[7], P. Van Duppen[4], M. J. Vermeulen[15], M. von Schmid[9], D. Voulot[6], N. Warr[7], K. Wimmer[19], K. Wrzosek-Lipska[4,16], C. Y. Wu[17], and M. Zielińska[16,21]

[1]School of Engineering, University of the West of Scotland, Paisley PA1 2BE, UK
[2]SUPA, Scottish Universities Physics Alliance, Glasgow G12 8QQ, UK
[3]Oliver Lodge Laboratory, University of Liverpool, Liverpool L69 7ZE, UK
[4]Instituut voor Kern- en Stralingsfysica, KU Leuven, Leuven B-3001, Belgium
[5]Dept. of Physics & Astronomy, University of Rochester, Rochester, NY 14627-0171, USA
[6]ISOLDE, CERN, Geneva CH-1211, Switzerland
[7]Institut fr Kernphysik, Universität zu Köln, Köln D-50937, Germany
[8]GANIL, Caen, BP 55027, F-14076, France
[9]Institut für Kernphysik, Technische Universität Darmstadt, Darmstadt D-64289, Germany
[10]Department of Nuclear Physics, Lund University, Lund PO Box 118, 221 00, Sweden
[11]Department of Physics, University of Michigan, Ann Arbor, Michigan 48104, USA
[12]School of Physics & Astronomy, University of Edinburgh, Edinburgh EH9 3JZ, UK
[13]Department of Physics, University of Jyväskylä, Jyväskylä FI-40014, Finland
[14]Helsinki Institute of Physics, PO Box 64, FI-00014 Helsinki, Finland
[15]Department of Physics, University of York, York YO10 5DD, UK
[16]Heavy Ion Laboratory, University of Warsaw, Warsaw 02-093, Poland
[17]Physics Division, Lawrence Livermore National Laboratory, Livermore, California 94551, USA
[18]Maier-Leibnitz-Laboratorium, Ludwig-Maximilians-Universität, Garching D-85748, Germany
[19]Physik Department E12, Technische Universität München, Garching D-85748, Germany
[20]Departamento de Fisica Teorica, Universidad Autónoma de Madrid, Madrid 28049, Spain
[21]DSM/IRFU/SPhN, CEA Saclay, Gif-sur-Yvette F-91191, France

E-mail: marcus.scheck@uws.ac.uk

Journal of Physics: Conference Series **533** (2014) 012007 doi:10.1088/1742-6596/533/1/012007

Abstract. The IS475 collaboration conducted Coulomb-excitation experiments with post-accelerated radioactive ^{220}Rn and ^{224}Ra beams at the REX-ISOLDE facility. The beam particles (E_{beam} ≈2.83 MeV/u) were Coulomb excited using ^{60}Ni, ^{114}Cd, and ^{120}Sn scattering targets. De-excitation γ-rays were detected employing the Miniball array and scattered particles were detected in a silicon detector. Exploiting the Coulomb-excitation code GOSIA for each nucleus several matrix elements could be obtained from the measured γ-ray yields. The extracted $\langle 3^- || \hat{E}3 || 0^+ \rangle$ matrix element allows for the conclusion that, while ^{220}Rn represents an octupole vibrational system, ^{224}Ra has already substantial octupole correlations in its ground state. An observation that has implications for the search of CP-violating Schiff moments in the atomic systems of the adjacent odd-mass nuclei.

1. Introduction

There is substantial experimental evidence for nuclei within specific regions of the nuclear landscape to have strong octupole correlations. Possibly, some nuclei possess these correlations even in their ground state. In a geometrical picture these correlations lead to a pear-shape distortion of the nuclear surface. The experimental evidence consists of low-lying 3^- states (see Fig. 1) in near-spherical and 1^- band heads of the K=0 octupole band in deformed nuclei, odd-even staggering of positive and negative parity yrast bands at comparably low spins (e.g., see Fig. 2), parity doublets in the neighboring odd-mass systems and, most importantly, enhanced $B(E3)$ strength for stretched E3 transitions [1].

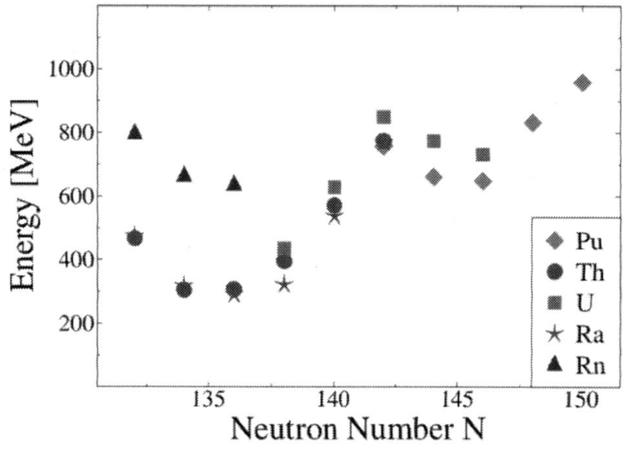

Figure 1. Excitation energy of the lowest-lying 3^- state in nuclei situated in the mass region north-east of ^{208}Pb.

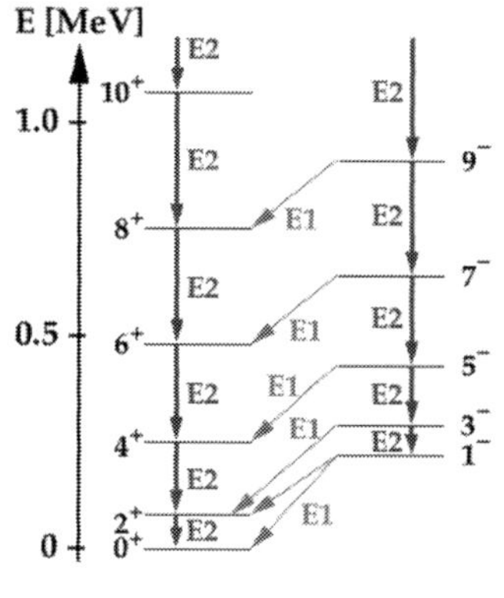

Figure 2. Low-energy level scheme of ^{224}Ra as relevant to sub-barrier Coulomb excitation. Data are taken from Ref. [2].

Strong octupole correlations occur in nuclei for which the Fermi level for protons as well as for neutrons is situated between the unique-parity subshell and the subshell having an orbital (l) as well as total (j) angular momentum difference of $\Delta l = 3\hbar$ and $\Delta j = 3\hbar$. These particular subshell combinations are realized near the neutron (N) and proton (Z) numbers $N, Z = 36, 56, 88, 136$. Supported by theoretical investigations using various approaches (e.g. see Ref. [3] and references therein), especially ^{224}Ra is a promising candidate for a nucleus with strong octupole correlations. Prior to this work a Coulomb excitation measurement on ^{226}Ra already exhibited strong $B(E3, J^+ \rightarrow (J+3)^-)$ transition probabilities in this nucleus [4]. Strong octupole correlations and the reflection asymmetric pear-shape associated with it

Journal of Physics: Conference Series **533** (2014) 012007 doi:10.1088/1742-6596/533/1/012007

lead to an asymmetric charge distribution in the nucleus. The latter results in an enhancement of a possible CP-violating nuclear Schiff moment [5] in the neighboring odd-mass nuclei. While it is unlikely to be measured in a nuclear system, it is predicted to be amplified in the atomic system and can at least be used for measuring upper limits for atomic electric-dipole moments and, therefore, constraining models which propose physics beyond the standard model.

2. Experiments

The experiments were performed at ISOLDE using the Radioactive ion beam EXperiment (REX) accelerator [6]. The REX-ISOLDE facility has the world-wide unique capability to produce and post-accelerate the radon and radium nuclei of interest with a sufficient intensity to perform sub-barrier Coulomb excitation measurements. Since the latter are sensitive to E2- and E3-matrix elements they represent a tool to measure E3-transition strength which otherwise cannot be observed due the presence of fast E1 transitions predominantly depopulating the level of interest.

In order to produce the nuclei of interest in a spallation reaction a primary UC_x target was irradiated with 1.4 GeV protons. The radioactive isotopes were mass separated, charge bred ($A/q \approx 4.1$), post-accelerated to 2.82 and 2.83 MeV/u, respectively, using REX and delivered to the secondary Coulomb-excitation target inside the Miniball setup. The Miniball setup consists of 24 high-purity germanium detectors grouped into eight triple clusters for γ-ray detection and a double-sided silicon strip detector for particle detection and identification. Particle-γ coincidences allow for a Doppler correction and background suppression. Exemplary spectra of ^{220}Rn Coulomb excited on ^{60}Ni and ^{120}Sn secondary targets are shown in Fig. 3. Furthermore, ^{114}Cd was used as secondary target.

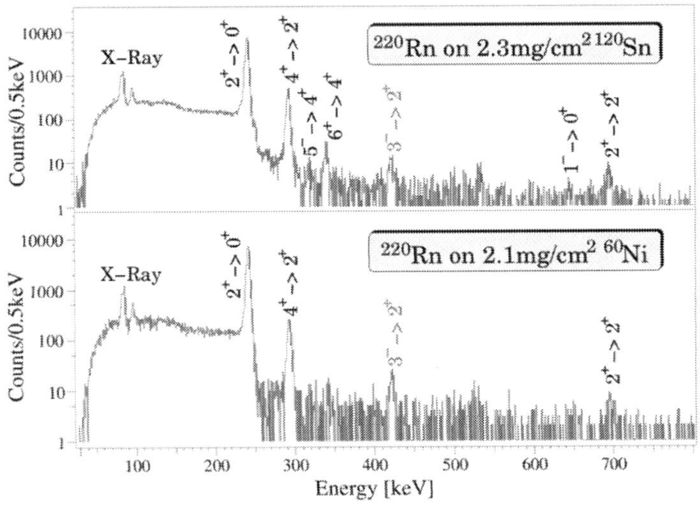

Figure 3. Particle-γ coincidence gated and Doppler-corrected γ-ray spectra of ^{220}Rn Coulomb excited using secondary ^{60}Ni (bottom) and ^{120}Sn (top) targets. Peaks are labeled corresponding to the transition they represent.

3. Results and Conclusions

The Coulomb-excitation code GOSIA was used to extract matrix elements from the observed γ-ray yields. This analysis resulted amongst others [8] in matrix elements corresponding to a $B(E3, 3^- \to 0^+)$ value of 33(4) W.u. for ^{220}Rn and 42(3) W.u. for ^{224}Ra. The corresponding values are shown in Fig. 4 as an inverse sum rule of the squared matrix element divided by the excitation energy of the corresponding 3^- level. The inverse sum rule is applied as it combines the two most important experimental observables and highlights the special role of the two radium isotopes (^{224}Ra and ^{226}Ra [4]) for which the B(E3) strength was measured so far. Here, it is worthwhile to point out, that the experimental trend of the two radium isotopes favours mean-field based models and opposes cluster models [8].

Assuming axial symmetry, the nuclear shape can be parametrized in terms of deformation parameters, β_λ,

$$R(\Theta) = c(\beta_\lambda)R_0 \left[1 + \sum_{\lambda=2}^{\infty} \sqrt{\frac{2\lambda+1}{4\pi}} \beta_\lambda P_{\lambda 0}(\cos\theta)\right] \quad (1)$$

and the corresponding Legendre Polynomials, $P_{\lambda 0}(\cos\theta)$. The deformation parameters can be calculated using the measured matrix elements [4]. The shape of ^{220}Rn and ^{224}Ra as resulting from the current experiment are illustrated in Fig. 5. Obviously, ^{224}Ra exhibits nicely the pear-shape associated with an octupole correlated nucleus.

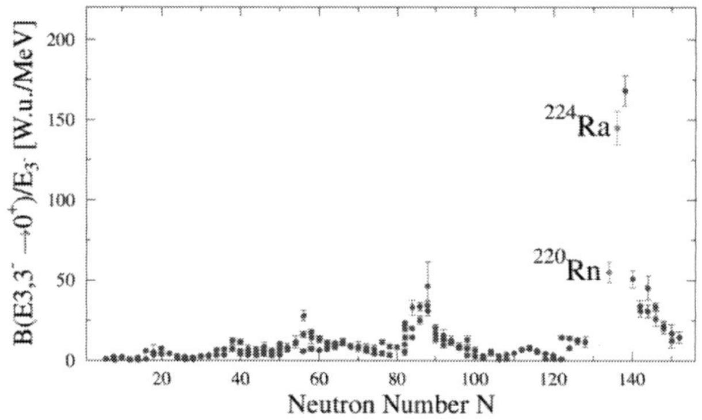

Figure 4. Inverse sum rule of the $B(E3, 3^- \rightarrow 0^+)$ strength divided by the energy of the 3_1^- level plotted over the neutron number, N.

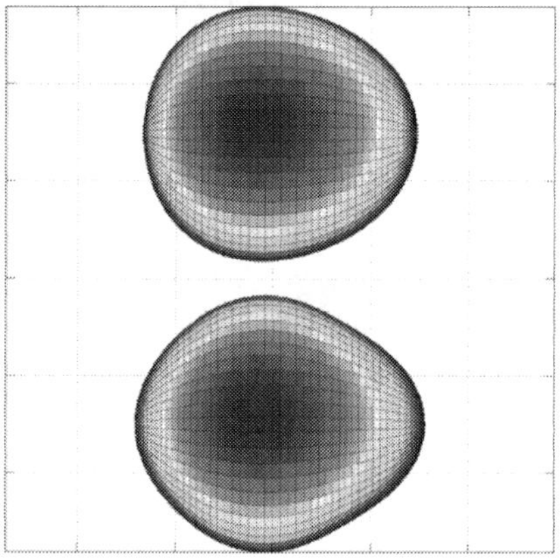

Figure 5. Surface contour plot of ^{220}Rn (top) and ^{224}Ra (bottom).

In conclusion the presented Coulomb-excitation experiment provided strong evidence that ^{220}Rn is rather an octupole vibrational system, while ^{224}Ra possesses an octupole deformed ground state. Therefore, at present, atoms of the odd-mass radium isotopes are favourable cases in the search for CP-violating physics. The collaboration has already further approved experiments at HIE-ISOLDE which aim to investigate even-even and odd-mass isotopes near $A = 224$ [9] and also the mass region near ^{144}Ba [10].

References

[1] Butler P A and Nazarewicz W 1996 *Rev. Mod. Phys.* **68** 349.
[2] Cocks J F *et al.* 1997 *Phys. Rev. Lett.* **78** 2920
[3] Robledo L M, Baldo M, Schuck P and Viñas X 2010 *Phys. Rev. C* **81** 034315
[4] Wollersheim H J *et al.* 1993 *Nucl. Phys.* **A556** 261
[5] Engel J, Ramsey-Musolf M J and van Kolck U 2013
[6] Voulot D *et al.* 2008 *Nucl. Instrum. Methods B* **266** 4103
[7] Van de Walle J *et al.* 2013 *Euro. Phys. Jour. A* **49** 40
[8] Gaffney L P *et al.* 2013 *Nature* **497** 199
[9] Butler P A, Joss D T and Scheck M *et al.* 2012 http://cds.cern.ch/record/1482416/files/INTC-P-347.pdf
[10] Scheck M and Joss D T et al. 2012 http://cds.cern.ch/record/1387489/files/INTC-P-305.pdf

XX International School on Nuclear Physics, Neutron Physics and Applications (Varna2013) IOP Publishing
Journal of Physics: Conference Series **533** (2014) 012008 doi:10.1088/1742-6596/533/1/012008

News on the Scissors Mode

N. Pietralla[1], J. Beller[1], T. Beck[1], V. Derya[2], B. Löher[3,4], C. Romig[1], D. Savran[3,4], M. Scheck[1,5], W. Tornow[6], and M. Zweidinger[1]

[1] Institut für Kernphysik, TU Darmstadt, Schlossgartenstraße 9, D-64289 Darmstadt, Germany
[2] Institut für Kernphysik, Universität zu Köln, Zülpicher Straße 77, D-50937 Köln, Germany
[3] ExtreMe Matter Institute EMMI and Research Division, GSI Helmholtzzentrum für Schwerionenforschung, Planckstr. 1, D-64291 Darmstadt, Germany
[4] Frankfurt Institute for Advanced Studies FIAS, Ruth-Moufang-Str. 1, D-60438 Frankfurt am Main, Germany
[5] School of Engineering, University of the West of Scotland, Paisley PA1 2BE, UK
[6] Department of Physics, Duke University, Durham, NC 27708-0308, USA

E-mail: `pietralla@ikp.tu-darmstadt.de`

Abstract. We report on our recent nuclear resonance fluorescence experiments on ^{152,154,156}Gd. Decay branches of the scissors mode to intrinsic excitations are observed. They are interpreted as a new signature for a spherical-to-deformed nuclear shape phase transition.

1. Introduction

Atomic nuclei provide examples of strongly interacting two-component fermionic quantum systems. The nuclear proton-neutron degree of freedom and its fundamental impact on the formation of nuclear structures, e.g. [1] represent one of the central aspects of modern nuclear structure physics. The structure of collective nuclear excitations is determined by the nuclear shell structure that itself evolves as a function of proton and neutron numbers, e.g., [2]. Nuclear excited states that partly involve an out-phase coupling of valence protons and valence neutrons such as the class of mixed-symmetry states of vibrational nuclei [3, 4] or the scissors mode of deformed nuclei [5, 6] are particular sensitive to the details of the effective proton-neutron interaction in the nuclear valence shell.

The evolution of nuclear shells and the residual interaction between valence protons and neutrons dominate the formation of nuclear deformation and collectivity. Depending on particle number a rather sudden onset of nuclear deformation can occur at so-called shape phase transitional points in the nuclear landscape. Near the shape phase transitions the fluctuations of nuclear deformation are large and shape excitations occur at rather low excitation energies. Examples of such sudden shape changes as a function of nucleon number are the stable $N = 90$ isotones [7]. At these points the filling of nuclear orbitals and their relative single-particle energies due to the residual proton-neutron interactions are such that collective nuclear quantum states can form within the valence shell with wave functions that include many basis states with proton excitations and neutron excitations that coherently couple in phase.

Recently, experimental evidence has been provided that this in-phase coupling of proton and neutron valence-shell excitations simultaneously leads to the occurrence of nuclear valence-shell excitations with partial out-of-phase coupling of valence protons and neutrons [1]. This represents a closely related aspect of the nuclear proton-neutron degree of freedom. Indeed, nuclear excitations may be classified according to the proton-neutron symmetry of their wave functions, either in terms of isospin in the framework of models that use fermionic nucleons as the fundamental degree of freedom or in terms of F-spin [8] in approximate bosonic nuclear models [9]. While, except for $N = Z$ nuclei, excited nuclear states at low energy (*e.g.*, at considerably lower energy as compared to the particle separation threshold) all have isospin quantum number $T = T_< = |N - Z|/2$, there occur states at rather low excitation energies with F-spin quantum numbers that differ from the F-spin of the ground state $F_{\mathrm{gs}} = F_{\mathrm{max}} = (N_\pi + N_\nu)/2$. Here $N_{\pi(\nu)}$ is half the number of valence protons (neutrons) or corresponding valence holes. Nuclear excitations with F-spin quantum number $F < F_{\mathrm{max}}$ have previously been called [10] mixed-symmetry states (MSSs) within the framework of the interacting boson model (IBM). Prominent examples of MSSs are the $J^\pi = 1^+$ scissors mode [5, 6, 9, 11, 12, 13, 14] in deformed nuclei or the mixed-symmetry $2^+_{1,\mathrm{ms}}$ one-phonon vibration for spherical nuclei [3, 4, 10]. Information on the proton-neutron symmetry of the low-lying nuclear states, including the ground state, its collective excitations, and MSSs, in nuclei that are as close as possible to the shape phase transitional point is desirable for a detailed understanding of the role of the nuclear proton-neutron degree of freedom in the formation of collective nuclear structure. Such information is indispensable when nuclei are used as quantum laboratories for fundamental interactions, such as searches for neutrinoless double-beta ($0\nu\beta\beta$) decay or searches for signals from scattering reactions between nuclei and Weakly Interacting Massive Particles (WIMPs) that are expected as candidates for the Dark Matter in the universe.

We have recently demonstrated [15] that information on the decay pattern of the scissors mode can provide sensitive constrains for nuclear models used for estimating the nuclear matrix elements (NMEs) for $0\nu\beta\beta$-decay rates. For instance, the IBM is used to calculate nuclear matrix elements for neutrinoless double β ($0\nu\beta\beta$) decays [16]. Especially for the double-β emitters ^{154}Sm and ^{150}Nd, sensitive information on the proton-neutron symmetry of the low-energy eigenstates is needed for a reliable calculation of $0\nu\beta\beta$ matrix elements within the IBM. Details on the decay pattern of the scissors mode of ^{154}Gd had a significant impact on the estimates for the branching ratio of the $0\nu\beta\beta$ decay of ^{154}Sm to ^{154}Gd [15].

2. Experiments

Photon-scattering experiments were performed exploiting the Darmstadt High Intensity Photon Setup (DHIPS) [17] at the Superconducting DArmstadt electron LINear ACcelerator (S-DALINAC) and the High Intensity $\vec{\gamma}$-ray Source (HI$\vec{\gamma}$S) [18] at the Triangle Universities Nuclear Laboratories (TUNL).

At DHIPS bremsstrahlung-photon beams are produced by stopping the electron beam in a radiator. The resulting bremsstrahlung-photon beams were scattered off the targets of interest. The scattered photons were detected by two or three large-volume high-purity germanium detectors (HPGe) positioned at polar angles 90° and 130° with respect to the incident γ-ray beam. The resonant photon-scattering cross sections $I_{s,f} = g\pi^2(\hbar c/E_\gamma)^2\Gamma_0\Gamma_f/\Gamma$ were measured relative to the calibration lines stemming from ^{27}Al [19]. Here, $g = (2J+1)/(2J_0+1)$ is a statistical factor. The total level width $\Gamma = \hbar/\tau$ and the partial decay width to the ground (final) state Γ_0 (Γ_f) are only accessible if all decay intensity ratios Γ_f/Γ are determined. The partial decay widths $\Gamma_{f,\pi\lambda}$ are propertional to the reduced transition strengths $B(\pi\lambda; J^\pi \to J^\pi_f)$.

For the investigation of ^{152}Gd a target consisting of 2,525 mg Gd_2O_3 powder enriched to 32.5% in the isotope ^{152}Gd and 440 mg Al was used. Besides the transitions stemming from other stable Gd isotopes, the γ decays of dipole excited states of ^{152}Gd to the ground state, to the first excited 2^+ state, and for the state at 3.14 MeV also to the first excited 0^+ state could be observed. For the investigation of ^{154}Gd details about setup and data analysis can be found in Ref. [15].

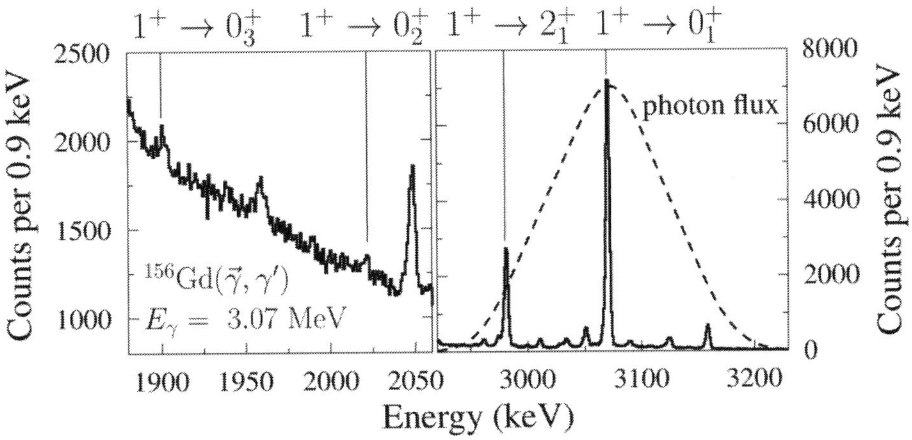

Figure 1. ^{156}Gd($\vec{\gamma},\gamma'$) spectrum taken at $\theta = 135°$, $\phi = 0°$ at HI$\vec{\gamma}$S. Deexciting transitions from the $J^\pi = 1^+$ scissors mode state at 3.07 MeV to the 0_1^+, 2_1^+, 0_2^+, and 0_3^+ states could be observed. The dashed curve shows schematically the photon-flux distribution.

HI$\vec{\gamma}$S provides quasi-monoenergetic linearly-polarized γ-ray beams with tunable energies well suited for NRF measurements [20]. The γ rays are produced in Compton-backscattering of internal FEL-photons on electrons. After collimation the spectral distribution is about 3% of the mean γ-ray energy. The γ-beam was scattered off a Gd_2O_3 target enriched in the isotope ^{156}Gd to 93.79% with a total weight of 11.62 g. The quasi-monochromatic beam was used to excite $J^\pi = 1^+$ states around 3.07 MeV in ^{156}Gd. Their γ decays were studied with four HPGe detectors under polar angles of 135° and azimuthal angles of 0°, 90°, 180°, and 270°. Figure 1 shows the ($\vec{\gamma},\gamma'$) spectrum. The γ decay lines of the 1^+ scissors mode at 3.07 MeV to the ground state, to the first excited 2^+ state, to the 0_2^+ state, and to the 0_3^+ state are clearly visible at 3069, 2980, 2020, and 1900 keV. From the peak areas the intensity ratios Γ_f/Γ were deduced.

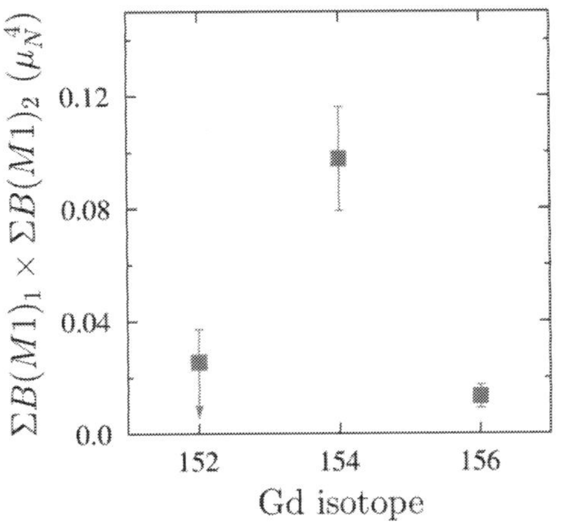

Figure 2. Product of the $M1$ transition strengths of scissors mode states to the 0_2^+ state and to the 0_1^+ ground state for ^{152,154,156}Gd. At ^{154}Gd the $M1$ product shows a distinct maximum.

3. Results

Total $B(M1)$ excitation strengths of $J = 1$ states in the energy region of the scissors mode was measured for ^{152,154}Gd. These yield $\sum B(M1) \uparrow = 0.45(9)\ \mu_N^2$ for 152Gd and $3.12(36)\ \mu_N^2$ for 154Gd. $M1$ branching ratios of those states to the first excited 0^+ state were observed in all three isotopes: ^{152,154,156}Gd. The total $M1$ transition strengths for these $1^+_{sc} \rightarrow 0^+_2$ transitions were deduced to $\sum B(M1; 1^+_{sc} \rightarrow 0^+_2) = 0.17(7)$, $0.094(14)$, and $0.013(4)\ \mu_N^2$ for ^{152,154,156}Gd, respectively.

The $M1$ product of the transitions strengths to the 0^+_2 state and to the ground state

$$B(M1)_1 \times B(M1)_2 \equiv B(M1; 1^+_{sc} \rightarrow 0^+_1) \times B(M1; 1^+_{sc} \rightarrow 0^+_2) \qquad (1)$$

is given in Fig. 2. The product exhibits a pronounced maximum at the $N = 90$ isotope ^{154}Gd and, thus, can be considered as a new signature for a shape phase transition. Work along these lines are in progress.

4. Conclusion

With the combination of different photon scattering experiments a novel decay branch (and its branching ratio) of $J^\pi = 1^+$ scissors mode states of ^{152,154,156}Gd to their first excited 0^+ state were observed. In addition, for the first time absolute excitation strengths of the scissors mode of ^{152,154}Gd were measured in photon scattering experiments. The $B(M1)_1 \times B(M1_2$ product exhibits a distinct maximum at the $N = 90$ shape phase transitional point. This quantity may represent a new signature for a spherical-to-deformed nuclear shape phase transition.

5. Acknowledgments

The authors thank the S-DALINAC accelerator crew and the TUNL staff for delivering excellent beams. This work was supported by the Deutsche Forschungsgemeinschaft under grant No. SFB 634 and No. ZI 510/4-2 and partly by the Alliance Programme of the Helmholtz Association (HA216/EMMI), BMBF-Verbundforschungsprojekt No. 06DA7047I, Fondecyt No. 1120462, and U.S. DOE Grant No. DE-FG-02-91ER-40608. N.P. thanks the Bulgarian Science Foundation and the DAAD for support.

[1] C. Walz *et al.*, Phys. Rev. Lett. **106**, 062501 (2011).
[2] K. Hebeler and A. Schwenk. Phys. Rev. C **82**, 014314 (2010).
[3] F. Iachello, Phys. Rev. Lett. **53**, 15 (1984).
[4] N. Pietralla, P. von Brentano, and A.F. Lisetskiy, Prog. Part. Nucl. Phys. 60, 225 (2008).
[5] D. Bohle, A. Richter *et al.*, Phys. Lett. **B137**, 27 (1984).
[6] N. Pietralla *et al.*, Phys. Rev. C **58**, 184 (1998).
[7] E. Kirchuk, P. Federman, and S. Pittel, Phys. Rev. C **47**, 2 (1993).
[8] P. Van Isacker *et al.*, Annals of Physics, 171, 253 (1986).
[9] F. Iachello and A. Arima, *"The Interacting Boson Model"* (Cambridge University Press, Cambridge, 1987).
[10] A. Arima *et al.*, Phys. Lett. **66B**, 205 (1977).
[11] N. Lo Iudice and F. Palumbo, Phys. Rev. Lett. **41**, 1532 (1978); Nucl. Phys. **A326**, 193 (1979).
[12] A. Richter, *Proc. of the Int. Conf. on Nuclear Physics*, Florence 1983, *ed.* P. Blasi and R. A. Ricci, (Tipografica Compositori Bologna, Vol. 2), p. 189.
[13] M.A. Caprio and F. Iachello, Annals of Physics 318, 454 (2005).
[14] K. Heyde, P. von Neumann-Cosel, and A. Richter, Rev. Mod. Phys. 82 (2010).
[15] J. Beller *et al.*, Phys. Rev. Lett. **111**, 172501 (2013).
[16] J. Barea, J. Kotila, and F. Iachello, Phys. Rev. Lett. **109**, 042501 (2012).
[17] K. Sonnabend *et al.*, Nucl. Instrum. Methods Phys. Res. A 640, 6 (2011).
[18] H. R. Weller *et al.*, Prog. Part. Nucl. Phys. **62**, 257 (2009).
[19] N. Pietralla *et al.*, Phys. Rev. C **51**, 1021 (1995).
[20] N. Pietralla *et al.*, Phys. Rev. Lett. **88**, 012502 (2002).

Development of a time resolution and position sensitive Multi-Gap Multi-Strip RPC for high counting rate experiments

Mariana Petriş and Mihai Petrovici

Horia Hulubei National Institute of Physics and Nuclear Engineering, Bucharest-Magurele, P.O.Box MG-6, RO-077125, Romania

E-mail: mpetris@nipne.ro

Abstract. A full differential strip readout multi-gap RPC was developed to meet high counting rate and high multiplicity requirements of a high collision rate experiments as Compressed Baryonic Matter at the future FAIR facility. The Multi-Gap, Multi-Strip Resistive Plate Chamber (MGMSRPC) is a completely symmetric two stack structure with time resolution around 50 ps and a detection efficiency better than 95%. A very good efficiency and time resolution at counting rates exceeding the CBM-TOF requirements is obtained using low resistivity ($\sim 10^{10}$ Ωcm) glass electrodes.

1. Introduction

Next generation experiments such as the Compressed Baryonic Matter (CBM) [1] at the future Facility for Antiproton and Ion Research (FAIR) [2] in Darmstadt will be confronted with the selection of rare probes in high multiplicity environment (up to 1000 tracks/event) at collision rates up to 10^7 events/s. Hadron identification in such a limiting environment is a real challenge. In the current design of the CBM experimental set-up the hadron identification is based on Time-of-Flight (ToF) measurements provided by a wall of Multi-gap Resistive Plate Chambers (MRPC). The low polar-angle region of a fixed target experiment like CBM will be exposed to high counting rates up to 25 kHz/cm^2 and high multiplicities. Therefore, a high counting rate and high granularity detector is required for this particular region.

2. Multi-Gap, Multi-Strip Resistive Plate Chamber

Fall of 1999 we proposed a new multi-gap resistive plate chamber architecture with readout on a multi-strip electrode - MGMSRPC (Multi-Gap, Multi-Strip Resistive Plate Chamber) [3]. The prototype was a double stack, multi-gap symmetric structure with 2 x 2 gas gaps of 300 μm thickness each. The resistive electrodes are made of commercial float glass with a resistivity of the order of $\sim 10^{13}$ Ωcm. The counter was based on a single ended standard readout with the signals picked up only from the central anode. The readout on 16 strip lines (2.54 mm strip pitch) placed on both sides of the anode allows in addition to a very good time resolution the access to position information on both orthogonal directions that define the readout electrode plane: along and across the strips. The time of flight is obtained as the mean of the time information measured at the both ends of a strip. In this way, the dependence of the time information on the position along the strip is canceled. The difference of the times measured at the two ends gives the position information along the strip. The transverse position is reconstructed applying the centre of gravity method to the signals induced on adjacent strips around the particle track. The functionality of the two designed and constructed prototype versions (300 mm and 900 mm strip length), was first tested with a ^{60}Co radioactive source in the detector laboratory. A typical ^{60}Co time resolution of about 90 ps was obtained [3]. The in-beam test performed with minimum ionizing particles at SIS18 accelerator of GSI Darmstadt demonstrated a very good performance: a time resolution of 70 ps for an efficiency larger than 95% [4]. Based on this new architecture, the upgrading of the FOPI ToF barrel was done, extending the kaon identification from 0.5 GeV/c to 1.3 GeV/c for beam energies up to 2 AGeV and 1 m flight path [5]. The successful commissioning of FOPI ToF barrel showed that large area detectors based on such an architecture is feasible.

3. High Counting Rate, High Multiplicity MGMSRPC

Our recent R&D activity has been focused on development of a MGMSRPC prototype for high counting rate environments, as it is anticipated to be in the most inner zone of the CBM-ToF wall. The classical timing multi-gap RPCs using float glass as resistive electrodes keep their performance up to about 1 kHz/cm^2 counting rate [6]. To extend the rate capability, the drop of the potential of the glass electrode should be restored faster. The solutions are the use of electrodes with lower resistivity, thinner resistive electrodes or build counters with smaller gap size in order to reduce the avalanche charge per gap.

Our first chambers developed for high counting rate experiments were based on resistive electrodes made from a special glass, with a resistivity of the order of 10^{10} Ωcm, known as Pestov glass [7]. The first prototype built using 300 mm length x 46 mm width glass of 2 mm thickness [8] had the same inner structure as the MGMSRPC prototype described in Section 2. A second prototype was built with a strip differential readout [9]: the signals are picked up from both the central anode and the outermost cathodes. This new architecture was developed being aware that the pick-up of external noise and the effect of internal cross talk between transmission lines corresponding to different strips are reduced once the counter has a differential structure and is correspondingly operated. All prototypes developed further have a differential read-out, the signals being processed by an amplifier/discriminator based NINO ASIC chip [10]. The in-beam test of both prototypes was performed with 28-30 MeV electron beam at ELBE facility in Dresden. The obtained time resolution for single-ended readout MGMSRPC was around 60 ps in conditions of a 1 kHz/cm^2 uniform illumination of the detector [8]. The results obtained in the in beam test for the differential architecture showed that the time resolution remains at about 80 ps up to 16 kHz/cm^2 (the maximum counting rate reached during these tests) [9].

The region of small polar angles of the CBM-ToF subsystem has to cope, besides of high counting rate, also with a high multiplicity which requires a high granularity detector. For this particular region of the ToF wall a new RPC prototype with a readout electrode with much shorter strips, of 46 mm length instead of 300 mm as the previous prototypes had, was designed and constructed. The counter has a 2 x 5 gap structure with a gap size reduced to 140 μm. A 40 ps time resolution was measured in the in-beam test at the SIS18 accelerator of GSI Darmsadt and reported in reference [11].

The prototypes developed so far have been supplied only with negative potential applied on cathodes. The next step was to build a completely symmetric structure (see Figure 1) with high voltage electrodes for both negative and positive polarities in order to reduce at half the potential applied on each high voltage electrode [12]. The strip geometry was maintained the same as in the case of previous high granularity prototype, (i.e. 2.54 mm pitch, 46 mm length). Three such prototypes were built: the first two (RPC1 and RPC2) are based on float glass resistive electrodes of 0.55 mm thickness and have 2x7 140 μm gas gaps. The difference between

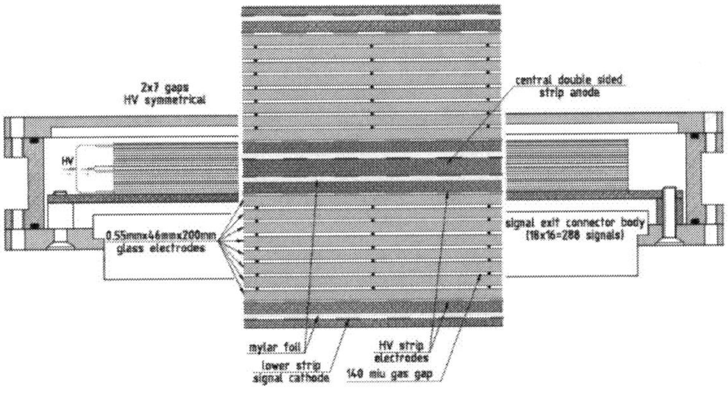

Figure 1. Schematic drawing of the full differential MGMSRPC structure.

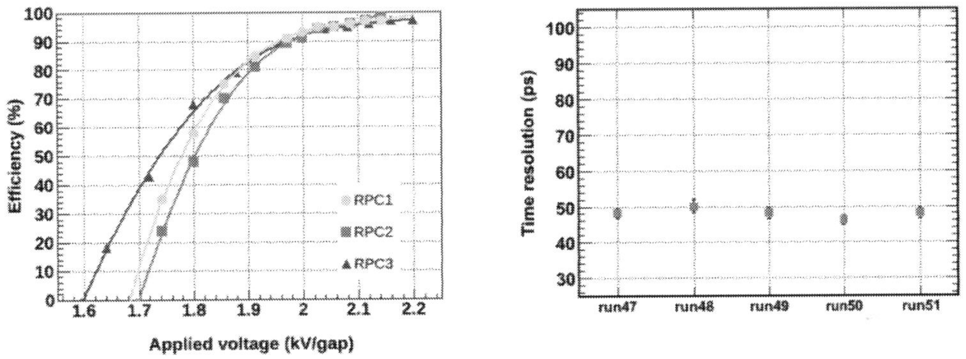

Figure 2. Efficiency for the three detectors as a function of applied voltage - left side. Time resolution using the time difference between RPC3 and RPC2 under the hypothesis that both counters have the same time resolution - right side.

them is that RPC1 has the strip structured high voltage electrode in contact with a resistive layer deposited on the last float glass electrode while in case of RPC2 the strip structured high voltage electrode is in direct contact with the last float glass electrode. The third one, RPC3 is based on low resistivity glass electrodes ($2.5\text{x}10^{10}$ Ωcm) [13] of 0.7 mm thickness and have 2x5 140 μm gas gaps. The in-beam test of the prototypes (RPC1, RPC2, RPC3) was performed at T10 beam line of CERN-PS, using pions of 6 GeV/c momentum. The measured efficiency at an applied electric field of 150 kV/cm was of 97% for all three detectors. A 50 ps time resolution for different runs measured over a time period of 6 hours is shown in Figure 2. This result confirms the stability of the counters, gas flow and electronics. A 4.5 mm position resolution along the strip lines was measured. The obtained transverse position resolution was of 400 μm [12].

For the regions where a lower granularity is required, a new prototype with the strip pitch equal with 7.1 mm and a strip length of 96 mm was designed and built. This new prototype has the active area defined by the low resistivity glass plate [13] size of 96 x 300 mm^2. Two identical detectors were mounted in a tight gas box in a staggered geometry along the strips (Figure 3-left side), a basic architecture of the disposal of the RPC cells similar with a proposed real one for the inner zone of the TOF wall. The detectors were tested in a high counting rate

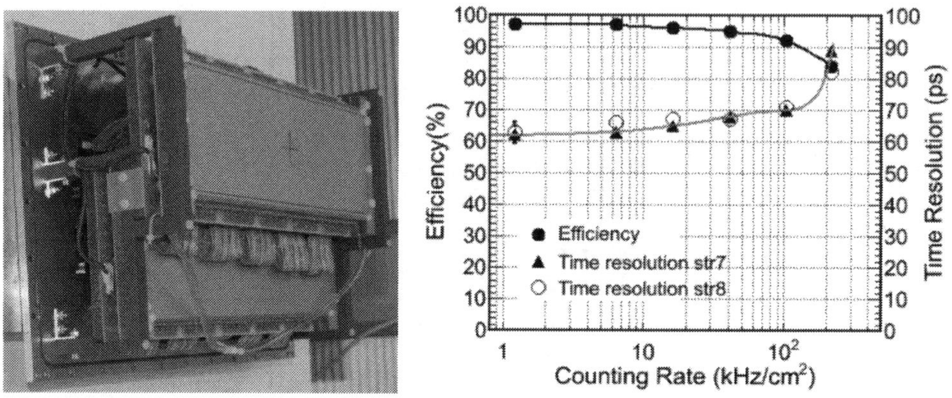

Figure 3. Photo of the two staggered RPC cells mounted on the back flange of the tight gas box - left side. Efficiency and time resolution as a function of counting rate - right side.

at the COSY facility in Jülich. As can be seen in Figure 3-right side, the efficiency is still larger than 90% and the time resolution is better than 70 ps at 100 kHz/cm^2 counting rate [12]. The next step is to complete the basic architecture of the inner zone with the staggering of the RPC cells across the strip direction for an uniform coverage of the full active area.

4. Conclusions

The obtained results in terms of efficiency, time and position resolutions and counting rate performance recommend the proposed high granularity MGMSRPC architecture based on low resistivity glass as a solution for large area detectors in high collision rate experiments.

Acknowledgments

We acknowledge V. Aprodu, D. Bartoş, G. Caragheorgheopol, F. Constantin, L. Prodan, A. Radu, L. Rădulescu, V. Simion, I. Deppner, N. Herrmann and P-A. Loizeau for their essential contribution in the design, construction and operation of the detectors. This work was supported by EU-FP7/HP2-WP18 grant no. 227431, EU-FP7/HP3-WP19 grant no.283286, NASR/CAPACITATI 179EU Project and NASR/NUCLEU Project.

References

[1] CBM Collaboration, 2005 *Technical Status Report* .
[2] http://www.fair-center.eu.
[3] M. Petrovici et al., 2002, *Nucl. Instr. and Meth. In Physics Research A***487**, 337.
[4] M. Petrovici et al., 2003, *Nucl. Instr. and Meth. In Physics Research A***508**, 75.
[5] M. Kis, 2012 *XI Workshop on Resistive Plate Chambers and Related Detectors*, 5-10 February, Frascati.
[6] ALICE Collaboration, Addendum to ALICE-TDR-8 (2002) ; CERN-LHCC-2002-016
[7] V.V. Parkhomchuck, Yu.N. Pestov, N.V. Petrovykh, 1971, *Nucl. Instr. and Meth. A* **93**, 269.
[8] M. Petriş et al., 2011 Rom. Journ. Phys. 56, 349.
[9] D. Bartoş et al., 2008 IEEE NSS/MIC, Book Series: IEEE Nuclear Science Symposium **Vol. 1-9**, 1933.
[10] F. Anghinolfi et al., 2004 *Nucl. Instr. and Meth. In Physics Research A* **533**, 183.
[11] M. Petriş et al., 2012 *Nucl. Instr. and Meth. In Physics Research A* **661**, Suppl.1, S129.
[12] M. Petrovici et al, 2012 *Journal Of Instrumentation* **7** (JINST 7 P11003).
[13] Wang Yi, 2012 *XI Workshop on Resistive Plate Chambers and Related Detectors*, 5-10 February, Frascati.

XX International School on Nuclear Physics, Neutron Physics and Applications (Varna2013) IOP Publishing
Journal of Physics: Conference Series **533** (2014) 012010 doi:10.1088/1742-6596/533/1/012010

Nuclear spin scissors – new type of collective motion

E B Balbutsev, I V Molodtsova

Joint Institute for Nuclear Research, 141980 Dubna, Moscow Region, Russia

E-mail: balbuts@theor.jinr.ru

Abstract. The coupled dynamics of the orbital and spin scissors modes is studied with the help of the Wigner Function Moments method on the basis of Time Dependent Hartree-Fock equations in the harmonic oscillator model including spin orbit potential plus quadrupole-quadrupole and spin-spin residual interactions. The relation between our results and the recent experimental data is discussed.

1. Introduction

The idea of the possible existence of the collective motion in deformed nuclei similar to the scissors motion continues to attract the attention of physicists who extend it on various kinds of objects, not necessary nuclei, (for example, magnetic traps, see the review by Heyde at al [1]) and invent new sorts of scissors, for example, the rotational oscillations of neutron skin against a proton-neutron core [2]. In a recent paper [3] the WFM method was applied for the first time to solve the TDHF equations including spin dynamics. As a first step, only the spin orbit interaction was included in the consideration, as the most important one among all possible spin dependent interactions because it enters into the mean field. The most remarkable result was the discovery of a new type of nuclear collective motion: rotational oscillations of "spin-up" nucleons with respect of "spin-down" nucleons (the spin scissors mode). It turns out that the experimentally observed group of peaks in the energy interval 2-4 MeV corresponds very likely to two different types of motion: the conventional (orbital) scissors mode and the spin scissors mode.

The aim of this work is to get a qualitative understanding of the influence of the spin-spin force on the new states analyzed in [3], as, for instance, the spin scissors mode. As a matter of fact we will find that the spin-spin interaction does not change the general picture of the positions of excitations described in [3]. The most interesting result concerns the B(M1) values of both scissors modes – the spin-spin interaction strongly redistributes M1 strength in the favour of the spin scissors mode, that allows us to give a tentative explanation of recent experimental findings [4, 5].

2. TDHF equation and model Hamiltonian

We consider the TDHF equation in coordinate space keeping all spin indices:

$$i\hbar\langle\mathbf{r}|\dot{\hat{\rho}}|\mathbf{r}''\rangle^{ss''} = \sum_{s'}\int d^3r' \left(\langle\mathbf{r}|\hat{h}|\mathbf{r}'\rangle^{ss'}\langle\mathbf{r}'|\hat{\rho}|\mathbf{r}''\rangle^{s's''} - \langle\mathbf{r}|\hat{\rho}|\mathbf{r}'\rangle^{ss'}\langle\mathbf{r}'|\hat{h}|\mathbf{r}''\rangle^{s's''}\right). \qquad (1)$$

We do not specify the isospin indices in order to make the formulae more transparent.

From the technical point of view it is more convenient to work with the Wigner function $f(\mathbf{r},\mathbf{p})$ instead of density matrix $\langle\mathbf{r}|\hat{\rho}|\mathbf{r}'\rangle$. To this end we rewrite the expression (1) with the help of the Wigner transformation [6]:

$$i\hbar\dot{f}^{+}=\frac{i\hbar}{2}\{h^{+},f^{+}\}+\frac{i\hbar}{2}\{h^{-},f^{-}\}+i\hbar\{h^{\uparrow\downarrow},f^{\downarrow\uparrow}\}+i\hbar\{h^{\downarrow\uparrow},f^{\uparrow\downarrow}\}+...$$

$$i\hbar\dot{f}^{-}=\frac{i\hbar}{2}\{h^{+},f^{-}\}+\frac{i\hbar}{2}\{h^{-},f^{+}\}-2h^{\downarrow\uparrow}f^{\uparrow\downarrow}+2h^{\uparrow\downarrow}f^{\downarrow\uparrow}+\frac{\hbar^2}{4}\{\{h^{\downarrow\uparrow},f^{\uparrow\downarrow}\}\}-\frac{\hbar^2}{4}\{\{h^{\uparrow\downarrow},f^{\downarrow\uparrow}\}\}+...$$

$$i\hbar\dot{f}^{\uparrow\downarrow}=-h^{\uparrow\downarrow}f^{-}+h^{-}f^{\uparrow\downarrow}+\frac{i\hbar}{2}\{h^{\uparrow\downarrow},f^{+}\}+\frac{i\hbar}{2}\{h^{+},f^{\uparrow\downarrow}\}+\frac{\hbar^2}{8}\{\{h^{\uparrow\downarrow},f^{-}\}\}-\frac{\hbar^2}{8}\{\{h^{-},f^{\uparrow\downarrow}\}\}+...$$

$$i\hbar\dot{f}^{\downarrow\uparrow}=h^{\downarrow\uparrow}f^{-}-h^{-}f^{\downarrow\uparrow}+\frac{i\hbar}{2}\{h^{\downarrow\uparrow},f^{+}\}+\frac{i\hbar}{2}\{h^{+},f^{\downarrow\uparrow}\}-\frac{\hbar^2}{8}\{\{h^{\downarrow\uparrow},f^{-}\}\}+\frac{\hbar^2}{8}\{\{h^{-},f^{\downarrow\uparrow}\}\}+... \quad (2)$$

where the functions h, f are the Wigner transforms of $\hat{h}$, $\hat{\rho}$ respectively, $\{f,h\}$ is the Poisson bracket of the functions $f(\mathbf{r},\mathbf{p})$ and $h(\mathbf{r},\mathbf{p})$ and $\{\{f,h\}\}$ is their double Poisson bracket, $f^{\pm}=f^{\uparrow\uparrow}\pm f^{\downarrow\downarrow}$, $h^{\pm}=h^{\uparrow\uparrow}\pm h^{\downarrow\downarrow}$; the dots stand for terms proportional to higher powers of $\hbar$. The conventional notation $\uparrow$ for $s=\frac{1}{2}$ and $\downarrow$ for $s=-\frac{1}{2}$ is used.

The microscopic Hamiltonian of the model, harmonic oscillator with spin orbit potential plus separable quadrupole-quadrupole and spin-spin residual interactions is given by

$$H=\sum_{i=1}^{A}\left[\frac{\hat{\mathbf{p}}_i^2}{2m}+\frac{1}{2}m\omega^2\mathbf{r}_i^2-\eta\hat{\mathbf{l}}_i\hat{\mathbf{S}}_i\right]+H_{qq}+H_{ss} \quad (3)$$

with

$$H_{qq}=\sum_{\mu=-2}^{2}(-1)^{\mu}\left\{\bar{\kappa}\sum_{i}^{Z}\sum_{j}^{N}+\frac{\kappa}{2}\left[\sum_{i,j(i\neq j)}^{Z}+\sum_{i,j(i\neq j)}^{N}\right]\right\}q_{2-\mu}(\mathbf{r}_i)q_{2\mu}(\mathbf{r}_j), \quad (4)$$

$$H_{ss}=\sum_{\mu=-1}^{1}(-1)^{\mu}\left\{\bar{\chi}\sum_{i}^{Z}\sum_{j}^{N}+\frac{\chi}{2}\left[\sum_{i,j(i\neq j)}^{Z}+\sum_{i,j(i\neq j)}^{N}\right]\right\}\hat{S}_{-\mu}(i)\hat{S}_{\mu}(j)\,\delta(\mathbf{r}_i-\mathbf{r}_j), \quad (5)$$

where N and Z are the numbers of neutrons and protons and $\hat{S}_\mu$ are spin matrices [7]. The quadrupole operator $q_{2\mu}$ can be written as the tensor product: $q_{2\mu}(\mathbf{r})=\sqrt{16\pi/5}\,r^2Y_{2\mu}(\theta,\phi)=\sqrt{6}\{r\otimes r\}_{2\mu}$, where $\{r\otimes r\}_{\lambda\mu}=\sum_{\sigma,\nu}C^{\lambda\mu}_{1\sigma,1\nu}r_\sigma r_\nu$, $C^{\lambda\mu}_{1\sigma,1\nu}$ is the Clebsch-Gordan coefficient and r_1,r_0,r_{-1} are cyclic coordinates [7].

3. Equations of motion

Integrating the set of equations (2) over phase space with the weights

$$W=\{r\otimes p\}_{\lambda\mu},\ \{r\otimes r\}_{\lambda\mu},\ \{p\otimes p\}_{\lambda\mu},\ \text{and } 1 \quad (6)$$

one gets dynamic equations for the following collective variables:

$$L^{\tau\varsigma}_{\lambda\mu}(t)=\int\!\!\int\frac{d^3r\,d^3p}{(2\pi\hbar)^3}\{r\otimes p\}_{\lambda\mu}f^{\tau\varsigma}(\mathbf{r},\mathbf{p},t),\quad R^{\tau\varsigma}_{\lambda\mu}(t)=\int\!\!\int\frac{d^3r\,d^3p}{(2\pi\hbar)^3}\{r\otimes r\}_{\lambda\mu}f^{\tau\varsigma}(\mathbf{r},\mathbf{p},t),$$

$$P^{\tau\varsigma}_{\lambda\mu}(t)=\int\!\!\int\frac{d^3r\,d^3p}{(2\pi\hbar)^3}\{p\otimes p\}_{\lambda\mu}f^{\tau\varsigma}(\mathbf{r},\mathbf{p},t),\quad F^{\tau\varsigma}(t)=\int\!\!\int\frac{d^3r\,d^3p}{(2\pi\hbar)^3}f^{\tau\varsigma}(\mathbf{r},\mathbf{p},t), \quad (7)$$

where $\varsigma=+,-,\uparrow\downarrow,\downarrow\uparrow$. By analogy with isoscalar $f^{\rm n}+f^{\rm p}$ and isovector $f^{\rm n}-f^{\rm p}$ functions we call the functions $f^{\tau+}$ and $f^{\tau-}$ and the corresponding collective variables $X^{\tau+}_{\lambda\mu}(t)$ and $X^{\tau-}_{\lambda\mu}(t)$ as spin-scalar and spin-vector ones. The required expressions for $h^{\pm}$, $h^{\uparrow\downarrow}$ and $h^{\downarrow\uparrow}$ are

$$h^{+}_{\tau}=\frac{p^2}{m}+m\,\omega^2r^2+12\sum_{\mu}(-1)^{\mu}Z^{\tau+}_{2\mu}(t)\{r\otimes r\}_{2-\mu}+V^{+}_{\tau}(\mathbf{r},t),$$

$$h_\tau^- = -\hbar\eta l_0 + V_\tau^-(\mathbf{r},t), \quad h_\tau^{\uparrow\downarrow} = -\frac{\hbar}{\sqrt{2}}\eta l_{-1} + V_\tau^{\uparrow\downarrow}(\mathbf{r},t), \quad h_\tau^{\downarrow\uparrow} = \frac{\hbar}{\sqrt{2}}\eta l_1 + V_\tau^{\downarrow\uparrow}(\mathbf{r},t),$$

where $Z_{2\mu}^{\mathrm{n}+} = \kappa R_{2\mu}^{\mathrm{n}+} + \bar{\kappa} R_{2\mu}^{\mathrm{p}+}$, $Z_{2\mu}^{\mathrm{p}+} = \kappa R_{2\mu}^{\mathrm{p}+} + \bar{\kappa} R_{2\mu}^{\mathrm{n}+}$ and

$$V_p^+(\mathbf{r},t) = -3\frac{\hbar^2}{8}\chi n_p^+(\mathbf{r},t), \quad V_p^-(\mathbf{r},t) = 3\frac{\hbar^2}{8}\chi n_p^-(\mathbf{r},t) + \frac{\hbar^2}{4}\bar{\chi} n_n^-(\mathbf{r},t),$$
$$V_p^{\uparrow\downarrow}(\mathbf{r},t) = 3\frac{\hbar^2}{8}\chi n_p^{\uparrow\downarrow}(\mathbf{r},t) + \frac{\hbar^2}{4}\bar{\chi} n_n^{\uparrow\downarrow}(\mathbf{r},t), \quad V_p^{\downarrow\uparrow}(\mathbf{r},t) = 3\frac{\hbar^2}{8}\chi n_p^{\downarrow\uparrow}(\mathbf{r},t) + \frac{\hbar^2}{4}\bar{\chi} n_n^{\downarrow\uparrow}(\mathbf{r},t). \quad (8)$$

The neutron potentials V_n^ς are obtained by the obvious change of indices $p \leftrightarrow n$.

We are interested in the scissors mode, the excitation with $K^\pi = 1^+$, therefore, we only need the part of dynamic equations with $\mu = 1$. These equations are nonlinear due to quadrupole-quadrupole and spin-spin interactions and will be solved in a small amplitude approximation.

Let us recall that all variables and equilibrium quantities $R_{\lambda 0}^+(eq)$ and $Z_{20}^+(eq)$ have isospin indices $\tau = n, p$. All the difference between neutron and proton systems is contained in the mean field quantities $Z_{20}^{\tau+}(eq)$ and V_τ^ς, which are different for neutrons and protons (see eq. (8)). It is convenient to rewrite the dynamical equations in terms of isovector and isoscalar variables $X_{\lambda\mu} = X_{\lambda\mu}^n + X_{\lambda\mu}^p$, $\bar{X}_{\lambda\mu} = X_{\lambda\mu}^n - X_{\lambda\mu}^p$ and isovector and isoscalar strength constants $\kappa_1 = (\kappa - \bar{\kappa})/2$ and $\kappa_0 = (\kappa + \bar{\kappa})/2$ connected by the relation $\kappa_1 = \alpha\kappa_0$. Then the equations for the neutron and proton systems are transformed into isovector and isoscalar ones. Supposing that all equilibrium characteristics of the proton system are equal to that of the neutron system one decouples isovector and isoscalar equations. Neglecting by the fourth order moments, generated by the terms $h^{\uparrow\downarrow} f^{\downarrow\uparrow} - f^{\uparrow\downarrow} h^{\downarrow\uparrow}$ of (2) we get the closed set of 19 equations.

4. Discussion and interpretation of the results

Imposing the time evolution via $e^{i\Omega t}$ for all variables one transforms dynamical equations into a set of algebraic equations. Eigenfrequencies are found as the zeros of its determinant. Excitation probabilities are calculated with the help of the linear response theory. The used spin-spin interaction is taken from the paper [8], where the notation $\chi = K_s/A$, $\bar{\chi} = q\chi$ was introduced. The results of calculations without spin-spin interaction (variant I) are compared with those performed with two sets of constants K_s, q (variants II, III). The strength of the spin-orbit interaction is taken from [9].

Table 1. Isovector energies and excitation probabilities of ^{164}Er. Deformation parameter $\delta = 0.25$, spin-orbit constant $\eta = 0.36$ MeV. Spin-spin interaction constants are: I – $K_s = 0$ MeV; II – $K_s = 92$ MeV, $q = -0.8$; III – $K_s = 200$ MeV, $q = -0.5$. Quantum numbers of variables responsible for the generation of the present level are shown in the first column.

$(\lambda, \mu)^\varsigma$	E, MeV			$B(M1)$, μ_N^2			$B(E2)$, B_W		
	I	II	III	I	II	III	I	II	III
$(1,1)^-$	1.61	2.02	2.34	3.54	5.44	7.91	0.12	0.36	0.82
$(1,1)^+$	2.18	2.45	2.76	5.33	4.48	2.98	1.02	1.23	1.26

One can see from table 1 that the spin-spin interaction does not change the qualitative picture of the positions of excitations described in [3]. It pushes all levels up proportionally to its strength (20-30% in the case II and 40-60% in the case III) without changing their order. The most interesting result concerns the relative B(M1) values of the spin scissors $(1,1)^-$ and the orbital scissors $(1,1)^+$. The spin-spin interaction strongly redistributes M1 strength in the favour of the spin scissors mode. We tentatively want to link this fact to the recent experimental finding in isotopes of Th and Pa [4]. The authors have studied deuteron and ^{3}He-induced reactions on ^{232}Th and found in the residual nuclei ^{231,232,233}Th and ^{232,233}Pa "an unexpectedly strong integrated strength of $B(M1) = 11-15\ \mu_N^2$ in the $E_\gamma = 1.0-3.5$ MeV region". The $B(M1)$ force in most nuclei in [4] shows evident splitting into two Lorentzians. "Typically, the experimental splitting is $\Delta\omega_{M1} \sim 0.7$ MeV, and the ratio of the strengths between the lower and upper resonance components is $B_L/B_U \sim 2$". The authors have tried to explain the splitting by a γ-deformation. To describe the observed value of $\Delta\omega_{M1}$ the deformation $\gamma \sim 15^\circ$ is required, that leads to the ratio $B_L/B_U \sim 0.7$ in an obvious contradiction with the experiment. The authors conclude that "the splitting may be due to other mechanisms". In this sense, we tentatively may argue as follows. On one side, theory [10] and experiment [11] give zero value of γ-deformation for ^{232}Th. On the other side, it is easy to see that our theory suggests the required mechanism. The calculations performed for ^{232}Th give $\Delta\omega_{M1} \sim 0.32$ MeV and $B_L/B_U \sim 1.6$ for the first variant of the spin-spin interaction and $\Delta\omega_{M1} \sim 0.28$ MeV and $B_L/B_U \sim 4.1$ for second one in reasonable agreement with experimental values.

Quite similar results were obtained one year earlier by A. S. Adekola et al [5]. They have studied the (γ,γ') reaction on ^{232}Th and found two groups of levels with energy centroids $E_1 = 2.1$ MeV and $E_2 = 2.9$ MeV having $B(M1)_1 = 2.52\ \mu_N^2$ and $B(M1)_2 = 1.74\ \mu_N^2$. The splitting $E_2 - E_1 = 0.8$ MeV and the ratio $B(M1)_1/B(M1)_2 = 1.45$ are in good agreement with the results of [4] and with our calculations.

5. Concluding remarks

In this work, we continued the investigation of spin modes [3] with the Wigner Function Method. The inclusion of spin-spin interaction does not change qualitatively the described picture concerning spin modes found in [3]. It pushes all levels up without changing their order. However, it strongly redistributes M1 strength between the conventional and spin scissors mode in the favour of the last one. We mentioned the recently appeared experimental works [4, 5], where for the two groups of low lying magnetic states a stronger B(M1) transition for the lower group with respect to the higher one was found. Our theory can naturally predict such a scenario with a non vanishing spin-spin force. It would indeed be very exciting, if the results of [4, 5] had already discovered the isovector spin scissors mode.

In the light of the above results, the study of spin excitations with pairing included, will be the natural continuation of this work. Pairing is important for a quantitative description of the orbital scissors mode. The same is expected for the novel spin scissors mode discussed here.

References

[1] Heyde K, von Neuman-Cosel P and Richter A 2010 *Rev. Mod. Phys.* **82** 2365
[2] Pena Arteaga D and Ring P, 2009 A new scissors mode on the skin of deformed neutron rich nuclei *Preprint* arXiv:0912.0908v1 [nucl-th]
[3] Balbutsev E B, Molodtsova I V and Schuck P 2011 *Nucl. Phys.* A **872** 42
[4] Guttormsen M, *et al* 2012 *Phys. Rev. Lett.* **109** 162503
[5] Adekola A S, *et al* 2011 *Phys. Rev.* C **83** 034615
[6] Ring P and Schuck P 1980 *The Nuclear Many-Body Problem* (Berlin: Springer)
[7] Varshalovich D A, Moskalev A N and Khersonskii V K 1988 *Quantum Theory of Angular Momentum* (Singapore: World Scientific)
[8] Sarriguren P, Moya de Guerra E and Nojarov R 1996 *Phys. Rev.* C **54** 690; 1997 *Z. Phys.* A **357** 143
[9] Soloviev V G 1976 *Theory of complex nuclei* (Oxford: Pergamon Press)
[10] Gogny D1S web mass table, http://www-phynu.cea.fr/HFB-Gogny.htm
[11] Korten W, *et al* 1993 *Phys. Lett.* B **317** 19

Journal of Physics: Conference Series **533** (2014) 012011 doi:10.1088/1742-6596/533/1/012011

Weak-decay rates of pf-shell nuclei in stellar scenarios

Pedro Sarriguren

Instituto de Estructura de la Materia, IEM-CSIC, Serrano 123, E-28006 Madrid, Spain

E-mail: p.sarriguren@csic.es

Abstract.

Weak decay rates under various stellar density and temperature conditions are studied in selected pf-shell nuclei of special relevance as constituents in presupernova formations. The rates are relevant to understand late stages of stellar evolution leading to supernova explosions and the nucleosynthesis of heavier nuclei. The nuclear structure part of the problem is described within a quasiparticle random-phase approximation. Results are compared to benchmark shell-model calculations and experimental data extracted from charge-exchange reactions.

1. Introduction

Weak β-decay and electron-capture (EC) processes are very important mechanisms to understand the late stages of the stellar evolution, playing a critical role to determine both the presupernova stellar structure and the nucleosynthesis of heavier nuclei. These processes are dominated by Gamow-Teller (GT) transitions and therefore, the GT properties of pf-shell nuclei are of special importance because these nuclei are the main constituents of the stellar core in presupernova formations leading to core-collapse (type II) or thermonuclear (type Ia) supernovae. An accurate understanding of those astrophysical processes requires input from nuclear physics that in most cases cannot be measured directly because of the extreme conditions of density (ρ) and temperature (T) that hold in stellar scenarios and therefore, the GT strength distributions must be estimated in many cases by model calculations. Obviously, nuclear physics uncertainties will finally affect the reliability of the description of those astrophysical processes.

A systematic evaluation of the ability to reproduce the measured GT strength distributions of various theoretical models based on shell model (SM) and QRPA was done recently in Ref. [1]. EC rates were also derived from those models at relevant ρ and T. While several sets of SM calculations using different effective interactions were compared, namely KB3G [2] and GXPF1a [3], in the case of QRPA only the formalism developed in [4] using deformations and masses obtained from the finite range droplet model [5] was considered in Ref. [1]. In this work, we study the dependence of the EC rates on both ρ and T with GT strength distributions calculated within a QRPA approach based on a selfconsistent deformed Hartree-Fock (HF) mean field with Skyrme interactions including pairing correlations and residual separable forces in both particle-hole (ph) and particle-particle (pp) channels. We compare our calculations with the benchmark calculations in Ref. [1]. This formalism represents an improvement over the QRPA-Möller approach in several aspects. First, instead of a phenomenological approach , the deformed mean field is obtained selfconsistently and secondly, a separable residual GT interaction in the pp channel is included. The present nuclear model has been tested successfully reproducing very reasonably the experimental information available on both bulk and decay properties of medium-mass nuclei [6, 7, 8, 9].

2. Theoretical approach

The decay rate corresponding to an initial nuclear parent state i is given by

$$\lambda_i = \sum_f \lambda_{if} = \frac{\ln 2}{D} \sum_f B_{if} \Phi_{if}(\rho, T) , \tag{1}$$

where the sum extends over all the states in the final nucleus reached in the process and $D = 6146$ s. This expression is decomposed into a phase space factor Φ_{if}, which is a function of ρ and T and a nuclear structure part B_{if} that contains the transition probabilities for allowed GT transitions.

In the astrophysical scenarios of our study, nuclei are fully ionized and continuum EC from the degenerate electron plasma are possible. The phase space factor for EC is given by

$$\Phi_{if}^{EC} = \int_{\omega_\ell}^{\infty} \omega p (Q_{if} + \omega)^2 F(Z, \omega) S_e(\omega) d\omega , \tag{2}$$

where ω is the total energy of the electron, $p = \sqrt{\omega^2 - 1}$ is the momentum, and Q_{if} is given by $Q_{if} = (Q_{EC} - m_e c^2 - E_f)/m_e c^2$. $F(Z, \omega)$ is the Fermi function that takes into account the distortion of the electron wave function due to the Coulomb interaction. S_e is the electron distribution function given by a Fermi-Dirac distribution. The lower integration limit is given by $\omega_\ell = 1$ if $Q_{if} > -1$, or $\omega_\ell = |Q_{if}|$ if $Q_{if} < -1$.

The nuclear structure involved in the weak rates is described within a microscopic deformed quasiparticle random-phase approximation (QRPA) based on a selfconsistent mean field obtained from Skyrme (SLy4 and SG2) Hartree-Fock + BCS calculations. A residual separable interaction in both *ph* and *pp* is added to the mean field and treated in a deformed proton-neutron QRPA. More details of the calculation can be found in Ref. [10, 11]. The GT strength for a transition from an initial state i to a final state f is given by

$$B_{if}(GT^{\pm}) = \frac{1}{2J_i + 1} \left(\frac{g_A}{g_V} \right)^2_{\text{eff}} \langle f || \sum_j^A \sigma_j t_j^{\pm} || i \rangle^2 , \tag{3}$$

$(g_A/g_V)_{\text{eff}} = 0.7 (g_A/g_V)_{\text{bare}}$ is the effective ratio of axial and vector coupling factors.

3. Results and discussion

As an example of the results obtained within this scheme [12], we can see in Fig. 1 the GT strength distributions corresponding to the ^{64}Zn$(0^+) \rightarrow ^{64}$Cu transition. Data are from $(d, ^2\text{He})$ [13] and $(t, ^3\text{He})$ [14] reactions achieving 115 keV and 280 keV energy resolution, respectively. The QRPA results show a bump centered at an excitation energy of 3 MeV (5 MeV) when using SG2 (SLy4) interaction. SG2 results reproduce better the experimental results, which show a strong peak at 3 MeV. The calculated total strength, including results from SM calculations, overestimates always the experiment. Results from SM-KB3G produce a strong peak at 5 MeV at variance with experiment.

In the next figure (Fig. 2) we present the EC rates for ^{64}Zn as a function of the temperature and for various densities. The range of T considered varies from $T_9 = 1$ up to $T_9 = 10$, whereas the range in ρY_e varies from $\rho Y_e = 10^6$ mol/cm^3 up to $\rho Y_e = 10^9$ mol/cm^3. This grid of ρ and T includes those ranges relevant for astrophysical scenarios related to the silicon-burning stage in a presupernovae star ($\rho Y_e = 10^7$ mol/cm^3 and $T_9 = 3$), as well as scenarios related to pre-collapse of the core and thermonuclear runaway type Ia supernovae ($\rho Y_e = 10^9$ mol/cm^3

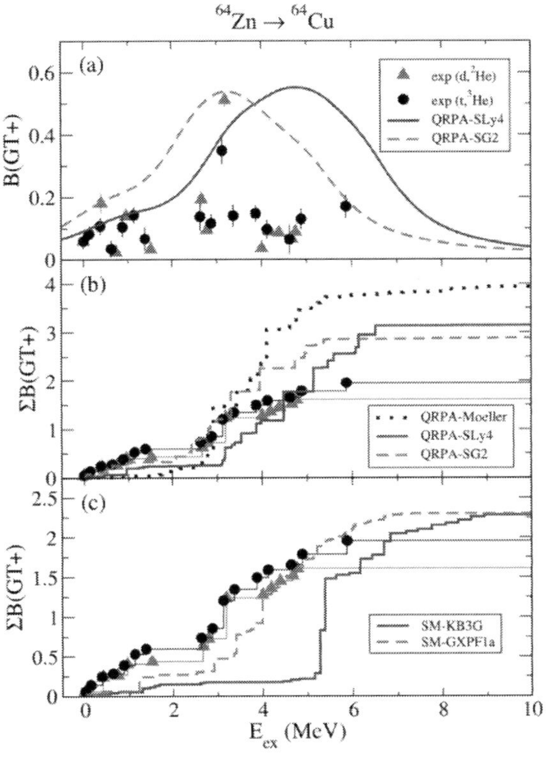

Figure 1. GT strength distributions for the transition ^{64}Zn to ^{64}Cu versus the excitation energy of the daughter nucleus. (a) Experimental data are compared to our QRPA results. Experimental accumulated strengths are compared to QRPA (a) and shell-model calculations (c).

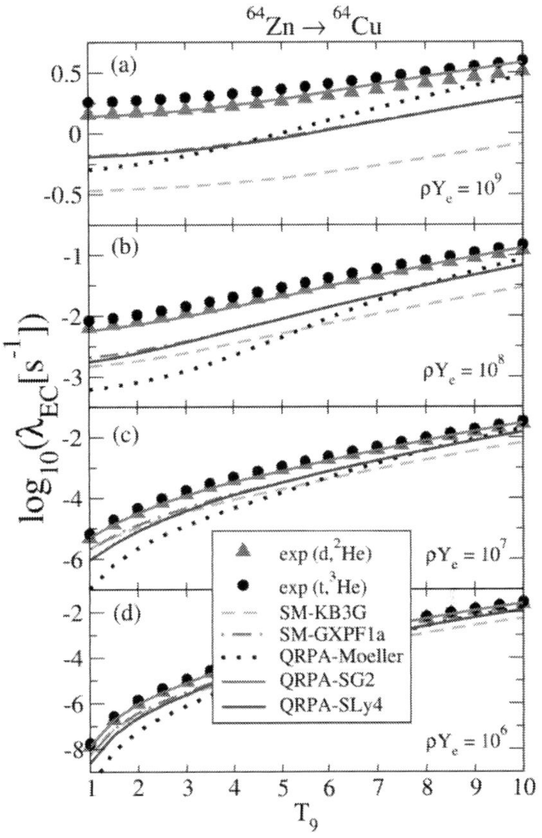

Figure 2. EC rates for ^{64}Zn obtained from experimental GT strength distributions and from different shell model and QRPA calculations for various densities.

and $T_9 = 10$). We show the EC rates obtained from the experimental GT strength distributions and from different SM (KB3G and GXPF1a) and QRPA (SLy4, SG2, and Möller) calculations.

Some general comments about the sensitivity of the EC rates to (ρ, T), to the Fermi and Q energies, and to the GT distribution are in order to understand better the EC rates. Since the Fermi energy increases with the electron density, it is expected that the EC rates at low densities are mainly sensitive to the GT strength of states at low excitation energies in the daughter nucleus. When ρ increases the rates also increase because the Fermi energy of the electrons is larger and larger allowing to reach higher excitation energies in the daughter nucleus and thus making the GT strength at these energies contribute to the rates in a more significant way. At low T, the shape of the electron energy distribution, S_e, has a sharp surface at the Fermi energy. When increasing T, the rates in general increase because the diffuseness of the electron energy distributions allows to reach more excited states at higher energy.

On the other hand the role of the Q_{if} (and Q_{EC}) energy is also very important. It determines the lower integration limit in the EC phase factor (2). Then, the energy of the available electrons must overcome this value to be captured or in other words, the Fermi energy has to be larger than Q_{if} at $T = 0$ where S_e changes abruptly from one to zero at the Fermi energy. When T increases, S_e is smeared out and ECs are possible even for Fermi energies lower than Q_{if}. In general, the larger the Q_{EC} energy (more negative in stable nuclei) the lower the EC rates. This effect is enhanced at low densities where the Fermi energy is small. Similarly, the rates of nuclei with large (negative) Q_{EC} values are mostly sensitive to the GT strength of the ground and lowest excited states whereas the opposite is true for the nuclei with small Q_{EC} values, such as ^{64}Zn ($Q_{EC} = -0.580$ MeV), where the rates are sensitive to high excitation energies that can contribute especially at high ρ values.

The agreement with experiment in Fig. 2 is quite reasonable, especially for QRPA-SG2 calculations. The rates are always underestimated

4. Conclusions

We have evaluated the continuum electron-capture rates in ^{64}Zn, as an example of a pf-shell nucleus, at different density and temperature conditions holding in stellar scenarios of particular interest. The nuclear structure involved in the calculation of the energy distribution of the Gamow-Teller strength is described within a selfconsistent deformed HF+BCS+QRPA formalism with density-dependent effective Skyrme interactions and spin-isospin residual interactions.

We find that the present QRPA calculation is able to reproduce the main features of the GT distributions extracted in these nuclei from charge-exchange reactions. Comparison of our results [12] to SM calculations and other QRPA results analyzed in Ref. [1] shows that on the overall the agreement with experiment of the present QRPA calculations is comparable to the agreement achieved by SM calculations and that QRPA based on Skyrme forces improve the agreement with experiment with respect to QRPA-Möller [4].

We have studied the sensitivity of the EC rates to both ρ and T. At low ρ (low Fermi energies) and low T (sharp shape of the energy distribution of the electrons), the rates are very sensitive to details of the GT strength of the low-lying excitations and therefore to model calculations. On the other hand, when the ρ and T are high enough, the EC rates are sensitive to all the spectrum. Then, the whole description of the GT strength distribution is more important than a detailed description of the low-lying spectrum. Since QRPA reproduces reasonably well the global behavior of the GT strength distributions, it is expected to be a good approach, especially for high ρ and T conditions.

Acknowledgments

This work was supported in part by MINECO (Spain) under Research Grant No. FIS2011–23565 and by Consolider-Ingenio 2010 Programs CPAN CSD2007-00042.

References

[1] Cole A L *et al.* 2012 *Phys. Rev.* C **86** 015809
[2] Poves A *et al.* 2001 *Nucl. Phys.* **A694** 157
[3] Honma M *et al.* 2002 *Phys. Rev.* C **69** 034335
[4] Krumlinde J and Möller P 1984 *Nucl. Phys.* **A417** 419 (1984)
[5] Möller P *et al.* 1995 *At. Data Nucl. Data Tables* **59** 185
[6] Sarriguren P *et al.* 2003 *Nucl. Phys.* **A716** 230; Sarriguren P *et al.* 2005 *Eur. Phys. J.* A **24** 193
[7] Nácher E *et al.* 2004 *Phys. Rev. Lett.* **92** 232501; Poirier E *et al.* 2004 *Phys. Rev.* C **69** 034307
[8] Sarriguren P 2009 *Phys. Rev.* C **79** 044315
[9] Sarriguren P and Pereira J 2010 *Phys. Rev.* C **81** 064314
[10] Sarriguren P *et al.* 1998 *Nucl. Phys.* **A635** 55; Sarriguren P *et al.* 1999 *Nucl. Phys.* **A658** 13; Sarriguren P *et al.* 2001 *Nucl. Phys.* **A691** 631; Sarriguren P *et al.* 2001 *Phys. Rev.* C **64** 064306
[11] Sarriguren P 2009 *Phys. Lett.* B **680** 438; Sarriguren P 2011 *Phys. Rev.* C **83** 025801
[12] Sarriguren P 2013 *Phys. Rev.* C **87** 045801
[13] Grewe EW *et al.* 2008 *Phys. Rev.* C **77** 064303
[14] Hitt GW *et al.* 2009 *Phys. Rev.* C **80** 014313

Pairing and specific heat in ^{161,162}Dy and ^{171,172}Yb isotopes

Danilo Gambacurta

GANIL, CEA and CNRS/IN2P3, Boîte Postale 55027, 14076 Caen Cedex, France

E-mail: gambacurta@ganil.fr

Denis Lacroix

Institut de Physique Nucléaire, IN2P3-CNRS, Université Paris-Sud, F-91406 Orsay Cedex, France

GANIL, CEA/DSM and CNRS/IN2P3, Boîte Postale 55027, 14076 Caen Cedex, France

N. Sandulescu

National Institute of Physics and Nuclear Engineering, P.O.Box MG-6, Magurele-Bucharest, Romania

Abstract. The finite-temperature variation after projection BCS approach (FT-VAP) is employed to study the pairing phase-transition for the ^{161,162}Dy and ^{171,172}Yb isotopes. Due to the restoration of particle number conservation, the pairing gap and the specific heat calculated in the FT-VAP approach vary smoothly with the temperature, indicating a gradual transition from the superfluid to the normal phase, as expected in finite systems. FT-VAP results are compared with those obtained in the exact canonical approach as well as with experimental data.

1. Introduction

The study of possible thermal signatures of pairing correlations in atomic nuclei has been the subject of intensive works, both from a theoretical and experimental point of view. The common description of the interplay between pairing correlations and thermal effects is generally done within the Bardeen-Cooper-Schrieffer (BCS) or the Hartree-Fock-Bogoliubov (HFB) theories [1]. However, these models are not well-suited to describe accurately pairing effects in hot nuclei, because of particle number fluctuation and quasiparticle parity mixing issues. Particle-number conserving approaches are indeed necessary and recently [2, 3] it has been shown that the particle-number projected BCS approximation extended to finite temperature (FT-VAP) [4] is an appropriate tool for this kind of studies.

The influence of pairing correlations on specific heat is here analyzed for the isotopes ^{161,162}Dy and ^{171,172}Yb, by using the FT-VAP and the exact canonical solution [5]. The comparison between the results obtained within these two different number-conserving approaches shows the validity of FT-VAP also in realistic cases. It is shown that the FT-VAP approach, applied with a level density provided by mean field calculations and supplemented, at high energies, by

Journal of Physics: Conference Series **533** (2014) 012012 doi:10.1088/1742-6596/533/1/012012

the level density of the back-shifted Fermi gas model, can approximate reasonably well the main properties of specific heat extracted from experimental data. However, the detailed shape of the calculated specific heat is rather sensitive to the assumption made for the mean field.

2. Description of the calculations: input and procedure

The hamiltonian used in the present calculations is

$$\hat{H} = \sum_i \epsilon_i a_i^+ a_i + G \sum_{i,j} a_i^+ a_{\bar{i}}^+ a_{\bar{j}} a_j, \tag{1}$$

containing a constant pairing interaction which scatters pairs among time-reversed single-particle states $(i, \bar{i})$ and where only like-particle pairing is considered. The single-particle (s.p.) states employed in the Hamiltonian (1) are taken from mean-field models. In particular, we have used two sets of s.p. states obtained, respectively, from Skyrme-Hartree-Fock (HF) and Relativistic Mean-Field (RMF) calculations allowing, in both cases, axial symmetry deformation. The SGII and PK1 force are used, respectively. As commonly done in BCS calculations at zero temperature, pairing interaction is considered acting in a limited energy window around the Fermi level.The pairing strength G is fixed at BCS level in order to give a pairing gap approximately equal to 0.8 MeV both for protons and neutrons. Excitations outside this energy window are incorporated assuming that pairing interaction affects the excitation energies generated by the s.p. energies taken from a restricted region around the Fermi energy while the rest of s.p. spectrum

3. Results

In the top panel of Fig. 1 the pairing gap dependence with respect to the the temperature in the isotopes ^{162}Dy and ^{172}Yb is shown. The results correspond to the RMF s.p. spectrum. The finite-temperature BCS(FT-BCS) predicts a sharp second order type transition from the superfluid to the normal phase, at variance with what it is expected for small finite systems and predicted by the FT-VAP approach where a smooth transition from the superfluid to the normal phase is found. In the lower panel, also the pairing gaps obtained for the odd nuclei are given. As expected, at low T, the gaps in odd systems are smaller than in the even systems. This difference tends to disappear as the temperature increases. For ^{161}Dy and to a lesser extend for ^{171}Yb, we also observe a small re-entrance effect around $T = 0.5$ MeV (resp. around 0.2-0.3 MeV).

In Fig. 2, we analyze the effect of pairing on specific heat separately for protons and neutrons for the isotopes ^{161,162}Dy, within both the FT-VAP approximation and the method of direct diagonalization. It can be seen that both treatments give similar results, confirming the good predictive power of the FT-VAP approximation.

How much depends the specific heat on the single-particle spectrum can be seen from Fig. 3, where are compared the results obtained with the energies provided by the Skyrme-HF and the RMF calculations, in the Yb isotopes. It can be seen that there are significant differences between the results obtained with the two mean fields. These differences are related to the distribution of the single particle levels around the Fermi energy, (see also Fig. 1 of [3]).

Finally, in Fig. 4, the specific heats for Dy and Yb extracted from the experimental level density in Refs [7, 8] are compared with the theoretical ones. The experimental heat capacities are obtained from the level density using specific assumptions. More precisely, in Refs. [7, 8] it is used the experimental level density for excitations below 8 MeV/nucleon while for higher excitation energies it is employed the level density provided by the Back Shifted Fermi Gas Model (BSFGM) (Eq. (5) of [8]). In the present calculations, we have used for all excitations energies

Journal of Physics: Conference Series **533** (2014) 012012 doi:10.1088/1742-6596/533/1/012012

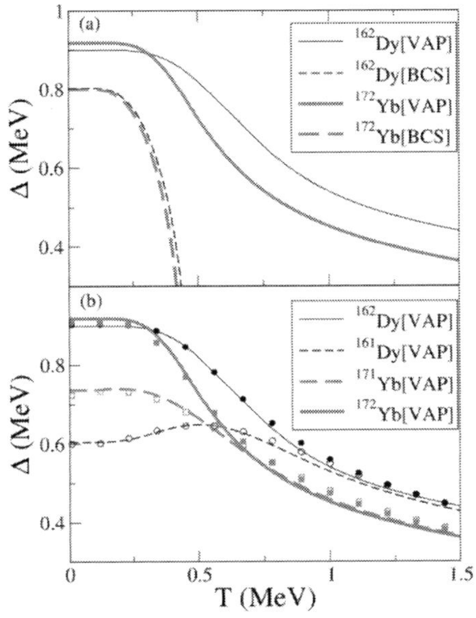

Figure 1. (color online) Top: neutron pairing gap as a function of temperature for ^{162}Dy and ^{172}Yb obtained in FT-VAP and FT-BCS approximations. Bottom: comparison between the pairing gaps in odd-even and even-even Dy and Yb isotopes obtained in FT-VAP approximation. By symbols are shown the gaps corresponding to exact diagonalization.

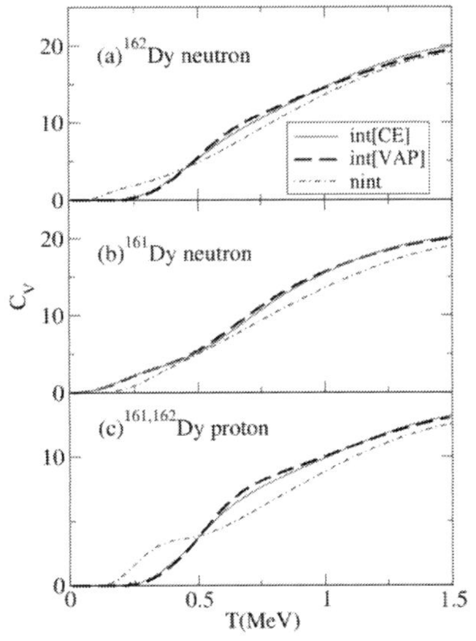

Figure 2. (color online) Neutron specific heat for ^{162}Dy (top), ^{161}Dy (middle) and the proton specific heat for ^{162,161}Dy (bottom) . The lines labeled by "int(VAP)" and "int(CE)" give the results corresponding to FT-VAP and to exact diagonalization, respectively, while "nint" are the results for non-interacting particles.

the level density generated by a discrete set of single-particle states. This assumption is expected to work reasonable well for low energy excitations but not for high energy excitations for which the contribution of the continuum become important. For this reason, a good agreement can only be obtained assuming that the free fermions gas limit is provided by the BSFGM as it was assumed in the experiment.For comparison, for even isotopes we show also the specific heats obtained in the FT-BCS approach which present the unphysical sharp transition between the superfluid and the normal phase. It can be observed that the agreement of the FT-VAP results with the experiments is reasonable good.

4. Conclusion

In the present work, the effect of pairing correlations in hot nuclei has been studied in the framework of a variation after particle-number projection BCS formalism extended to finite temperature. The calculation are done in canonical ensemble and with a Hamiltonian composed of a single-particle term, generated by Skyrme-HF or RMF calculations, and a pairing interaction of seniority type. The specific heat in the isotopes ^{161,162}Dy and ^{171,172}Yb have been evaluated. It is thus shown that the pairing correlations have a significant influence on the specific heat, especially for temperatures below $T = 0.5$ MeV. The comparison with the non-interacting

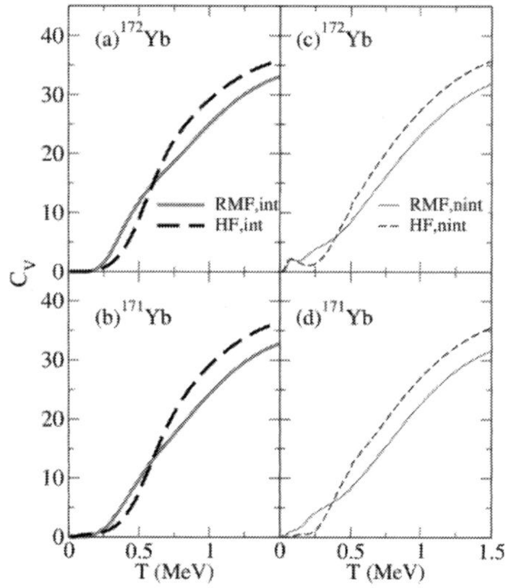

Figure 3. (color online) Heat capacities in ^{162}Dy and ^{161}Dy obtained using two different sets of single-particle energies corresponding to RMF and HF mean fields. In the left (right) panels are shown the results with (without) the pairing interaction included.

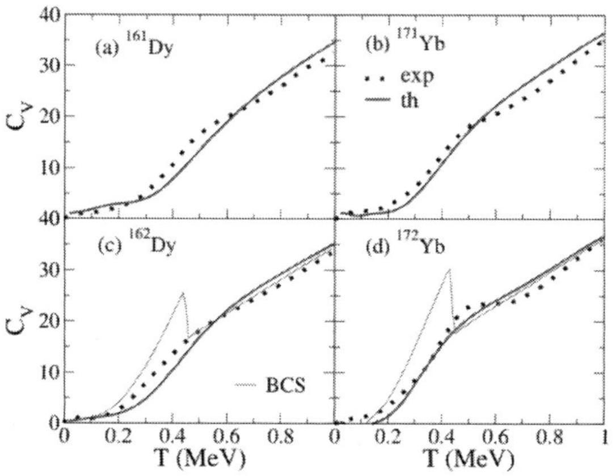

Figure 4. (color online) Comparison between the calculated FT-VAP specific heats and the specific heats extracted from experimental data [7, 8]. For comparison, we show also the results obtained within the FT-BCS approximation .

systems shows clearly that the pairing correlations contributes to the S-shape behaviour of the specific heat, as noticed in the data extracted from experiment.

5. Acknowledgments

This work was supported by the French-Romanian IN2P3-IFIN agreement and by the Romanian Ministry of Education and Research through the grant Idei nr 57.

References

[1] V. Martinez, J.L. Egido and L.M. Robledo 2003, Phys. Rev. C68, 034327 ; E. Khan, N. Van Giai and N. Sandulescu 2007, Nuclear Physics A 789, 94;
[2] D. Gambacurta, D. Lacroix 2012, Phys. Rev. C85, 044321;
[3] D. Gambacurta, D. Lacroix and N. Sandulescu 2013, Phys. Rev. C 88, 034324;
[4] C. Esebbag and J. L. Egido 1993, Nucl. Phys. A 552, 205;
[5] T. Sumaryada and Alexander Volya 2007, Phys. Rev. C76, 024319;
[6] Y. Alhassid, G.F. Bertsch, L. Fang 2003, Phys. Rev C 68, 044322;
[7] M. Guttormsen, A. Bjerve, M. Hjorth-Jensen, E. Melby, J. Rekstad, A. Schiller, S. Siem, and A. Belic 2000, Phys. Rev. C 62, 024306;
[8] A. Schiller, *et al.* 2001, Phys. Rev. C 63, 021306.

Journal of Physics: Conference Series **533** (2014) 012013 doi:10.1088/1742-6596/533/1/012013

The phonon vacuum state in a Lipkin model

S. Mishev

Joint Institute for Nuclear Research, 6 Joliot-Curie str., Dubna 141980, Russia

E-mail: mishev@theor.jinr.ru

Abstract. The action of the long-range residual force on the expectation value of observables in the nuclear ground states is evaluated by finding optimal values for the coefficients of the canonical transformation which connects the phonon vacuum state with the independent-particle ground-state using the two-level Lipkin-Meshkov-Glick (LMG) model. After estimating the improvements over the predictions of the independent-particle approximation we compare the ground and first excited states wave functions, obtained using the presented approach, with those, obtained using the conventional random phase approximation (RPA) and its extended version. The problem with overbinding of the nuclear ground state calculated using the RPA is shown to be removed if one sticks to the prescriptions of the present approach. The reason being that the latter conforms to the original variational formulation.

1. Introduction

The purpose of this note is to perform a comparative study for the binding energies and the wave functions of the nuclear ground states approximated by the phonon vacuum state [1, 2, 3] using different methods for evaluating average values of quantum mechanical operators in a many-body wave function. All considerations are performed employing the LMG model [4] for reasons of simplicity. In doing this comparison we subject a variational approach to a test against the true LMG ground state as well as against well established approximations such as RPA[5] and one of its extended variants (ERPA)[6, 7, 8]. A special attention is given to the cases when the interaction between the particles exceeds the inherent to the RPA critical point.

The importance of this development to nuclear structure physics is that it provides a blueprint for a variational theory which is able to account for the effects of the long-range part of the residual nuclear forces in the ground states. In the spirit of the variational method we examine the convergence for some of the attributes of the ground states, with its energy being the most important one, when expanding the configurational space. Building over the obtained ground state using the featured method we head off to construct an one particle-hole excited state and examine its qualities. It is argued that the discrepancies for the first excited states in the strong interaction regime are due to the fact that configurations with more than one particle and one hole give major contribution to this state.

2. Model

2.1. LMG model basics

In order to access the utility of different approaches and to prove the usefulness of the proposed scheme we limit the configuration space and simplify the inter-nucleon interaction to monopole-monopole one as suggested by Lipkin, Meshkov and Glick. This setting permits comparisons between the rates of convergence of different approximation methods, including the hereby described, to the exact solution.

In this model the interaction of N particles on 2 quantum levels is presented by the following Hamiltonian

$$H = H_0 + V;\ H_0 = \varepsilon J_0;\ V = \frac{G}{2}(J_+ + J_-)^2, \tag{1}$$

where

$$J_+ = \sum_i a_{1i}^\dagger a_{-1i};\ J_- = \sum_i a_{-1i}^\dagger a_{1i};\ J_0 = \frac{1}{2}\sum_i (a_{1i}^\dagger a_{1i} - a_{-1i}^\dagger a_{-1i}) \tag{2}$$

are analogous to the raising, lowering and angular momentum' z-component of the quasi-spin algebra respectively, $a^\dagger$ represents the particle creation operator, the suffix ± 1 denotes the upper or lower level, ε is the energy gap between the two levels and G is the interaction strength.

The exact solution for this Hamiltonian is readily obtained after performing a simple diagonalization [9].

2.2. Phonon vacuum solution

The method that we examine is expressed in the language of the LMG model in the following way. The phonon vacuum wave function is taken in the form

$$|) = N_0 e^{\frac{1}{2}SJ_+^2}\rangle. \tag{3}$$

Up to arbitrary order $n \le N/2$ the ground-state energy is obtained to be

$$(|H|) = N_0^2 \sum_n \frac{1}{(n!)^2}\left(\frac{S}{2}\right)^{2n} \langle J_-^{2n} H J_+^{2n}\rangle + 2N_0^2 \sum_n \frac{n}{(n!)^2}\left(\frac{S}{2}\right)^{2n-1} \langle J_-^{2n-2} H J_+^{2n}\rangle$$

with

$$N_0^2 = \left[\sum_n \frac{1}{(n!)^2}\left(\frac{S}{2}\right)^{2n} \langle J_-^{2n} J_+^{2n}\rangle\right]^{-1} \tag{4}$$

The variational equation $\partial_S(|H|) = 0$ then yields the following problem

$$\begin{aligned} & N_0^2 \sum_n \frac{n}{(n!)^2}\left(\frac{S}{2}\right)^{2n-2} \left[\frac{S}{2}\langle J_-^{2n} H J_+^{2n}\rangle + (2n-1)\langle J_-^{2n-2} H J_+^{2n}\rangle\right] + \\ & \partial_S N_0^2 \sum_n \frac{1}{(n!)^2}\left(\frac{S}{2}\right)^{2n-1} \left[\frac{S}{2}\langle J_-^{2n} H J_+^{2n}\rangle + 2n\langle J_-^{2n-2} H J_+^{2n}\rangle\right] = 0. \end{aligned} \tag{5}$$

Respectively, the energy of the $1p - 1h$ excited state is evaluated as

$$\omega = \psi^2(|J_- H J_+|) - 2\psi\varphi(|J_- H J_-|) + \varphi^2(|J_+ H J_-|). \tag{6}$$

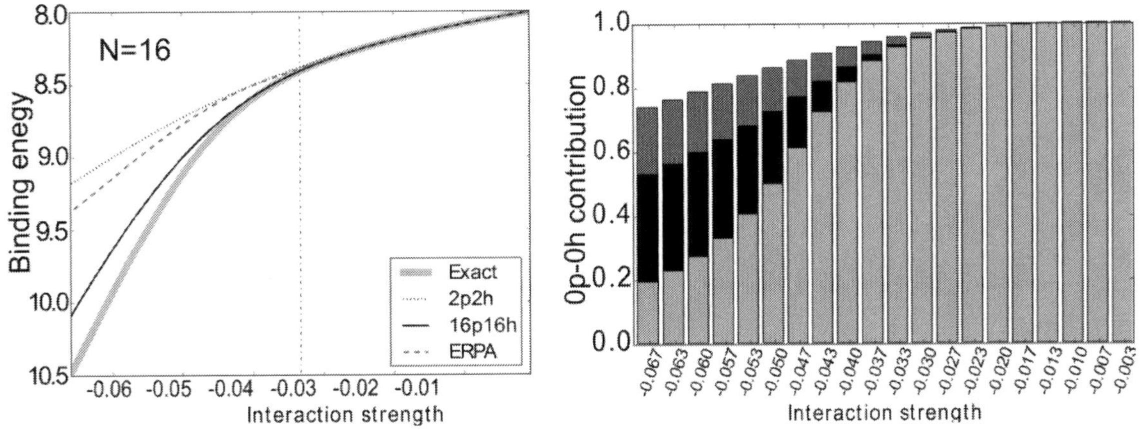

Figure 1. Binding energy (left panel) and contribution of the independent particle state to the correlated ground state (right panel) for a two-level LMG model system with $N = 16$ particles. The energy gap ε between the two LMG levels is set to 1 MeV. *Left panel.* Energies obtained using the phonon vacuum explicit variation with $2p - 2h$ and up $16p - 16h$ admixtures into the ground-state wave function along with the ERPA solution are compared to the exact one. The vertical dash-dotted line indicates the strength at which the RPA experiences a collapse. *Right panel.* The $0p - 0h$ contribution to the exact ground-state wave function (light gray), the phonon vacuum solution from Eq.(3) with $2p - 2h$ admixtures (dark gray) and with correlations of order $N/2$ (black) as function of the interaction strength G.

The expressions for the Hamiltonian average values in Eqs.(5) and (6) are given in Ref. [10]. The forward and backward phonon amplitudes in Eq.(6) are obtained by applying the normalization condition for the one-phonon state

$$N_0^2\left(\psi^2-\varphi^2\right)\sum_n\left(\frac{1}{n!}\right)^2\left(\frac{S}{2}\right)^{2n}(N-4n)\left\langle J_-^{2n}J_+^{2n}\right\rangle=1, \tag{7}$$

along with the definitive equation

$$Q|\rangle=0, \tag{8}$$

resulting in the relation

$$[\varphi-(N-1)\psi S]^2+6\psi^2S^4(N-1)(N-2)=0. \tag{9}$$

Note that the latter relation is independent of the expansion order n.

3. Results

In Figs. 1 and 2 are present results for the energies and the contributions to the wave functions of the ground and first excited states respectively.

From the analysis of the plot in the left panel of Fig. 1 it is concluded that the explicit variation of the phonon vacuum with $2p-2h$ admixtures only in the ground state yields solution at any value of G. However, increasing the interaction strength beyond the RPA critical point (G_{crit}) causes progressive divergence of this solution from the exact one. Adding higher order terms to the energy functional, which account for further correlations effects, greatly improves the results bringing the energy of the phonon vacuum much closer to the exact value. It is also

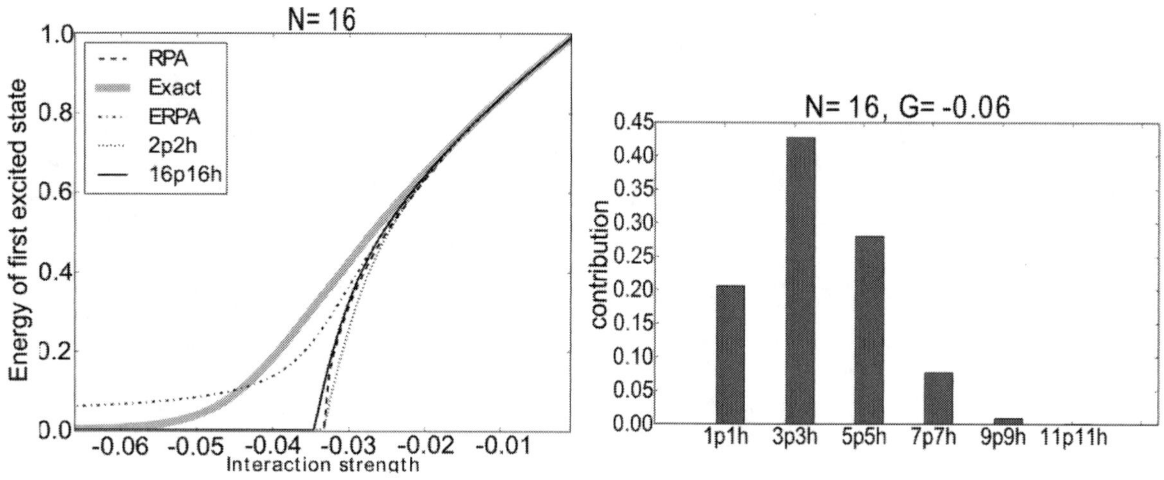

Figure 2. *Left panel.* Same as the left panel of Fig. 1 but for the first excited state with the RPA solution added. *Right panel.* The distribution of the contributions (squared) from different multi-particle-hole configurations to the first excited state in the strong interaction regime.

shown that the ERPA energy strongly underbinds the ground state beyond G_{crit}. However, as seen from the right panel of this figure, despite the reasonable agreement for the ground state energies in this regime the wave function of the phonon vacuum yields a substantially less correlated state than the true ground state.

In contrast, as seen in the left panel of Fig. 2, for the first excited state's energy the ERPA gives rather good results in the strong interaction regime whereas the phonon vacuum variational solution experiences a collapse. The reason for the behavior can be tracked in the right panel of Fig. 2 where it is seen that the higher order particle-hole admixtures contributing in the structure of the first excited state are giving a dominant contribution. Although the reproduction of the energy within the ERPA is good, the wave function of the first excited state happens to be almost orthogonal to the true ground state.

4. Conclusions

The explicit variational approach to the ground states using the phonon vacuum as a trial wave function proves a tractable method for reproducing the binding energies but it underestimates the degree of correlation in the strong interaction regime. Building a $1p1h$ excitations on top of the correlated ground state in order to reproduce the first excited state renders unsatisfactory results in this regime since configurations with a higher number of particles and holes give the major contribution to this state.

References

[1] R.A. Sorensen, Nucl. Phys. 25, 674 (1961).
[2] A. Goswami, M.K. Pal, Nucl. Phys. 44, 294 (1963).
[3] E.A. Sanderson, Phys Lett. 19(2), 141 (1965).
[4] H. Lipkin, N. Meshkov, A. Glick, Nucl. Phys. 62, 188 (1965).
[5] M. Baranger, Phys. Rev. 120, 957 (1960).
[6] K. Hara, Progr. Theor. Phys. 32, 88 (1964).
[7] R. V. Jolos, W. Rybarska, Z. Phys. A 296, 73 (1980).
[8] D. Karadjov, V. V. Voronov, and F. Catara, Phys. Lett. B 306, 197 (1993)
[9] J.A Maruhn, P.-G. Reinhard, and E. Suraud, Simple Models of Many-Fermion systems (Springer, 2010).
[10] S. Mishev, Phys. Rev. C 87, 064310 (2013)

XX International School on Nuclear Physics, Neutron Physics and Applications (Varna2013) IOP Publishing
Journal of Physics: Conference Series **533** (2014) 012014 doi:10.1088/1742-6596/533/1/012014

Selfconsistent mean field calculations of the nuclear response using a realistic nucleon-nucleon interaction with a density dependent corrective term

P Veselý[1], G De Gregorio[1,2], F Knapp[3], N Lo Iudice[1,2], D Bianco[1,2]‡, F Andreozzi[1,2], A Porrino[1,2]

[1] Istituto Nazionale di Fisica Nucleare (INFN), Sezione di Napoli, Monte S. Angelo, via Cintia, I-80126 Napoli, Italy

[2] Dipartimento di Fisica, Università di Napoli Federico II, Monte S. Angelo, via Cintia, I-80126 Napoli, Italy

[3] Faculty of Mathematics and Physics, Charles University, Prague, Czech Republic

Abstract. Tamm-Dancoff and random-phase approximations are formulated in the canonical Hartree-Fock-Bogoliubov quasi-particle basis and adopted to compute the $E1$ response in Ca isotopes using an intrinsic Hamiltonian composed of a V_{lowk} potential, deduced from the CD-Bonn nucleon-nucleon interaction, corrected with phenomenological density dependent and spin-orbit terms. Attention is focused on the evolution of the dipole strength distribution, including the low-lying transitions associated to the pygmy resonance, in going from the $N = Z$ ^{40}Ca to Ca isotopes with neutron excess.

1. Introduction

Following the procedure proposed in Ref. [1], we have solved the QRPA and quasi-particle Tamm-Dancoff approximation (QTDA) eigenvalue equations in a canonical Hartree-Fock-Bogoliubov (HFB) basis starting from a V_{lowk} potential [2], derived from the CD-Bonn NN interaction [3]. This approach may represent an alternative to the generally adopted QRPA calculations rooted in energy density functionals derived either from Skyrme forces [4] or from relativistic classical meson nucleon Lagrangians [5]. As in Ref. [1], we added a phenomenological density dependent potential which reduces drastically the too large spacing between levels produced by the V_{lowk} potential. Moreover, we added a phenomenological spin-orbit term in order to enhance the spin-orbit splitting.

We applied the formalism to neutron rich oxygen and tin isotopes [6] obtaining dipole cross sections in fair agreement with experiments. In nuclei with neutron excess, an appreciable low-energy dipole strength, comparable to the one measured in recent

‡ present address : Ecole Normale Superieur (ENS) de Cachan, 61 av. du Président Wilson, Cachan - France

Journal of Physics: Conference Series **533** (2014) 012014 doi:10.1088/1742-6596/533/1/012014

experiments [7], was obtained. The states in this region were shown to be excited by isovector and isoscalar probes, consistently with recent experimental findings [8]. We have applied here the same method to Ca isotopes and focus our attention at the low-lying states which are supposed to describe the pygmy resonance.

2. QTDA and QRPA $E1$ response in Ca isotopes

The method is illustrated in Ref. [6]. Here we just point that we have solved the QTDA and QRPA eigenvalue equations in the HFB canonical basis. In such a basis, the HFB transformation coefficients u and v from particle to quasi-particle operators assume the simple BCS form. We pay a price for that. The one-body quasi-particle Hamiltonian is no longer diagonal in the quasi-particle operators, which makes more involved the expression of the diagonal (QTDA) block of the QRPA matrix.

We consider an intrinsic Hamiltonian obtained by subtracting the CM kinetic energy T_{CM} from the shell model kinetic operator. It includes a modified one-body kinetic term and a potential $V = V_2 + V_\rho + V_{so}$ composed of three pieces. The first one $V_2 = V_{lowk} + T_2$ is the sum of the CM two-body kinetic piece $T_2 = -1/(2mA)\sum_{i\neq j}\vec{p}_i \cdot \vec{p}_j$ plus a V_{lowk} potential [2] deduced from the NN CD-Bonn force [3] with a cut-off $\Lambda = 1.9\ fm^{-1}$. The second term is a density dependent potential $V_\rho = \sum_{i<j} v_\rho(ij)$, where

$$v_\rho = \frac{C_\rho}{6}(1 + P_\sigma)\rho(\frac{\vec{r}_1 + \vec{r}_2}{2})\delta(\vec{r}_1 - \vec{r}_2). \qquad (1)$$

It was introduced in Ref. [1] and was shown [9] to give to the ground state energy the same contribution of a contact three-body interaction. Our potential V includes also a spin-orbit term $V_{so} = C_{so}\sum_i \vec{l}_i \cdot \vec{s}_i$. The parameter of V_ρ was determined by the request that the peak of giant-dipole-resonance (GDR) be reproduced. The spin-orbit parameter was fitted on the empirical neutron spin-orbit separation energies. Our space, which includes up to 12 major oscillator shells, is not sufficient to decouple completely the 1^- and 0^+ spurious states from the intrinsic ones, as it should be in a fully self-consistent QRPA. On the other hand, we eliminate completely and exactly the spurious admixtures in QTDA by resorting to the Gramm-Schmidt orthogonalization procedure.

One can see from Figure 1 that V_ρ shifts the peaks down by $\sim$ 20MeV into the experimental region. The spin-orbit term induces only a redistribution of the strength. As shown in Figures 2 and 3, the QTDA to QRPA cross sections are almost identical. Both over-exhaust the TRK sum rule by a factor ~ 1.4 for all isotopes. Experimentally, the enhancement factor is ~ 1.4 in the doubly magic nuclei ^{40}Ca and ^{48}Ca and ~ 1.3 in the open shell isotopes. This enhancement is not surprising. In fact, the TRK sum rule holds only if the Hamiltonian does not contain momentum dependent terms. This is not the case for a realistic potential like the one we have adopted. The cross sections, computed with a width $\Delta = 3$ MeV, reproduce the experimental data in ^{40}Ca but underestimate the peak in ^{48}Ca. In the open shell isotopes, the data can be reproduced if a larger width ($\Delta = 6$ MeV) is used. In the isotopes with neutron excess, non negligible peaks appear in the region < 13 MeV. Their summed strengths are

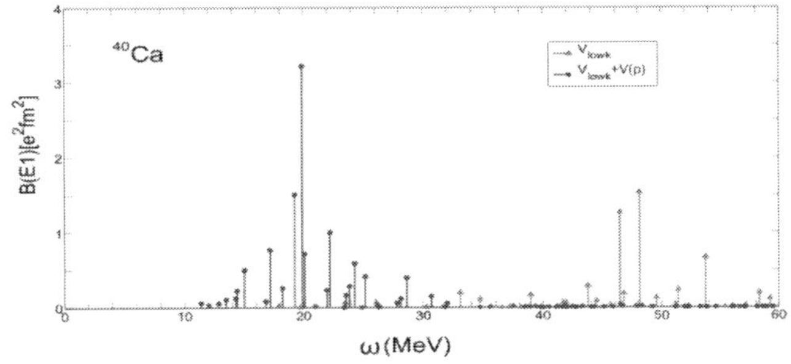

Figure 1. TDA B(E1) in ^{40}Ca with V_{lowk} and with $V_{lowk} + V_\rho$.

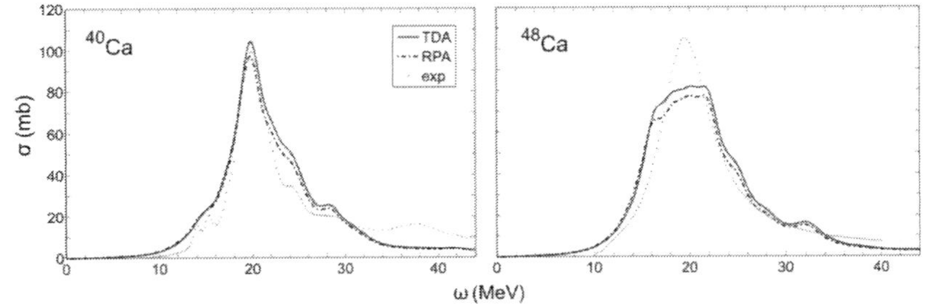

Figure 2. Comparison of the QTDA and QRPA cross sections with the experimental data for ^{40}Ca and ^{48}Ca

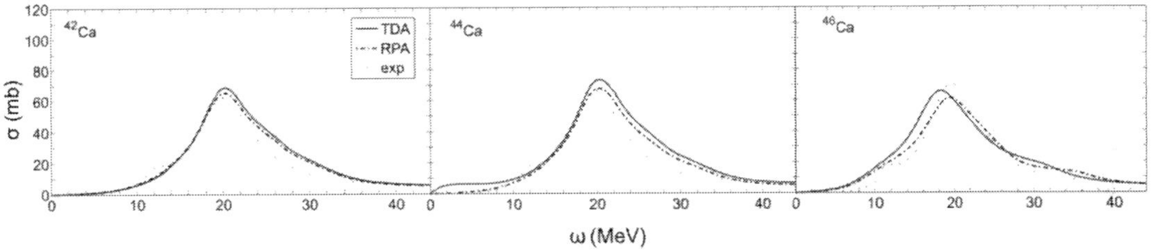

Figure 3. Comparison of the QTDA and QRPA cross sections with the experimental data for ^{42}Ca, ^{44}Ca and ^{46}Ca with $\Delta = 6$ MeV

$\sum B(E1) \sim 0.33\ e^2fm^2$ in ^{42}Ca, $\sum B(E1) \sim 0.13\ e^2fm^2$ in ^{44}Ca, $\sum B(E1) \sim 2\ e^2fm^2$ in ^{46}Ca. In ^{48}Ca only one peak appears at $\omega = 11.4$ MeV and gets a strength $B(E1) = 0.13$ e^2fm. As shown for ^{42}Ca in Figure 4, the 1^- states are excited by both isovector and isoscalar probes. The isoscalar strength is concentrated below and above the region of the $E1$ GDR. The high energy strength is induced by the $\sim 3\hbar\omega_0$ excitations and is associated to the compressional squeezed mode (see [10] and references therein). The low-lying isoscalar and isovector dipole spectra have a specular structure. They are both composed of two fairly high peaks and a small one. Only the lowest 1^- can be associated to the pygmy resonance due to the overwhelming dominance of its neutron component (98.5%). Moreover its dominant quasi-particle component $\mid (1d_{5/2})_n(0f_{7/2})_n\rangle$, with a weight 59%, describes valence neutrons excitations and, therefore, emphasizes the role of neutron skin.

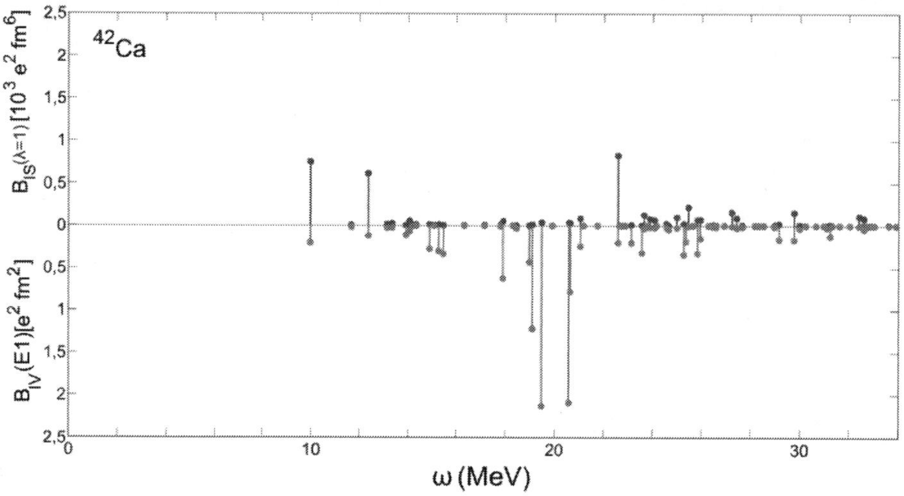

Figure 4. Comparison between isoscalar and isovector B(E1) for ^{42} Ca

In summary, the present calculation has confirmed that it is crucial to add a density dependent two-body potential V_ρ simulating a three-body contact force to V_{lowk}. The spin-orbit term improves the spectra by enhancing the spin-orbit splitting but has only a fine tuning effect on the $E1$ strength distribution. This distribution changes little in going from QTDA to QRPA for all the Ca isotopes and is compatible with the experimental data. Low-lying states collecting a non negligible low-energy dipole strength appear in nuclei with neutron excess. They are excited by both isovector and isoscalar operators and describe excitations of the valence neutrons, two signatures of the pygmy resonance.

References

[1] Hergert H, Papakonstantinou P and Roth R 2011 *Phys. Rev. C* **83** 064317
[2] Bogner S K, Kuo T T S and Coraggio L 2001 *Nucl. Phys. A* **684** 432c
[3] Machleidt R 2001 *Phys. Rev. C* **63** 024001
[4] Bender M, Heenen P-H and Reinhard P-G 2003 *Rev. Mod. Phys.* **75** 121 and references therein
[5] Vretenar D, Afanasjev A V, Lalazissis G A and Ring P 2005 *Phys. Rep.* **409** 101 and references therein
[6] Bianco D, Knapp F, Lo Iudice N, Veselý P, Andreozzi F, De Gregorio G, Porrino A, submitted to J. Phys. G
[7] Adrich P *et al.* 2005 *Phys. Rev. Lett.* **95** 132501
[8] Endres J *et al.* 2010 *Phys. Rev. Lett.* **105** 212503
[9] Waroquier M, Heyde K and Vincx H 1976 *Phys. Rev. C* **13** 1664
[10] Kvasil J, Lo Iudice N, Stoyanov Ch, and Alexa P 2003 *J. Phys. G: Nucl. Phys* **29** 1

Journal of Physics: Conference Series **533** (2014) 012015 doi:10.1088/1742-6596/533/1/012015

Negative parity states in the IVBM

H. G. Ganev

Joint Institute for Nuclear Research, 141980 Dubna, Russia

Abstract. The low-lying spectra including the first few excited positive and negative parity bands of some heavy even-even nuclei from the rare earth and actinide mass regions are investigated within the framework of the symplectic Interacting Vector Boson Model with $Sp(12, R)$ dynamical symmetry group. Symplectic dynamical symmetries allow the change of the number of excitation quanta or phonons building the collective states providing for larger representation spaces and richer subalgebraic structures to incorporate more complex nuclear spectra. The theoretical predictions for the energy levels and transition probabilities between the collective states are compared with experiment and some other collective models incorporating octupole degrees of freedom.

1. Introduction

It is well known [1],[2] that in some mass regions several bands of negative parity are observed in the low-lying nuclear spectra, like $K^{\pi} = 0^-$, 1^- and 2^- bands. The most well-studied of them is the $K^{\pi} = 0^-$ band, usually interpreted as an octupole vibrational band, connected to the ground state band (GSB) by enhanced $E1$ transitions.

Negative parity states have been described within different approaches mainly by inclusion of octupole or/and dipole degrees of freedom. The bands of negative parity states are often associated with the reflection asymmetry in the intrinsic frame of reference. In the geometrical approach this is achieved by including the $\alpha_{30} \equiv \beta_3$ deformation [3]. In the Interacting Boson Model (IBM) the description of negative states requires the introduction of p or/and f bosons with negative parity in addition to the standard s and d bosons (*spdf*-IBM) [4],[5]. An alternative interpretation of the low-lying negative parity states has been provided in different cluster models [6]-[7] in which the dipole degrees of freedom are related with the relative motion of the clusters. Based on the Bohr Hamiltonian different critical point symmetries (CPS) including quadrupole and octupole deformations have been proposed [8]-[10], [11] extending the concept of CPS introduced for the description of positive parity states.

In the present paper we give a unified description of the first few excited low-lying positive and negative parity bands in the even-even nuclei from the the rare earth and actinide mass regions within the framework of the symplectic Interacting Vector Boson Model (IVBM) with $Sp(12, R)$ dynamical symmetry group [12]. It is shown that the negative parity bands arise along with the positive bands (ground state, β and γ) without the introduction of any additional collective degrees of freedom.

2. The IVBM

It was suggested by Bargmann and Moshinsky [13] that two types of bosons are needed for the description of nuclear dynamics. It was shown there that the consideration of only two-body system consisting of two different interacting vector particles will suffice to give a complete description of N three-dimensional oscillators with a quadrupole-quadrupole interaction. The latter can be considered as the underlying basis in the algebraic construction of the *phenomenological* IVBM [12].

The algebraic structure of the IVBM is realized in terms of creation and annihilation operators of two kinds of vector bosons $u_m^\dagger(\alpha)$, $u_m(\alpha)$ $(m = 0, \pm 1)$, which differ in an additional quantum number $\alpha = \pm 1/2$ (or $\alpha = p$ and n)−the projection of the T−spin (an analogue to the F−spin of IBM-2 or the I−spin of the particle-hole IBM). All bilinear combinations of the creation and annihilation operators of the two vector bosons generate the boson representations of the non-compact symplectic group $Sp(12, R)$. Its irreducible representations are infinite dimensional.

We consider the following chain [12],[14]

$$\begin{array}{ccccccc} Sp(12,R) \supset U(6) \supset & SU(3) & \otimes & U(2) & \supset SO(3) & \otimes & U(1), \\ [N]_6 & (\lambda,\mu) & \Longleftrightarrow & (N,T) & K \quad L & & T_0 \end{array} \tag{1}$$

where below the different subgroups the quantum numbers characterizing their irreducible representations are given. We note that within the symmetric irreducible representation $[N]_6$ of $U(6)$ the groups $SU(3)$ and $U(2)$ are mutually complementary [15], i.e. the quantum numbers of the two groups are related: $N = \lambda + 2\mu$ and $T = \lambda/2$. Making use of the latter we can write the basis as

$$\mid [N]_6; (\lambda, \mu); K, L; T_0\rangle = \mid (N, T); K, L; T_0\rangle \tag{2}$$

The basis states associated with the even irreducible representation of the $Sp(12, R)$ can be constructed by the application of powers of raising generators $F_M^L(\alpha, \beta) = \sum_{k,m} C_{1k1m}^{LM} u_k^+(\alpha) u_m^+(\beta)$ of the same group on the vacuum. Each raising operator will increase the number of bosons N by two. The resulting infinite set of basis states so obtained is denoted as (2) and is shown in Table 1.

Table 1. Classification of the basis states.

N	T	$T_0 \cdots$	-3	-2	-1	0	1	2	3	$\cdots$
0	0					$(0,0)$				
2	1				$(2,0)$	$(2,0)$	$(2,0)$			
	0					$(0,1)$				
	2			$(4,0)$	$(4,0)$	$(4,0)$	$(4,0)$	$(4,0)$		
4	1				$(2,1)$	$(2,1)$	$(2,1)$			
	0					$(0,2)$				
$\cdots$	$\cdots$	$\cdots$	$\cdots$	$\cdots$	$\cdots$	$\cdots$	$\cdots$	$\cdots$	$\cdots$	$\cdots$

We use the following Hamiltonian [14]:

$$H = aN + bN^2 + \alpha_3 T^2 + \beta_3 L^2 + \alpha_1 T_0^2 \tag{3}$$

expressed in terms of the first and second order invariant operators of the different subgroups in the chain (1). It is obviously diagonal in the basis (2) and its eigenvalues are just the energies of the nuclear system:

$$E(N, L, T, T_0) = aN + bN^2 + \alpha_3 T(T+1) + \beta_3 L(L+1) + \alpha_1 T_0^2. \tag{4}$$

3. Application

In our application, the most important point is the identification of the experimentally observed states with a certain subset of the basis states (2). In this regard, the following two points are of importance. As we noted, the irreducible representations of $Sp(12, R)$ are infinite dimensional. Then the natural question arises -how to select a certain subset of the basis states given in Table 1? It turns out that an appropriate set of states is given by the so called "stretched states" [16], which are defined as the $SU(3)$ states of the type $(\lambda, \mu) = (\lambda_0, \mu_0 + k)$, where $k = 0, 2, 4, \ldots$. The latter represent dominant $SU(3)$ multiplets in the low-lying collective states. In the symplectic IVBM the change of the number k, which is related in the applications to the angular momentum L of the states, gives rise to the collective bands. The second point concerns the parity of the state. We assume that the one type of two vector bosons, say p-boson, transforms under space reflections as a pseudovector, while the other - n-boson - transforms as a vector. So, we define the parity of the considered collective state as $\pi = (-1)^{N_n}$ which generalizes our previous definition of the parity $\pi = (-1)^T$ given in Ref.[14]. This allows us to describe both positive and negative parity states in the IVBM on the same footing without introducing of any additional collective degrees of freedom.

We consider the first few excited low-lying positive (ground state, β, γ) and negative ($K^\pi = 0^-, 1^-$) parity bands. We take two examples for which there is enough experimental data on $E1$ transitions. In Fig.1(a) we compare our theoretical predictions for the excitation energies in ^{152}Sm with experiment [1] and the results obtained by the diagonalization of the *spdf*-IBM Hamiltonian [17]. This nucleus in the positive parity part of the spectrum is considered as an example of $X(5)$ CPS [18]. The results obtained by the CPS approach [10] are also shown in Fig.1(a). The next example is ^{226}Ra considered in the literature as possessing a stable octupole shape. In Fig.1(b) we compare the results obtained by us in the IVBM and the pure $SU(3)$ dynamical limit of the *spdf*-IBM for this nucleus with experiment [19]. Calculations in the $SU(3)$ limit of *spdf*-IBM are performed using the Hamiltonian and matrix elements given in [21]. For comparison, the predictions of the diagonalized *spdf*-IBM Hamiltonian [20] and CPS approach [9] are also shown. One sees that the IVBM describes reasonably well the low-lying excitation energies of the two nuclei under consideration. Note that in the case of ^{226}Ra, the experimental data show large deviations from the rotational $L(L+1)$ rule ($SU(3)$ limit of the spdf-IBM) for both the ground state and $K^\pi = 0^-$ bands despite the fact that $R_{4/2} = 3.13$.

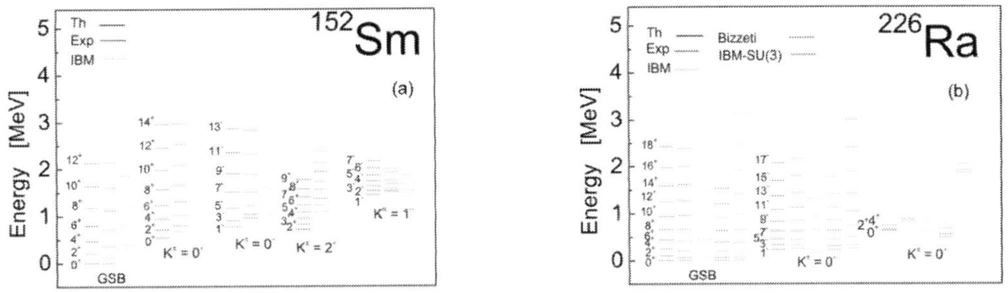

Figure 1. (Color online) Comparison of the theoretical energies for the first few low-lying positive and negative parity bands in ^{152}Sm and ^{226}Ra with experiment and some other collective models incorporating octupole degrees of freedom.

It is well known that the transition probabilities are more sensitive test for each model.

Negative parity states of the $K^\pi = 0^-$ band are characterized by the enhanced $E1$ transitions to the GSB. Thus, we consider only the $E1$ transitions characteristic for the octupole/dipole collectivity. In Fig.2 we compare our theoretical results for the matrix elements of the interband $E1$ transitions between the states of the GSB and $K^\pi = 0^-$ band in ^{226}Ra and ^{148}Nd with experiment [19], [1] and some other collective models. We see that IVBM describes well the experimental data.

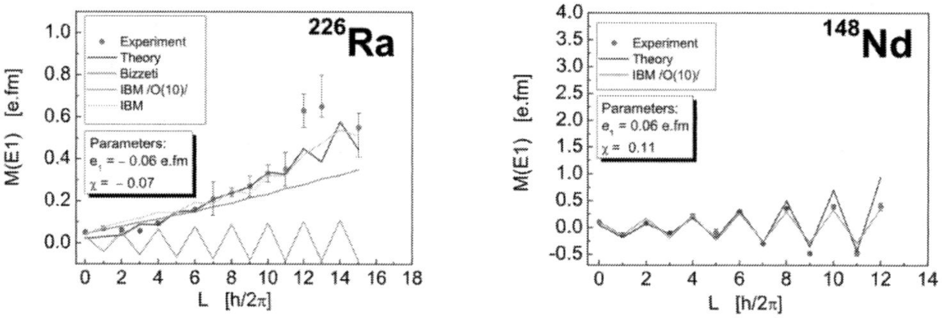

Figure 2. (Color online) Theoretical and experimental values for the matrix elements of the interband $E1$ transitions between the states of the GSB and $K^\pi = 0^-$ band in ^{226}Ra and ^{148}Nd. For comparison, the theoretical predictions of some other collective models incorporating octupole degrees of freedom are also shown.

4. Conclusions

In the present work the low-lying spectra including the first few excited positive and negative parity bands of some heavy even-even nuclei from the rare earth and actinide mass regions are investigated within the framework of the symplectic IVBM with $Sp(12, R)$ dynamical symmetry group. Symplectic dynamical symmetries allow the change of the number of excitation quanta or phonons building the collective states providing for larger representation spaces and richer subalgebraic structures to incorporate more complex nuclear spectra. The theoretical predictions for the energy levels and matrix elements of the $E1$ transitions between the states of the GSB and $K^\pi = 0^-$ band are compared with experiment and some other collective models incorporating octupole degrees of freedom. The IVBM describes well the experimental data. The results obtained for the energy levels and the $E1$ transitions of the considered nuclei prove the correct mapping of the basis states to the experimentally observed ones and reveal the relevance of the used dynamical symmetry of the IVBM in the description of negative parity states.

References

[1] National Nuclear Data Center (NNDC), http://www.nndc.bnl.gov/
[2] P. A. Butler and W. Nazarewicz, Rev. Mod. Phys. **68**, 349 (1996).
[3] A. Bohr and B. R. Mottelson, Nucl. Phys. **4**, 529 (1957).
[4] C. S. Han *et al.*, Phys. Lett. **B163**, 295 (1985).
[5] J. Engel and F. Iachello, Phys. Rev. Lett. **54**, 1126 (1985); Nucl. Phys. **A472**, 61 (1987).
[6] F. Iachello and A. D. Jackson, Phys. Lett. **B108**, 151 (1982); F. Iachello, Nucl. Phys. **A396**, 233c (1983); H. Daley and F. Iachello, Phys. Lett. **B131**, 281 (1983); Ann. Phys. **167**, 73 (1986);
[7] T. M. Shneidman *et al.*, Phys. Lett. **B526**, 322 (2002).
[8] P. G. Bizzeti and A. M. Bizzeti-Sona, Phys. Rev. **C70**, 064319 (2004).
[9] P. G. Bizzeti and A. M. Bizzeti-Sona, Phys. Rev. **C77**, 024320 (2008).
[10] P. G. Bizzeti and A. M. Bizzeti-Sona, Phys. Rev. **C81**, 034320 (2010).
[11] D. Bonatsos *et al.*, Phys. Rev. **C71**, 064309 (2005); D. Lenis and D. Bonatsos, Phys. Lett. **B633**, 474 (2006).

[12] A. Georgieva, P. Raychev, and R. Roussev, J. Phys. **G8**, 1377 (1982).
[13] V. Bargmann and M. Moshynsky, Nucl. Phys. **18**, 697 (1960); Nucl. Phys. **23**, 177 (1961).
[14] H. Ganev, V. P. Garistov, and A. I. Georgieva, Phys. Rev. **C69**, 014305 (2004).
[15] M. Moshinsky and C. Quesne, J. Math. Phys. 11, 1631 (1970).
[16] D. J. Rowe, Rep. Prog. Phys. **48**, 1419 (1985).
[17] M. Babilon *et al.*, Phys. Rev. **C72**, 064302 (2005).
[18] R. F. Casten and N. V. Zamfir, Phys. Rev. Lett. **87**, 052503 (2001).
[19] H. J. Wollersheim *et al.*, Nucl. Phys. **A556**, 261(1993).
[20] N. V. Zamfir and D. Kusnezov, Phys. Rev. **C 63**, 054306 (2001).
[21] H. Y. Ji *et al.*, Nucl. Phys. **A 658**, 197 (1999); G. L. Long *et al.*, Phys. Rev. **C57**, 2301 (1998).

Journal of Physics: Conference Series **533** (2014) 012016 doi:10.1088/1742-6596/533/1/012016

Symmetry Energy and Structure of Exotic Nuclei

M K Gaidarov[1], P Sarriguren[2], A N Antonov[1], E Moya de Guerra[3]
[1]Institute for Nuclear Research and Nuclear Energy, Bulgarian Academy of Sciences, Sofia 1784, Bulgaria
[2]Instituto de Estructura de la Materia, IEM-CSIC, Serrano 123, E-28006 Madrid, Spain
[3]Departamento de Fisica Atomica, Molecular y Nuclear, Facultad de Ciencias Fisicas, Universidad Complutense de Madrid, E-28040 Madrid, Spain

E-mail: gaidarov@inrne.bas.bg

Abstract. The symmetry energy, the neutron pressure and the asymmetric compressibility of spherical Ni, Sn, Pb and deformed Kr, Sm neutron-rich even-even nuclei are calculated within the coherent density fluctuation model using the symmetry energy as a function of density within the Brueckner energy-density functional. The correlation between the thickness of the neutron skin and the characteristics related with the density dependence of the nuclear symmetry energy is investigated for isotopic chains of these nuclei in the framework of the deformed self-consistent mean-field Skyrme HF+BCS method. The mass dependence of the nuclear symmetry energy and the neutron skin thickness are also studied together with the role of the neutron-proton asymmetry. The studied correlations reveal a smoother behavior in the case of spherical nuclei than for deformed ones. We also notice that the neutron skin thickness obtained for ^{208}Pb with SLy4 force is found to be in a good agreement with the recent data. In addition to the interest that this study may have by itself, we give some numerical arguments in proof of the existence of peculiarities of the studied quantities in Ni and Sn isotopic chains that are not present in the Pb chain.

1. Introduction

Great attention has been paid to the nuclear equation of state of isospin asymmetric nuclear matter (ANM), in particular the nuclear matter symmetry energy. The density-dependent symmetry energy governs numerous isospin-dependent properties of nuclei such as the binding energy, the location of the drip lines, the density distributions, as well as the reactions: giant resonances, heavy ion collisions, isospin diffusion, and multifragmentation. At the same time, the nuclear symmetry energy is crucial in the astrophysical calculations of neutron stars, supernova explosions and stellar nucleosynthesis. The neutron skin thickness is one of the observables where symmetry energy shows up in the ground state of nuclei [1]. Neutron skin defined through the rms radii of protons and neutrons depends on the properties of the nuclear surface. The relative differences of the neutron and the proton distributions in this region are sensitive to the symmetry energy at the subsaturation densities.

The main aim of this work is to investigate the relation between the neutron skin thickness and some nuclear matter properties in finite nuclei, such as the symmetry energy at the saturation point s, symmetry pressure p_0 (proportional to the slope of the bulk symmetry energy), and asymmetric compressibility ΔK, considering nuclei in given isotopic chains and within a certain theoretical approach. For this purpose, the Brueckner energy density functional for infinite nuclear matter [2, 3] was applied to calculate s, p_0, and ΔK of medium-heavy and heavy spherical Ni ($A = 74 - 84$), Sn ($A = 124 - 152$), Pb ($A = 202 - 214$) nuclei [4] and deformed Kr ($A = 82 - 120$), Sm ($A = 140 - 156$) isotopes [5] that include surface effects. A theoretical approach that combines the deformed HF+BCS method with Skyrme-type density-dependent effective interactions (e.g., [6]) and the coherent density fluctuation model (CDFM) [7] was used. We would like to note the capability of the CDFM to be applied as an alternative way to make a transition from the properties of nuclear matter to the properties of finite nuclei. Finally, we have analyzed in details the existence of kinks in Ni and Sn isotopic chains and the lack of such a kink for the Pb chain. The kinks are produced because of the sensitivity of the symmetry energy and neutron pressure to the shell structure (see, for instance, the discussion in Refs. [4, 8]).

2. Theoretical framework

The quantity $s^{ANM}(\rho)$, which refers to the infinite system and therefore neglects surface effects, is related to the second derivative of the energy per particle $E(\rho,\delta)$ using its Taylor series expansion in terms of the isospin asymmetry $\delta = (\rho_n - \rho_p)/\rho$ (ρ, ρ_n and ρ_p being the baryon, neutron and proton densities, respectively) (see, e.g., [4]):

$$s^{ANM}(\rho) = \frac{1}{2}\left.\frac{\partial^2 E(\rho,\delta)}{\partial\delta^2}\right|_{\delta=0} = a_4 + \frac{p_0^{ANM}}{\rho_0^2}(\rho-\rho_0) + \frac{\Delta K^{ANM}}{18\rho_0^2}(\rho-\rho_0)^2 + \cdots . \quad (1)$$

In Eq. (1) the parameter a_4 is the symmetry energy at equilibrium ($\rho = \rho_0$). In ANM the pressure p_0^{ANM}, the "slope" parameter L^{ANM} and the curvature ΔK^{ANM} are:

$$p_0^{ANM} = \rho_0^2 \left.\frac{\partial s^{ANM}(\rho)}{\partial\rho}\right|_{\rho=\rho_0}, \quad L^{ANM} = \frac{3p_0^{ANM}}{\rho_0}, \quad \Delta K^{ANM} = 9\rho_0^2 \left.\frac{\partial^2 s^{ANM}(\rho)}{\partial\rho^2}\right|_{\rho=\rho_0}. \quad (2)$$

For instance, using the Brueckner theory, the symmetry energy $s^{ANM}(x)$ has the form:

$$s^{ANM}(x) = 41.7\rho_0^{2/3}(x) + b_4\rho_0(x) + b_5\rho_0^{4/3}(x) + b_6\rho_0^{5/3}(x). \quad (3)$$

Under some approximation the symmetry energy for *finite nuclei* and the related quantities can be written within the CDFM as infinite superpositions of the corresponding ANM quantities weighted by $|f(x)|^2$:

$$s = \int_0^\infty dx|f(x)|^2 s^{ANM}(x), \quad (4)$$

$$p_0 = \int_0^\infty dx|f(x)|^2 p_0^{ANM}(x), \qquad \Delta K = \int_0^\infty dx|f(x)|^2 \Delta K^{ANM}(x). \quad (5)$$

The neutron skin thickness is usually estimated as the difference of the rms radii of neutrons and protons:

$$\Delta R = <r_{\rm n}^2>^{1/2} - <r_{\rm p}^2>^{1/2} . \quad (6)$$

In our calculations the following Skyrme force parametrizations are used: SLy4, Sk3, SGII, and LNS (see, e.g., [4]). Also, we use the proton and neutron densities obtained from self-consistent deformed Hartree-Fock calculations with density-dependent Skyrme interactions [6] and pairing correlations.

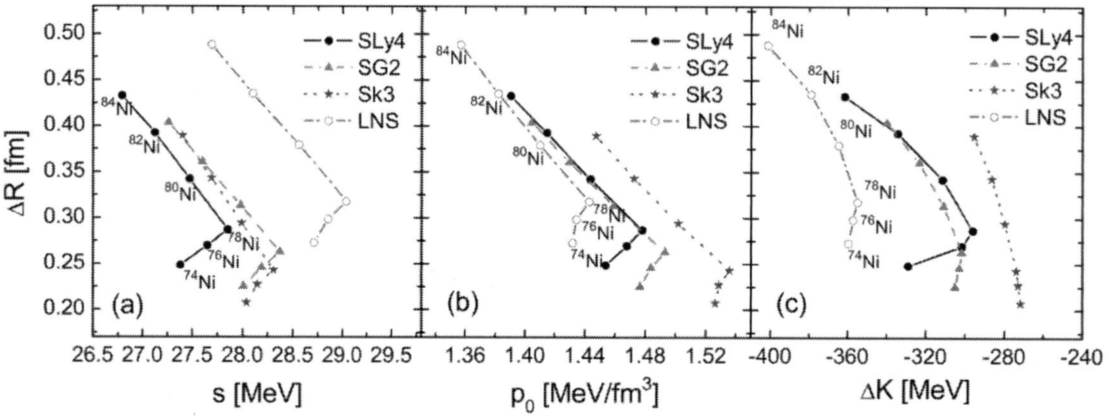

Figure 1. HF+BCS neutron skin thicknesses ΔR for Ni isotopes as a function of the symmetry energy s (a), pressure p_0 (b), and asymmetric compressibility ΔK (c) calculated with SLy4, SG2, Sk3, and LNS forces.

3. Results and discussion

The correlation of the neutron skin thickness ΔR in the Ni isotopic chain with the s, p_0 and ΔK parameters is shown in Fig. 1. It is seen from Fig. 1(a) that there exists an approximate linear dependence between ΔR and s for the even-even Ni isotopes with $A = 74 - 84$. We observe a smooth growth of the symmetry energy till the double-magic nucleus ^{78}Ni ($N = 50$) and then a linear decrease of s while the neutron skin thickness of the isotopes increases. This behavior is valid for all Skyrme parametrizations used in the calculations, in particular, the average slope of ΔR for various forces is almost the same. We also find a similar approximate linear correlation for Ni isotopes between ΔR and p_0 [Fig. 1(b)] and less strong correlation between ΔR and ΔK [Fig. 1(c)]. As in the symmetry energy case, the behavior of the curves drawn in these plots shows the same tendency, namely the inflexion point transition at the double-magic ^{78}Ni nucleus. We would like to note that the predictions for the difference ΔR between the rms radii of neutrons and protons with Skyrme forces obtained in Ref. [9] exhibited a steep change at the same place in which the number of neutrons starts to increase in the chain of nickel isotopes.

In our work we consider a chain of deformed neutron-rich even-even Kr isotopes (including, as well, the case of some extreme neutron-rich nuclei up to ^{120}Kr). The results for the symmetry energy s as a function of the mass number A for the whole Kr isotopic chain are presented in Fig. 2. We observe peaks of the symmetry energy at specific Kr isotopes, namely at semi-magic ^{86}Kr ($N = 50$) and ^{118}Kr ($N = 82$) nuclei. In addition, a flat area is found surrounded by transitional regions $A = 88 - 96$ and $A = 110 - 116$. The results shown in Fig. 2 are closely related to the evolution of the quadrupole parameter β as a function of the mass number A that is presented in Fig. 3. First, one can see from Fig. 3 that the semi-magic $A = 86$ and $A = 118$ Kr isotopes are spherical, while the open-shell Kr isotopes within this chain possess two equilibrium shapes, oblate and prolate. In the case of open-shell isotopes, the oblate and prolate minima are very close in energy and the energy difference is always less than 1 MeV. Nevertheless, we specify in Fig. 3 which shape corresponds to the ground state of each isotope by encircling them. Thus, the trend that the evolution of the symmetry energy shown in Fig. 2 follows can be clearly understood. The peaks of the symmetry energy correspond to the closed-shell nuclei that are spherical. Mid-shell nuclei ($A = 96 - 110$) are well deformed and exhibit a stabilized behavior with small values of s. The transitional regions from spherical to well deformed shapes correspond to transitions from the peaks to the valley in the symmetry energy.

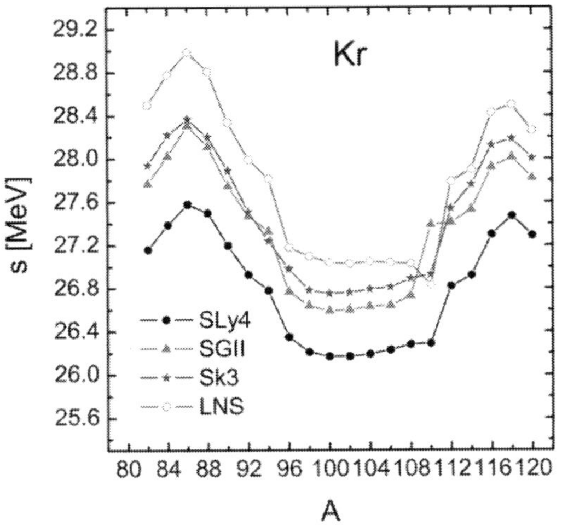

Figure 2. The symmetry energies s for Kr isotopes ($A = 82 - 120$) calculated with SLy4, SGII, Sk3, and LNS forces.

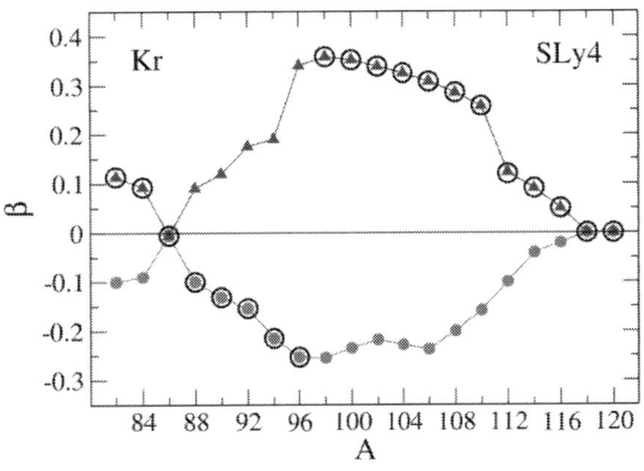

Figure 3. The quadrupole parameter β as a function of the mass number A for the even-even Kr isotopes ($A = 82 - 120$) in the case of SLy4 force.

4. Conclusions

The results of the present work can be summarized as follows:

i) There exists an approximate linear correlation between ΔR and s, as well as p_0, while the relation between ΔR and ΔK is less pronounced. A behavior containing an inflexion point transition at specific shell closure is observed for these correlations (^{78}Ni, ^{132}Sn, ^{86}Kr, and ^{144}Sm).

ii) Our HF+BCS calculations lead to s in the range of 27–30 MeV, which is in agreement with the empirical value of 30 ± 4 MeV. The calculated values of p_0=1.36–1.68 MeV/fm^3 lead to values of the slope parameter L=26–32 MeV, in agreement with other theoretical predictions.

iii) The kinks displayed by the Ni and Sn can be understood as consequences of particular differences in the structure of these nuclei and the resulting densities and weight functions. It is shown that for the Pb isotopes the different signs of the relative deviations corresponding to the range of integration on x that contains the peak of $|f(x)|^2$ (being sensitive to the different density profiles of the considered isotopes around double-magic ones) is in favor of the absence of kink in the Pb chain.

Concluding, we would like to note that the capability of the present method can be further demonstrated by taking into consideration Skyrme-type and relativistic nuclear energy density functionals.

References

[1] Brown B A 2000 *Phys. Rev. Lett.* **85** 5296
[2] Brueckner K A, Buchler J R, Jorna S and Lombard R J 1968 *Phys. Rev.* **171** 1188
[3] Brueckner K A, Buchler J R, Clark R C and Lombard R J 1969 *Phys. Rev.* **181** 1543
[4] Gaidarov M K, Antonov A N, Sarriguren P and Moya de Guerra E 2011 *Phys. Rev.* C **84** 034316
[5] Gaidarov M K, Antonov A N, Sarriguren P and Moya de Guerra E 2012 *Phys. Rev.* C **85** 064319
[6] Moya de Guerra E, Sarriguren P, Caballero J A, Casas M and Sprung D W L 1991 *Nucl. Phys.* A **529** 68
[7] Antonov A N, Hodgson P E and Petkov I Zh 1988 *Nucleon Momentum and Density Distributions in Nuclei* (Clarendon Press, Oxford); 1993 *Nucleon Correlations in Nuclei* (Springer-Verlag, Berlin-Heidelberg-New York)
[8] Warda M, Viñas X, Roca-Maza X and Centelles M 2010 *Phys. Rev.* C **81** 054309
[9] Sarriguren P, Gaidarov M K, Moya de Guerra E and Antonov A N 2007 *Phys. Rev.* C **76** 044322

Journal of Physics: Conference Series **533** (2014) 012017 doi:10.1088/1742-6596/533/1/012017

Description of pairing in finite like-particle systems in a formalism of quartets

M Sambataro[1] **and N Sandulescu**[2]

[1] Istituto Nazionale di Fisica Nucleare - Sezione di Catania, Via S. Sofia 64, I-95123 Catania, Italy

[2] National Institute of Physics and Nuclear Engineering, P.O. Box MG-6, Magurele, Bucharest, Romania

E-mail: michelangelo.sambataro@ct.infn.it, sandulescu@theory.nipne.ro

Abstract. We provide a description of the ground state correlations induced by a general pairing Hamiltonian in a finite system of like fermions in terms of four-body correlated structures (quartets). These are real superpositions of products of two pairs of particles in time-reversed states. Quartets are determined variationally through an iterative sequence of diagonalizations of the Hamiltonian in restricted model spaces and are, in principle, all distinct from one another. The ground state is represented as a product of quartets to which, depending on the number of particles (supposed to be even, in any case), an extra collective pair is added. The extra pair is also determined variationally. We show some realistic applications of the quartet formalism in the case of Sn isotopes.

1. Introduction

Pairing plays a crucial role in the comprehension of a large variety of quantum many-body systems ranging from very small, like nuclei, to very large, like stars. In a mean-field treatment of nuclei, pairing mimics the short range part of the effective nucleon-nucleon interaction. Its action basically consists in scattering the nucleons between pairs of time-reversed single-particle levels. The eigenstates of the pairing Hamiltonian for a finite system of like fermions can be constructed in a (semi)analytical form in the simplifying hypothesis that the interaction strength is independent of the single-particle levels [1, 2]. One finds in this case that, for an even number of particles ($2N$), these eigenstates are a product of distinct, collective pairs of particles in time-reversed states. These pairs can be either real or complex, with reference to their mixing amplitudes, depending on the value of the pairing strength. Complex pairs, in particular, always occur in a complex-conjugate form. It is possible to show [3], that the product of two such complex-conjugate pairs forms a four-body correlated structure with real mixing amplitudes (quartet) irrespective of the pairing strength. The exact ground state of the constant pairing Hamiltonian can therefore be equivalently formulated as a product of real quartets. Quartets are expected to be all distinct from one another and, for odd values of N, they are joined in by an extra collective pair also characterized by real amplitudes. Such a representation of the ground state in terms of quartets holds exactly only for a constant pairing Hamiltonian. We have extended this representation in an approximate form to the case of a general pairing Hamiltonian by resorting to a numerical procedure [4]. In the following we will shortly illustrate how this procedure has been implemented and we will present some of the results that have been obtained. A comparison with a representation of the pairing ground state in terms of independent pairs [5] will also be discussed.

2. The quartet formalism

The general pairing Hamiltonian has the form

$$H = \sum_i \epsilon_i \mathcal{N}_i - \sum_{ii'} g_{ii'} L_i^\dagger L_{i'}, \tag{1}$$

where

$$\mathcal{N}_i = \sum_{m_i} a^\dagger_{im_i} a_{im_i}, \qquad L_i^\dagger = \sum_{m_i>0} P^\dagger_{im_i}, \qquad P^\dagger_{im_i} = a^\dagger_{im_i}\tilde{a}^\dagger_{im_i}. \tag{2}$$

The operator $a^\dagger_{im_i}$ creates a particle on the single particle level i, with energy ϵ_i and degeneracy Ω_i, $\tilde{a}^\dagger_{im_i}$ being the time reversed operator. In the case of pairing in a spherically symmetric mean field, $i \equiv n_i, l_i, j_i$ (adopting the standard notation), $\Omega_i = 2j_i + 1$ and $\tilde{a}^\dagger_{im_i} = (-1)^{j_i - m_i} a^\dagger_{i-m_i}$, m_i being the projection of j_i. In this case, the operator $L_i^\dagger$ creates a pair of particles on the level i with total angular momentum $J = 0$. In the absence of spherical symmetry, $L_i^\dagger$ simply creates a pair of particles with opposite projections of the intrinsic spin on the doubly degenerate level i ($m_i = \pm$ and $\Omega_i = 2$ in this case).

We begin by considering a system with an even number N of pairs and we assume a ground state of the form

$$|\Psi_{gs}\rangle = \prod_{\nu=1}^{N/2} Q^\dagger_\nu |0\rangle, \qquad Q^\dagger_\nu = \sum_{k<k'}^{(1,\Omega)} q^{(\nu)}_{kk'} P^\dagger_k P^\dagger_{k'}. \tag{3}$$

The operator $Q^\dagger_\nu$ is a linear superposition of products of two $P^\dagger_k$ operators (each operator creating a pair of particles in time-reversed states) and the amplitudes $q^{(\nu)}_{kk'}$ are assumed to be real. In the following, we will refer to a four body correlated structure with these general properties as a quartet. $|\Psi_{gs}\rangle$ is therefore a product of quartets which are, in principle, all distinct from one another. In order to search for the most appropriate $q^{(\nu)}_{kk'}$'s we make use of an iterative variational procedure that resorts to diagonalizations of the Hamiltonian in spaces of a rather limited size. This procedure draws inspiration from an analogous technique previously developed for a treatment of pairing correlations in terms of a set of independent pairs [5] and works as follows. Let us suppose that, at a given stage of the iterative process, one knows the state (3) and let us construct the space

$$F^{(\rho)} = \left\{ P^\dagger_k P^\dagger_{k'} \prod_{\nu=1(\nu\neq\rho)}^{N/2} Q^\dagger_\nu |0\rangle \right\}_{1\leq k<k'\leq\Omega}. \tag{4}$$

The states of $F^{(\rho)}$ are generated by acting with the operators $P^\dagger_k P^\dagger_{k'}$ $(k < k')$ on the product of all the quartets $Q^\dagger_\nu$ but the ρ-th one. The dimension of each space (4) is therefore $\Omega(\Omega-1)/2$ and one can form at most $N/2$ such spaces. By diagonalizing the Hamiltonian in $F^{(\rho)}$ and searching for the lowest eigenstate, one constructs the state

$$|\Psi^{(new)}_{gs}\rangle = Q^{\dagger(new)}_\rho \prod_{\nu=1(\nu\neq\rho)}^{N/2} Q^\dagger_\nu |0\rangle. \tag{5}$$

This differs from $|\Psi_{gs}\rangle$ only for the new quartet $Q^{\dagger(new)}_\rho$ and its energy is by construction lower than (or, at worst, equal to) that of $|\Psi_{gs}\rangle$. As a result of this operation, then, the quartet $Q^{\dagger(new)}_\rho$

has updated $Q_\rho^\dagger$ while all the other quartets have remained unchanged. At the same time the energy of $|\Psi_{gs}\rangle$ has been driven towards its minimum. Performing a series of diagonalizations of H in $F^{(\rho)}$ for all possible ρ values ($1 \leq \rho \leq N/2$) exhausts what we define an iterative cycle. At the end of a cycle all the quartets $Q_\nu^\dagger$ have been updated and a new cycle can then start. The sequence of iterative cycles goes on until the difference between the ground state energy at the end of two successive cycles becomes vanishingly small.

The case of an odd number N of pairs proceeds along the same path but one now assumes that the ground state is formed by $(N-1)/2$ quartets and one extra pair. One has, then,

$$|\Psi_{gs}\rangle = B^\dagger \prod_{\nu=1}^{(N-1)/2} Q_\nu^\dagger|0\rangle, \qquad B^\dagger = \sum_{k=1}^{\Omega} \beta_k P_k^\dagger, \tag{6}$$

the quartets $Q_\nu^\dagger$ being defined as in equation (3) and the amplitudes β_k being supposed to be real. In this case, the iterative cycle consists of two different steps. In the first one, one performs a series of diagonalizations in the spaces

$$R^{(\rho)} = \left\{ P_k^\dagger P_{k'}^\dagger B^\dagger \prod_{\nu=1(\nu\neq\rho)}^{(N-1)/2} Q_\nu^\dagger|0\rangle \right\}_{1\leq k<k'\leq\Omega} \tag{7}$$

for all possible ρ values. These diagonalizations update the $(N-1)/2$ quartets while leaving unchanged the pair. Once these have been completed, as a second step, one performs a diagonalization in the space

$$S = \left\{ P_k^\dagger \prod_{\nu=1}^{(N-1)/2} Q_\nu^\dagger|0\rangle \right\}_{1\leq k\leq\Omega} \tag{8}$$

which updates the pair $B^\dagger$ with the quartets acting as spectators. Also in this case, the sequence of iterative cycles goes on up to the convergence of the ground state energy.

3. Ground state results for Sn isotopes

As an application of the procedure just described, we have considered a case of pairing in a spherically symmetric mean field. This application has concerned, in particular, the Sn isotopes with the valence neutrons in the 50-82 neutron shell. As customary for these nuclei, the model space has been restricted to the five neutron orbitals $g_{7/2}$, $d_{5/2}$, $d_{3/2}$, $s_{1/2}$, and $h_{11/2}$. Single-particle energies ϵ_j and pairing strengths $g_{ii'}$ have been taken from the work of Zelevinsky and Volya [6].

In figure 1, we have plotted the relative error in the ground state correlation energy (i.e., the energy of the ground state relative to the lowest uncorrelated many-body configuration) as a function of the number N of pairs. Results obtained with the procedure of section 2 are shown at various levels of approximation, each level corresponding to a different number of iterative cycles (indicated by the number next to each solid line). Already at the lowest level of approximation (1 cycle), the approach is seen to provide a remarkably good agreement with the exact results. This agreement progressively improves with increasing the number of iterative cycles up to becoming basically exact. We remark that the quartets $Q_\nu^\dagger$ of equation (3) are defined in an uncoupled formalism and so they are not characterized by a good angular momentum J. The same is true for the pair $B^\dagger$ of equation (6). In spite of that, $J=0$ pair and quartets naturally arise from the calculations due to the assumed spherical symmetry of the mean field.

It is of interest to compare these results with those which can be obtained in a formalism of independent $J=0$ pairs [5]. The dot-dashed line in figure 1 refers to a calculation in which the pairing ground state has been assumed to be a product of real, collective, distinct, $J=0$ pairs. The approach based on quartets appears to be considerably more effective than the corresponding one with $J=0$ pairs.

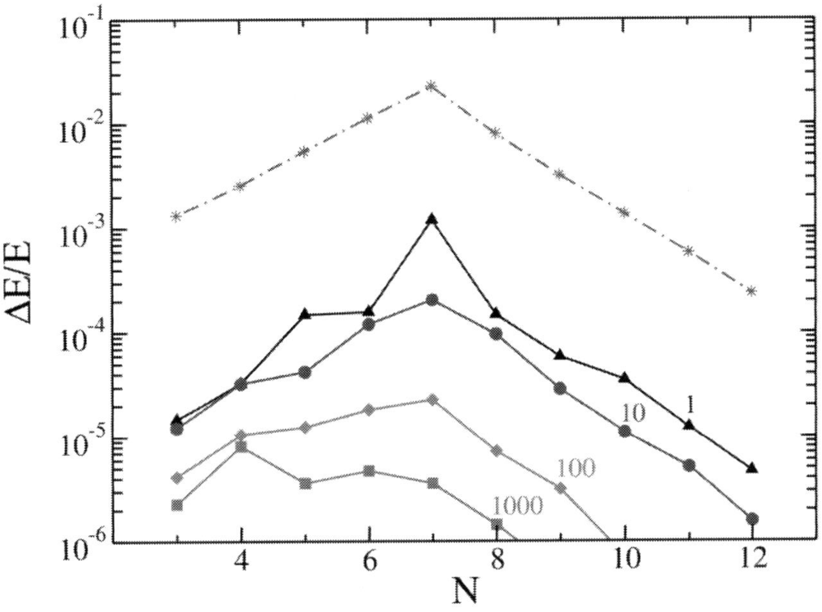

Figure 1. Relative errors in the ground state correlation energy of Sn isotopes with $2N$ active neutrons. Solid lines refer to the procedure of section 2 at various levels of approximations: the number next to each line indicates how many iterative cycles have been carried out in each calculation (see text). The dot-dashed line shows the same quantity calculated with the approach of [5] based on independent $J = 0$ pairs.

4. Conclusions

We have provided a description of the ground state of a general pairing Hamiltonian for a finite system of like particles in a formalism of quartets. We have shown that a product of distinct $J = 0$ quartets (and, in case, an extra $J = 0$ pair) can provide a basically exact representation of the pairing ground state. The quartet formalism has also been extended to excited states [4].

It has been recently observed that the ground state of the proton-neutron isovector pairing Hamiltonian can be very effectively described as a condensate of isospin T=1 quartets [7, 8]. An extension of this quartet condensation model in a formalism of independent quartets constructed with a procedure of the type discussed in section 2 is under way.

Acknowledgments

N.S acknowledges the support from UEFISCDI of Romanian Ministry of Education and Research through Grant IDEI No 57

References

[1] Richardson R W 1963 Phys. Lett. **3** 277
[2] Richardson R W and Sherman N 1964 Nucl. Phys. **52** 221
[3] Sambataro M 2007 Phys. Rev. C **75** 054314
[4] Sambataro M and Sandulescu N 2013 J. Phys. G: Nucl. Part. Phys. **40** 055107
[5] Sambataro M 2012 Phys. Rev. C **85** 064326
[6] Zelevinsky V and Volya A 2003 Phys. of Atomic Nuclei **66** 1781
[7] Sandulescu N, Negrea D, Dukelsky J and Johnson C W 2012 Phys. Rev. C **85** 061303(R)
[8] Sandulescu N, Negrea D and Johnson C W 2012 Phys. Rev. C **86** 041302(R)

XX International School on Nuclear Physics, Neutron Physics and Applications (Varna2013) IOP Publishing
Journal of Physics: Conference Series **533** (2014) 012018 doi:10.1088/1742-6596/533/1/012018

Proton-neutron pairing and alpha-type quartet condensation in nuclei

N. Sandulescu[1], D. Negrea[1], J. Dukelsky[2] and C. W. Johnson[3]

[1] National Institute of Physics and Nuclear Engineering, P.O. Box MG-6, RO-077125, Magurele-Bucharest, Romania

[2] Instituto de Estructura de la Materia, CSIC, Serrano 123, 28006 Madrid, Spain

[3] Department of Physics, San Diego State University, 5500 Campanile Drive, San Diego, CA 92182-1233

Abstract. Proton-neutron pairing in nuclei is commonly described by generalized BCS/HFB models. By construction, these models do not conserve the particle number and the isospin. Here we shall review an alternative approach, based on quartets instead of Cooper pairs, in which both the particle number and the isospin are exactly conserved. In this approach the ground state of N=Z nuclei is approximated by a condensate of alpha-like quartets built by two neutrons and two protons coupled to the isospin T=0. For nuclei with N > Z the ground state is taken as a superposition of a quartet condensate and a pair condensate formed by the neutrons in excess relative to the isotope with N=Z.

Pairing in fermionic systems is associated to the formation of collective Cooper pairs which are supposed to form a pair condensate of BCS-type [1]. In nuclei the BCS approach was first applied to describe like-particle pairing, specific to semi-magic nuclei or heavy nuclei with different valence shells for neutrons and protons [2]. To treat proton-neutron pairing, important in nuclei with protons and neutrons moving in the same major shell, the BCS approach was extended by introducing quasiparticle states which mix proton and neutron degrees of freedom [3]. The main drawback of quasiparticle approximations of BCS-type is the non-conservation of particle number and isospin. Here we shall present an alternative approach to treat the proton-neutron pairing in which the particle number and the isospin are exactly conserved from the outset. In contrast to BCS-type models, this approach is based on a condensate of alpha-like quartets instead of a condensate of Cooper pairs. The fact that in N=Z nuclei the proton-neutron pairing and isospin restoration can induce alpha-like correlations has been discused long ago [4]. In the literature there are also many studies related to the alpha-type condensation in N=Z nuclei [5, 6, 7, 8, 9, 10, 11]. However, majority of these studies have been done with schematic forces or using approximations justified only for a limited number of quartets. A model valid for any number of quartets and for a general charge-independent pairing force has been proposed recently in Refs. [12, 13]. In the present contribution we summarize the main assumptions of the quartet model and we shall show how this model can be applied in Skyrme-HF mean-field models.

The study is focused on a system formed by N neutrons and Z protons moving outside a self-conjugate core and interacting via an isovector pairing force. The Hamiltonian describing this system is

$$\hat{H} = \sum_{i,\tau=\pm 1/2} \varepsilon_{i\tau} N_{i\tau} + \sum_{i,j,t=-1,0,1} V_{ij} P^{+}_{i,t} P_{j,t} \tag{1}$$

where $\varepsilon_{i\tau}$ are the single-particle energies associated to the mean fields of neutrons and protons, supposed invariant to time reversal. The isovector interaction is expressed in terms of the isovector pair operators $P^{+}_{i,1} = \nu^{+}_{i}\nu^{+}_{\bar{i}}$, $P^{+}_{i,-1} = \pi^{+}_{i}\pi^{+}_{\bar{i}}$ and $P^{+}_{i,0} = (\nu^{+}_{i}\pi^{+}_{\bar{i}} + \pi^{+}_{i}\nu^{+}_{\bar{i}})/\sqrt{2}$; the operators ν^{+}_{i} and π^{+}_{i} create, respectively, a neutron and a proton in the state i while $\bar{i}$ denotes the time conjugate of the state i.

The Hamiltonian (1) is commonly treated in the generalised BCS approximation. The alternative explored here is to construct the ground state not in terms of independent T=1 pairs, as in BCS-type models, but with quartets formed by two pairs coupled to the isospin T=0, i.e.,

$$A^{+}_{ij} = [P^{+}_{i} P^{+}_{j}]^{T=0} = \frac{1}{\sqrt{3}}(P^{+}_{i,1}P^{+}_{j,-1} + P^{+}_{i,-1}P^{+}_{j,1} - P^{+}_{i,0}P^{+}_{j,0}). \tag{2}$$

With the non-collective quartets we construct the collective quartet

$$A^{+} = \sum_{i,j} x_{ij} A^{+}_{ij}. \tag{3}$$

Supposing that the amplitudes x_{ij} are separable, i.e., $x_{ij} = x_i x_j$, the collective four-nucleon operator can be written as

$$A^{+} = 2\Gamma^{+}_{1}\Gamma^{+}_{-1} - (\Gamma^{+}_{0})^{2}, \tag{4}$$

where $\Gamma^{+}_{t} = \sum_i x_i P^{+}_{i,t}$ denote, for t=0,1,-1, the collective pair operators for the pn, nn and pp pairs.

Following the analogy with PBCS models, it is supposed that the ground state of even-even N=Z nuclei can be approximated by a condensate of collective quartes

$$|\Psi\rangle = (A^{+})^{n_q}|0\rangle, \tag{5}$$

where $n_q = (N+Z)/4$. Using for the collective quartet the approximation (5), the state (6) can be written as

$$\begin{aligned} |\Psi\rangle &= (2\Gamma^{+}_{1}\Gamma^{+}_{-1} - \Gamma^{+2}_{0})^{n_q}|0\rangle \\ &= \sum_{k} \binom{n_q}{k} (-1)^{n_q-k} 2^{k} (\Gamma^{+}_{1}\Gamma^{+}_{-1})^{k} \Gamma^{+2(n_q-k)}_{0} |0\rangle \end{aligned} \tag{6}$$

From the equation above it can be seen that the quartet condensate is a particular superposition of nn, pp and pn pair condensates.

In Ref. [13] the quartet condensation model (QCM) presented above has been extended to describe isovector pairing correlations in N > Z nuclei. For these nuclei it is supposed that the neutrons in excess form a PBCS pair condensate which is appended to the quartet condensate. Thus, the ground state of N > Z nuclei is approximated by

$$|\Psi\rangle = (\tilde{\Gamma}^{+}_{1})^{n_N}(A^{+})^{n_q}|0\rangle = (\tilde{\Gamma}^{+}_{1})^{n_N}(2\Gamma^{+}_{1}\Gamma^{+}_{-1} - \Gamma^{+2}_{0})^{n_q}|0\rangle \tag{7}$$

where $n_N = (N-Z)/2$ is the number of neutron pairs in excess and $n_q = (N - 2n_N + Z)/4$ is the maximum number of alpha-like quartets which can be formed by the neutrons and protons. Since the quartets A^{+} have zero isospin, the state (7) has a well-defined total isospin given by the excess neutrons, i.e., T=n_N. The neutron pairs in excess are described by the collective pair operator $\tilde{\Gamma}^{+}_{1} = \sum_i y_i P^{+}_{i1}$. It can be seen that the collective pair describing the excess neutrons

Journal of Physics: Conference Series **533** (2014) 012018 doi:10.1088/1742-6596/533/1/012018

is taken of different structure from the collective neutron pair entering in the collective quartet. This is a requirement imposed by the Pauli principle in the HF limit.

One observes that in the state (7) one can identify two terms which play the role of PBCS approximations for N>Z systems interacting with charge-independent pairing forces, i.e.,

$$|PBCS0\rangle = (\tilde{\Gamma}_1^+)^{n_N}(\Gamma_0^+)^{2n_q}|0\rangle \tag{8}$$

$$|PBCS1\rangle = (\tilde{\Gamma}_1^+)^{N/2}(\Gamma_{-1}^+)^{Z/2}|0\rangle. \tag{9}$$

The state (8) is a product between a condensate of proton-neutron pairs and a condensate of neutron-neutron pairs while the state (9) is a product of a condensate of neutron-neutron pairs with a condensate of proton-proton pairs. Both states have the right number of protons and neutrons but have not a well-defined total isospin.

The mixing amplitudes x_i and y_i which define the ground state (7) are determind from the minimization of $\langle\Psi|H|\Psi\rangle$ under the normalization condition $\langle\Psi|\Psi\rangle = 1$. To calculate the average of the hamiltonian and the norm it was used the recurrence relations method. For details see Ref.[13].

Table 1. Pairing correlations energies for isotopes having as core ^{40}Ca. The results correspond to exact diagonalisation (Exact), quartet condensation model (QCM), and the PBCS1 approximation (9). Numbers in the brackets are the errors relative to the exact diagonalisation. The calculations are done with an isovector pairing force of seniority type and with axially-deformed single-particle states.

	Exact	QCM	PBCS1		Exact	QCM	PBCS1
^{44}Ti	3.147	3.142 (0.16%)	2.750 (12.61%)	^{48}Cr	4.248	4.227 (0.49%)	3.854 (9.27%)
^{46}Ti	3.526	3.509 (0.48%)	3.308 (6.18%)	^{50}Cr	4.461	4.444 (0.38%)	4.230 (5.18%)
^{48}Ti	3.882	3.853 (0.75%)	3.735 (3.79%)	^{52}Cr	4.743	4.721 (0.46%)	4.582 (3.39%)
^{50}Ti	3.973	3.956 (0.43%)	3.889 (2.11%)	^{54}Cr	4.869	4.855 (0.29%)	4.772 (1.99%)

To illustrate the accuracy of QCM in Table I are shown the pairing correlations energies for the nuclei with the valence nucleons moving outside the double-magic core ^{40}Ca. One starts with the N=Z isotopes and then add extra neutrons, keeping only those nuclei for which the ground state energy can be calculated exactly by diagonalisation. The results are obtained with a charge-independent pairing interaction of seniority type, with the strength $g = 24/A$, and with protons and neutrons moving in deformed mean fields obtained from axially-deformed Hartree-Fock (HF) calculations [15]. In the calculations we have neglected the Coulomb interaction and we have used for N>Z nuclei the same single-particle energies as for the corresponding N=Z isotope. The QCM results are compared to the exact SM calculations and with the results of PBCS1 approximation (9). It can be noticed that QCM describes with very high precision (errors below 1%) the pairing correlation energies for all calculated isotopes. One can also see that PBCS1 is less accurate, much less than the PBCS approximation for like-particle pairing [14].

In the example presented above, the QCM was applied on a given single-particle spectrum. However, QCM can be also applied for treating the isovector pairing correlations in self-consistent mean field calculations. Thus, in analogy with the HF+BCS calculations the isovector pairing force can be used as a residual interaction acting on the HF single-particle states from the vecinity of Fermi level. The residual pairing interaction is treated by QCM and is iterated with the mean field up to the convergence.

As an application of HF+QCM calculations we discuss here briefly the contribution of isovector pairing correlations to the Wigner energy. To see that we suppose that for an isobaric chain the binding energy, from which are extracted the Coulomb part, can be written as

$$B(N,Z) = B(N=Z) + a_S \frac{|N-Z|^2}{A} + a_w \frac{|N-Z|}{A} \tag{10}$$

Here we consider the binding energies for the even-even isotopes which in the ground state have the isospin $T = T_z$, which is the case for the nuclei with $A > 40$. The equation above can be equivalenty written as

$$B(T) = B(T=0) + \frac{T(T+x)}{2\Phi} \tag{11}$$

The parameter x quantifies the contribution of the linear term, associated to the Wigner energy. The fit with experimental data indicates that for many nuclei $a_S \approx a_W$, which correspond to the value $x \approx 1$. A more systematic analyses show that x is in fact fluctuating quite a lot with the mass region (e.g., see Ref. [16]).

As example we give here the values of x obtained for the isobar chains A=52,68. In the calculations we have used an isovector pairing force with the strength g=24/A and the Skyrme force SLy4. In HF+BCS one gets, respectively, x=0.002 and x=0.06. Thus, in the BCS approximation the isovector pairing gives a negligible contribution to the Wigner energy. On the other hand, in HF+QCM one gets for the two isobaric chains the values x=0.589 and, respectively, x=1.472. These values should be compared to the experimental values, equal to x=1.0 and, respectively, 1.26. One can thus see that in HF+QCM the isovector pairing contributes significantly to the Wigner energy.

In conclusion, in this contribution we have summarized the results of Refs.[12, 13], which show that the ground state of N=Z nuclei can be described with high accuracy by a condensate of alpha-like quartets. In N > Z nuclei with neutrons and protons in the same major shell, the best approximation is obtained when the excess neutrons form a PBCS condensate which is appended to the alpha-like condensate. The comparison with the exact shell model calculations indicates that the quartets and the associated proton-neutron pairing correlations persist far away of N=Z line. Finally, we have shown that the isovector pairing, when treated properly, gives a significant contribution to the Wigner energy.

Acknowledgments

This work was supported by the Romanian Ministry of Education and Research through the grant Idei nr 57 and by the U.S. Department of Energy through the grant DE-FG02-96ER40985. D. N. acknowledges the support from the PhD grant POSDRU/107/1.5/S/82514.

References

[1] J. Bardeen, L. N. Cooper, J. R. Schrieffer, Phys. Rev. **110**, 936 (1958)
[2] A. Bohr, B. Motelson, D. Pines, Phys. Rev. **110** (1958) 936
[3] A. L. Goodman, Adv. Nucl. Phys. **11**, 263 (1979)
[4] A. M. Lane, Nuclear Theory (W. A. Benjamin Inc., 1964)
[5] B. H. Flowers and M. Vujicic, Nucl. Phys. **49**, 585 (1963)
[6] R. Senkov and V. Zelevinsky, Phys. Atom. Nucl. **74**, 1267 (2011)
[7] J. Eichler and M. Yamamura, Nucl. Phys. **A182**, 33 (1972)
[8] Y. K. Gambhir, P. Ring, P. Schuck, Phys. Rev. Lett. **51**, 1235 (1983)
[9] M. Hasegawa, S. Tazaki, R. Okamoto, Nucl. Phys. **A592**, 45 (1995)
[10] J. Dobes and S. Pittel, Phys. Rev **C57**, 688 (1998)
[11] Igal Talmi, Nucl. Phys. **A686** (2001)217
[12] N. Sandulescu, D. Negrea, J. Dukelsky, C. W. Johnson, Phys. Rev. **C85**, 061303(R) (2012)
[13] N. Sandulescu, D. Negrea, C. W. Johnson, Phys. Rev. **C86**, 041302(R) (2012)
[14] N. Sandulescu and G. F. Bertsch, Phys. Rev. **C78**, 064318 (2008)
[15] P. Bonche, H. Flocard, P.H. Heenen, Comp. Phys. Comm. **171** (2005) 49
[16] I. Bentley, S. Frauendorf, Phys. Rev. C88, 014322 (2013)

Journal of Physics: Conference Series **533** (2014) 012019 doi:10.1088/1742-6596/533/1/012019

Pairing properties of hot nuclei within the finite temperature Hartree Fock+BCS approximation

E Yüksel[1], E Khan[2] and K Bozkurt[1]

[1]Physics Department, Yildiz Technical University, 34210 Esenler, Istanbul, Turkey
[2]Institut de Physique Nucléaire, Universite Paris-Sud, IN2P3-CNRS, F-91406 Orsay Cedex, France

E-mail: eyuksel@yildiz.edu.tr

Abstract. In this work, finite temperature Hartree Fock BCS (FT+HFBCS) calculations are performed in Sn isotopes with the zero-range Skyrme type interactions. Vanishing of pairing correlations due to the increase in temperature are investigated with various types of Skyrme interactions. In addition, dependence of the critical temperature with respect to the several properties of these energy density functionals are analysed.

1. Introduction

Pairing correlations constitutes short-range part of the nucleon-nucleon interaction and have been investigated over the years within the mean field models. Pairing correlations are included either with BCS or Bogoliubov approximations at zero temperature [1].

One of the most remarkable results in finite nuclei was obtained with the disappearance of pairing properties at critical temperatures as in the metal superconductors. Experimental signatures for the phase transition in hot nuclei were firstly investigated with the formation of S-shape in the heat capacities of ^{161,162}Dy ^{171,172}Yb [2] and ^{166,167}Er [3] nuclei using level density measurements. The formation of the S-shape in the heat capacities is related to the phase transition of nuclei from superfluid state to the normal state. Also, the critical temperature for quenching of pairing correlations was found at around 0.5 MeV.

Theoretically, investigations about the effect of temperature on finite nuclei have also been carried out within mean-field theories. First studies in hot nuclei were performed with Hartree Fock-BCS model [4, 5]. It has been shown that pairing correlations disappears and nuclei go through a sharp transition from superfluid state to the normal state at a critical temperature. In addition, critical temperature is related to the pairing gap value at zero temperature and follows the $T_c = 0.567\Delta(T = 0)$ equation according to the BCS model of nuclei. Thereafter, finite temperature HFB model has been developed which gave same results as in the BCS model [6-12].

The aim of this proceeding is to investigate the relation between pairing and temperature with various Skyrme energy density functionals. In addition, we investigate the relation between effective nucleon mass and K_{inf} (incompressibility) Skyrme parameters and critical temperature. In Sec 2. We briefly give the main ingredients of the finite temperature HFBCS model (FT+HFBCS). In Sec. 3 the results are presented for Tin isotopic chain and interpreted. The main outcome is summarized in Sec.4.

Journal of Physics: Conference Series **533** (2014) 012019 doi:10.1088/1742-6596/533/1/012019

2. Finite temperature Hartree Fock BCS model

In this work, we used finite temperature HFBCS approach for the study of temperature and pairing correlations in Sn isotopes. BCS approach of nuclei is well known [1]. For the finite temperature extension of BCS model, we used temperature dependent Fermi-Dirac distribution function which is given by,

$$f_k = [1 + \exp(E_k/k_B T)]^{-1} \tag{1}$$

where f_k is the Fermi function, E_k is the quasiparticle energy, k_B is the Boltzmann constant and T is the temperature. For the FT+HFBCS model, all equations keep the same structure except for particle number, gap equation, normal and abnormal densities [7]. Thermal averaged densities: particle density, spin density and kinetic energy density are given as [7, 14]:

$$\rho_T(r) = \frac{1}{4\pi}\sum_k (2j_k + 1)[v_k^*(r)v_k(r)(1 - f_k) + u_k^*(r)u_k(r)f_k], \tag{2}$$

$$J_T(r) = \frac{1}{4\pi}\sum_k (2j_k + 1)\left[j_k(j_k + 1) - l_k(l_k + 1) - \frac{3}{4}\right] \times \{v_k^2(1 - f_k) + u_k^2 f_k\}, \tag{3}$$

$$\tau_T(r) = \frac{1}{4\pi}\sum_k (2j_k + 1)\left\{\left[\left(\frac{dv_k}{dr} - \frac{v_k}{r}\right)^2 + \frac{l_k(l_k+1)}{r^2}v_k^2\right](1 - f_k) + \left[\left(\frac{du_k}{dr} - \frac{u_k}{r}\right)^2 + \frac{l_k(l_k + 1)}{r^2}u_k^2\right] f_k\right\}. \tag{4}$$

The thermal averaged pairing field is given by:

$$\Delta_T(r) = \frac{V_0}{2}\left(1 - \eta\left(\frac{\rho(r)}{\rho_0}\right)\right)\frac{1}{4\pi}\sum_k (2j_k + 1)u_k^*(r)v_k(r)(1 - 2f_i). \tag{5}$$

We used zero-range Skyrme interactions in the calculations. Calculations are performed in a 20 fm box with 0.1 fm mesh point. In the HFBCS framework, the pairing correlations are taken into account with,

$$V_{pair}(\boldsymbol{r}_1, \boldsymbol{r}_2) = V_0\left[1 - \eta\left(\frac{\rho(r)}{\rho_0}\right)\right]\delta(\boldsymbol{r}_1 - \boldsymbol{r}_2) \tag{6}$$

where $\rho(r)$ is the particle density, $\rho_0 = 0.16$ fm^3 is the nuclear saturation density and $\eta=1$ for the surface pairing interaction. The pairing strength V_0 is fixed for each nuclei according to $12/\sqrt{A}$.

3. Numerical results for Tin nuclei

In this work, we employed FT+HFBCS approximation for the selected Sn isotopes. Since we deal with magic proton number Z=50, calculations are performed with average neutron pairing gap value which is defined as [11],

$$< \Delta_n >_k = \frac{\int dr\kappa(r)\Delta_{T,n}(r)}{\int dr\kappa(r)} \tag{7}$$

In here, $\Delta(r)$ and $\kappa(r)$ are the pairing field and pairing tensor which are obtained from the solution of the FT+ HFBCS model.

In Figure 1, we present the change of average neutron pairing gaps as a function of temperature with different Skyrme energy density functionals: SLy5, SGII, SIII and SkM*. According to the BCS theory, critical temperature is related with the pairing gap as; $T_c \approx 0.57\Delta_n(T = 0)$ [4, 5]. In our calculations, disappearance of pairing correlations is obtained around 0.65 MeV for all Skyrme interactions. Phase transition is obtained from superfluid to normal state and critical temperature follows the generally admitted rule: $T_c \approx 0.57\Delta_n(T = 0)$ for all Sn isotopes and Skyrme interactions.

Because of the high temperature, cooper pairs around Fermi level are broken and phase transition from superfluid state to normal state occurred [6].

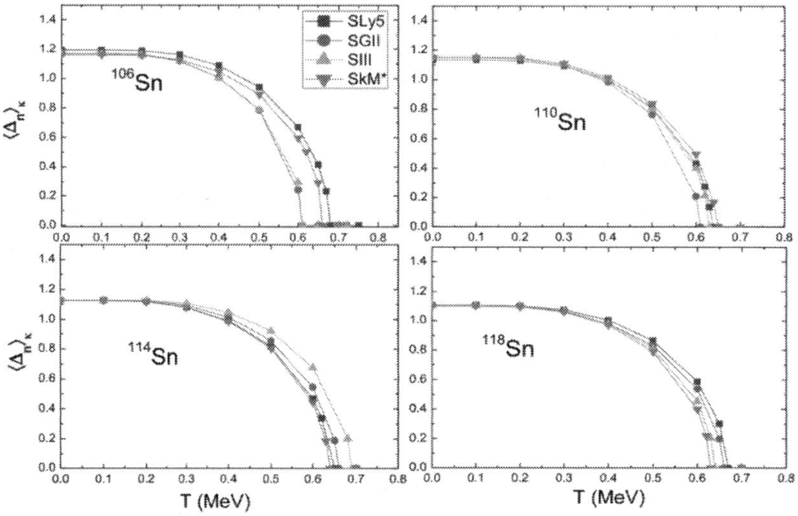

Figure 1. Mean value of the neutron pairing gap as a function of temperature in ^{106,110,114,118}Sn nuclei, calculated with different Skyrme interactions.

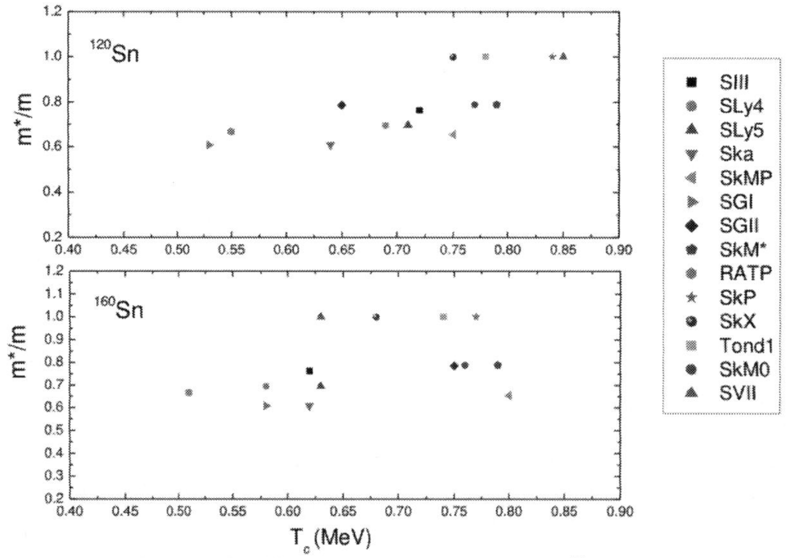

Figure 2. Change of critical temperature with the effective nucleon mass values of the used Skyrme interactions.

To see the effect of various Skyrme energy density functionals on the critical temperature, the relation between the effective mass and K_{inf} parameters of Skyrme interactions are investigated as a function of critical temperatures of ^{120}Sn and ^{160}Sn nuclei. The aim is to check whether the nuclear incompressibility and/or the effective mass of a given interaction has any impact on the critical temperature of nuclei. Since we want to investigate the effect of effective nucleon mass and Kinf parameters, we used constant pairing strength which is $V_0 = 680\ MeV\ fm^3$. In Figure 2, we display the results in terms of the effective nucleon mass and critical temperature of nuclei with 14 different Skyrme interactions. Critical temperature values are obtained between 0.5 MeV and 0.9 MeV for all

Skyrme interactions. A weak relationship is obtained between the effective mass and critical temperature. The effective mass is known to impact the level density which in turn may have an effect on the critical temperature. In Figure 3, we present the results for K_{inf} and critical temperature values of ^{120}Sn and ^{160}Sn nuclei. However no relation is found between K_{inf} and critical temperatures.

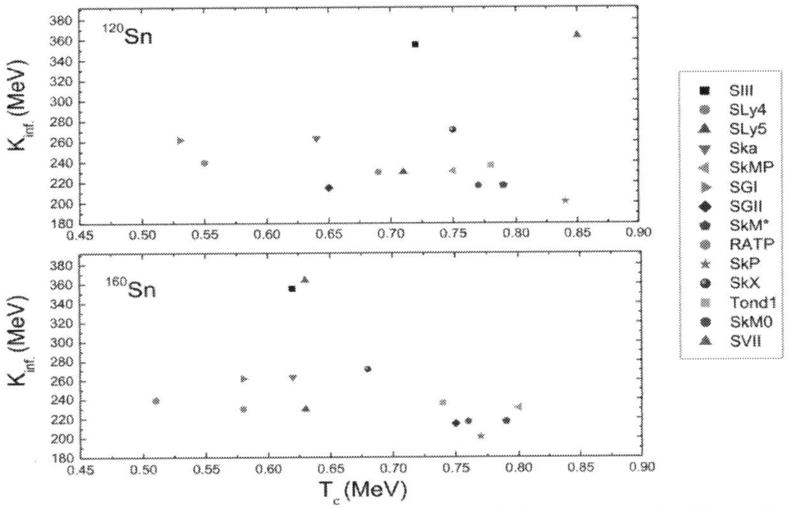

Figure 3. Change of critical temperature with respect to the K_{inf} values of the used Skyrme interactions.

4. Conclusions

Quenching of pairing properties of Tin isotopes with temperature are investigated within the framework of the finite temperature HFBCS approximation by using zero-range Skyrme interactions. All calculations are carried out with surface pairing interaction. The relation between critical temperature and average neutron pairing gap obeys the generally admitted rule in the BCS theory and found around $T_c \approx 0.57\Delta(T = 0)$ with all Skyrme interactions. To see the effect of different Skyrme interactions on the critical temperature values of ^{120}Sn and ^{160}Sn, we investigated the effective nucleon mass and K_{inf} parameters of Skyrme interactions as a function of critical temperature. Although, no meaningful relation is obtained between incompressibility and critical temperature, we obtained a weak correlation in the case of the effective mass.

References

[1] Ring P, Schuck P 1980 *The Nuclear Many-Body Problem* (Verlag, Heidelberg: Springer).
[2] Schiller A, Bjerve A, Guttormsen M, Hjorth-Jensen M, Ingebretsen F, Melby E, Messelt S, Rekstad J, Siem S, and Ødegård S W 2001 *Phys. Rev. C* **63** 021306
[3] Melby E, Guttormsen M, Rekstad J, Schiller A, Siem S, and Voinov A 2001 *Phys. Rev. C* **63** 044309
[4] Sano M, Yamasaki S 1963 *Prog. Theor. Phys.* **29** 397
[5] Bonche P, Levit S and Vautherin V 1984, *Nucl. Phys. A* **428** 95
[6] Egido J L, Robledo L M and Martin V 2000 *Phys. Rev. Lett.* **85** 26
[7] Goodman A L 1981, *Nucl. Phys. A* **352** 30
[8] Goodman A L 1986 *Phys. Rev. C* **34** 1942
[9] Egido J L, Ring P 1993 *J. Phys. G: Nucl. Part. Phys.* **19** 1
[10] Martin V, Egido J L and Robledo L M 2003 *Phys. Rev. C* **68** 034327
[11] Khan E, Van Giai N , Sandulescu N 2007 *Nucl. Phys. A* **789** 94
[12] Khan E, Giai N V, Grasso M 2004 *Nucl. Phys. A* **731** 311

Journal of Physics: Conference Series 533 (2014) 012020 doi:10.1088/1742-6596/533/1/012020

Pseudospin symmetry: Recent progress with supersymmetric quantum mechanics

Haozhao Liang[1,2,*], Jie Meng[2,3], Shihang Shen[2], Nguyen Van Giai[4], Shuangquan Zhang[2], Ying Zhang[2,5], and Pengwei Zhao[2]

[1]RIKEN Nishina Center, Wako 351-0198, Japan
[2]School of Physics, Peking University, Beijing 100871, China
[3]School of Physics and Nuclear Energy Engineering, Beihang University, Beijing 100191, China
[4]Institut de Physique Nucléaire, IN2P3-CNRS, Université Paris-Sud, F-91406 Orsay, France
[5]Graduate School of Science and Technology, Niigata University, Niigata 950-2181, Japan

E-mail: haozhao.liang@riken.jp

Abstract. It is an interesting and open problem to trace the origin of the pseudospin symmetry in nuclear single-particle spectra and its symmetry breaking mechanism in actual nuclei. In this report, we mainly focus on our recent progress on this topic by combining the similarity renormalization group technique, supersymmetric quantum mechanics, and perturbation theory. We found that it is a promising direction to understand the pseudospin symmetry in a quantitative way.

1. Introduction

Although the concept of pseudospin symmetry (PSS) in nuclear single-particle spectra has been introduced for more than 40 years, the origin of PSS and its breaking mechanism in realistic nuclei have not been fully understood. Basically, whether or not its nature is perturbative remains an open problem. In this report, we mainly focus on our recent progress on this topic by combining the similarity renormalization group (SRG) technique, supersymmetric (SUSY) quantum mechanics, and perturbation theory [1, 2].

The PSS was introduced in 1969 to explain the near degeneracy between pairs of single-particle states with the quantum numbers $(n-1, l+2, j=l+3/2)$ and $(n, l, j=l+1/2)$ [3, 4]. They are regarded as the pseudospin doublets with modified quantum numbers $(\tilde{n}=n-1, \tilde{l}=l+1, j=\tilde{l}\pm 1/2)$. Since the suggestion of PSS, there have been comprehensive efforts to understand its origin. In 1997, the PSS was shown to be a relativistic symmetry of the Dirac Hamiltonian, and the equality in magnitude but difference in sign of the scalar $\mathcal{S}(\boldsymbol{r})$ and vector $\mathcal{V}(\boldsymbol{r})$ potentials was suggested as the exact PSS limit [5]. A more general condition $d(\mathcal{S}+\mathcal{V})/dr=0$ [6, 7] can be better satisfied in exotic nuclei with diffuse potentials [8]. On the other hand, since there exist no bound nuclei within such PSS limit, the non-perturbative or dynamical nature of PSS in realistic nuclei was suggested [9]. A recent overview on the PSS investigation can be found in Ref. [1], and the readers are referred to Refs. [10, 11, 12] for some recent progress.

Recently, the perturbation theory was used to investigate the symmetries of the Dirac Hamiltonian and their breaking in realistic nuclei [13, 14], which provides a clear and quantitative

Journal of Physics: Conference Series **533** (2014) 012020 doi:10.1088/1742-6596/533/1/012020

way for investigating the perturbative nature of PSS. On the other hand, the SUSY quantum mechanics can provide a PSS-breaking potential without singularity [15], and naturally interpret the unique feature that all states with $\tilde{l} > 0$ have their own pseudospin partners except for the intruder states [15, 16]. Furthermore, the SRG technique fills the gap between the perturbation calculations and the SUSY descriptions by transforming the Dirac Hamiltonian into a diagonal form which keeps every operator Hermitian [17, 18]. Therefore, we deem it promising to understand the PSS and its breaking mechanism in a quantitative way by combining the SRG technique, SUSY quantum mechanics, and perturbation theory [1, 2].

2. SRG, SUSY quantum mechanics, and perturbation theory

Within the relativistic scheme, our starting point is the Dirac Hamiltonian for nucleons, $H_D = \boldsymbol{\alpha} \cdot \boldsymbol{p} + \beta(M + \mathcal{S}) + \mathcal{V}$, where $\boldsymbol{\alpha}$ and β are the Dirac matrices, M is the nucleon mass, $\mathcal{S}$ and $\mathcal{V}$ are the scalar and vector potentials, respectively.

Using the SRG technique [17], the Dirac Hamiltonian can be transformed into a series of $1/M$ having a diagonal form. By keeping the leading-order terms of the kinetic energy, the central and spin-orbit (SO) potentials, respectively, the eigenequations for nucleons in the Fermi sea read

$$\left[-\frac{1}{2M}\frac{d^2}{dr^2} + \frac{\kappa(\kappa+1)}{2Mr^2} + V(r) + \frac{\kappa}{Mr}U(r)\right] R(r) = ER(r) \tag{1}$$

with $V(r) = \mathcal{V} + \mathcal{S}$ and $U(r) = -(\mathcal{V} - \mathcal{S})'/(4M)$. Here the spherical symmetry is adopted, the symbol $'$ means the derivative with respect to r, and the good quantum number κ is defined as $\kappa = \mp(j + 1/2)$ for $j = l \pm 1/2$. One finds that Eq. (1) is nothing but the Schrödinger equation including a SO term.

In the SUSY framework [19], a couple of Hermitian conjugate first-order differential operators are defined as

$$B_\kappa^+ = \left[Q_\kappa(r) - \frac{d}{dr}\right]\frac{1}{\sqrt{2M}}, \qquad B_\kappa^- = \frac{1}{\sqrt{2M}}\left[Q_\kappa(r) + \frac{d}{dr}\right], \tag{2}$$

where the $Q_\kappa(r)$ are the so-called superpotentials to be determined. In order to explicitly identify the $\kappa(\kappa+1)$ structure and the SO term shown in Eq. (1), one can introduce the reduced superpotentials $q_\kappa(r) = Q_\kappa(r) - \kappa/r - U(r)$. In such a way, the SUSY partner Hamiltonians H_1 and H_2 can be expressed as

$$H_1(\kappa) = B_\kappa^+ B_\kappa^- = \frac{1}{2M}\left[-\frac{d^2}{dr^2} + \frac{\kappa(\kappa+1)}{r^2} + q_\kappa^2 + 2q_\kappa U + U^2 + \frac{2\kappa}{r}q_\kappa - q_\kappa' - U'\right] + \frac{\kappa}{Mr}U, \tag{3a}$$

$$H_2(\kappa) = B_\kappa^- B_\kappa^+ = \frac{1}{2M}\left[-\frac{d^2}{dr^2} + \frac{\kappa(\kappa-1)}{r^2} + q_\kappa^2 + 2q_\kappa U + U^2 + \frac{2\kappa}{r}q_\kappa + q_\kappa' + U'\right] + \frac{\kappa}{Mr}U. \tag{3b}$$

Note that the Hamiltonian $H(\kappa)$ in Eq. (1) and its SUSY partner Hamiltonian $\tilde{H}(\kappa)$ differ from H_1 and H_2 by a common constant $e(\kappa)$, the so-called energy shift [1, 15, 19]. By combining Eqs. (1) and (3a), one obtains the first-order differential equations for the reduced superpotentials $q_\kappa(r)$, and then obtains $\tilde{H}(\kappa)$ with Eq. (3b). Now it is important to see that not only does the $\kappa(\kappa+1)$ structure appear in H (H_1) but also the $\kappa(\kappa-1)$ structure explicitly appears in the SUSY partner Hamiltonian $\tilde{H}$ (H_2). The so-called pseudo-centrifugal barrier (PCB) terms $\kappa(\kappa-1)/(2Mr^2)$ are identical for the pseudospin doublets a and b with $\kappa_a + \kappa_b = 1$.

For the perturbation analysis [13], the Hamiltonian $\tilde{H}$ is further expressed as $\tilde{H} = \tilde{H}_0^{\rm PSS} + \tilde{W}^{\rm PSS}$, where $\tilde{H}_0^{\rm PSS}$ and $\tilde{W}^{\rm PSS}$ are the PSS-conserving and PSS-breaking terms, respectively. By

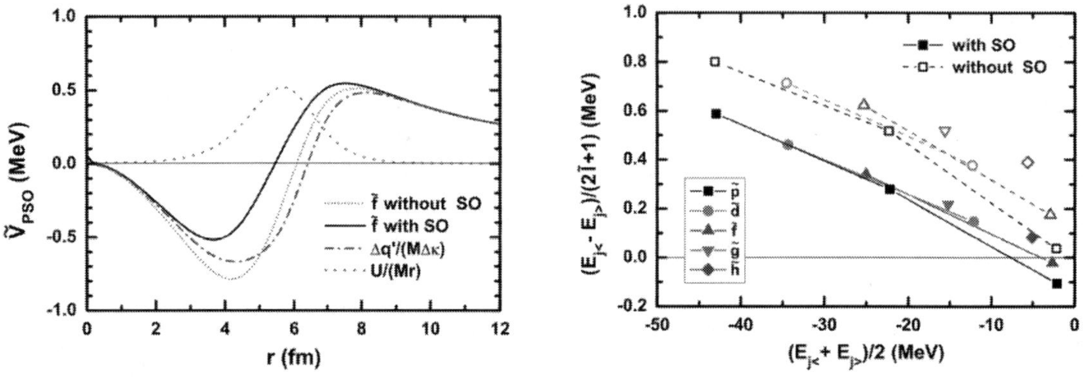

Figure 1. (Color online) Left: Pseudospin-orbit potentials $\tilde{V}_{\rm PSO}(r)$ for the $\tilde{f}$ block. The symmetry-breaking potential obtained with the SO term (solid line) is decomposed into the contributions from the first (dash-dotted line) and second (dotted line) terms on the right-hand side of Eq. (5). The symmetry-breaking potential obtained without SO term is shown with short dotted line for comparison. Right: Reduced pseudospin-orbit splittings $(E_{j_<} - E_{j_>})/(2\tilde{l}+1)$ vs the average single-particle energies $(E_{j_<} + E_{j_>})/2$. The results obtained with and without SO term are shown with filled and open symbols, respectively. Taken from Ref. [2].

requiring that $\tilde{W}^{\rm PSS}$ should be proportional to κ [1, 2], which is similar to the case of the SO term in the normal scheme, one has

$$\tilde{H}_0^{\rm PSS} = \frac{1}{2M}\left[-\frac{d^2}{dr^2} + \frac{\kappa(\kappa-1)}{r^2}\right] + \tilde{V}_{\rm PSS}(r), \qquad \tilde{W}^{\rm PSS} = \kappa\tilde{V}_{\rm PSO}(r). \tag{4}$$

Finally, for a given pair of pseudospin doublets, the pseudospin-orbit (PSO) potential $\tilde{V}_{\rm PSO}(r)$ can be uniquely determined as

$$\tilde{V}_{\rm PSO}(r) = \frac{1}{M}\frac{q'_{\kappa_a}(r) - q'_{\kappa_b}(r)}{\kappa_a - \kappa_b} + \frac{1}{Mr}U(r). \tag{5}$$

3. An example

In this report, we take the scalar and vector potentials of the Woods-Saxon form from Ref. [20], and take the neutron potentials in the nucleus ^{132}Sn as an example.

First of all, the PSS-breaking potentials $\tilde{V}_{\rm PSO}(r)$ for the $\tilde{f}$ block obtained without and with the SO term are shown in the left panel of Fig. 1. These potentials show several special features [1, 2], which are crucial for understanding the PSS in a quantitative way: (1) they are regular functions of r; (2) their amplitudes directly determine the sizes of reduced PSO splittings $\Delta E_{\rm PSO} \equiv (E_{j_<} - E_{j_>})/(2\tilde{l}+1)$ according to the perturbation theory; (3) their shape, being negative at small radius but positive at large radius with a node at the surface region, can explain a general tendency that the PSO splittings become smaller with increasing single-particle energies, and even reverse as approaching the single-particle threshold.

In order to identify the SO effects, the $\tilde{V}_{\rm PSO}(r)$ obtained with the SO term is decomposed into the contributions from the first and second terms on the right-hand side of Eq. (5), denoted as $\Delta q'/M\Delta\kappa$ and U/Mr, respectively. These two terms can be regarded as the indirect and direct effects of the SO term, respectively, because the former one represents the SO effects on $\tilde{V}_{\rm PSO}(r)$ via the reduced superpotentials $q_\kappa(r)$, while the latter is nothing but the SO potential itself appearing in Eq. (1). Comparing to the result obtained without SO term, it is found that the indirect effect on $\tilde{V}_{\rm PSO}(r)$ is only around $0.1 \sim 0.2$ MeV, and eventually results in less influence on $\Delta E_{\rm PSO}$ due to the cancellation between $r < 5$ fm and $r > 5$ fm regions. On the other hand, the SO potential $U(r)/Mr$ is always positive with a surface-peak shape. It substantially raises the $\tilde{V}_{\rm PSO}(r)$, in particular for the surface region. This systematically reduces the PSO splittings $\Delta E_{\rm PSO}$.

All of these properties are confirmed in the right panel of Fig. 1, in which $\Delta E_{\rm PSO}$ for all bound pseudospin doublets are shown as a function of the average single-particle energies $E_{\rm av} = (E_{j_<} + E_{j_>})/2$. The results obtained with and without SO term are shown with filled and open symbols, respectively. It is found that the sizes of $\Delta E_{\rm PSO}$ match the amplitudes of $\tilde{V}_{\rm PSO}(r)$. The decreasing of the PSO splittings with increasing single-particle energies is due to the special shape of $\tilde{V}_{\rm PSO}(r)$. Last but not least, the SO term reduces the $\Delta E_{\rm PSO}$ systematically by $0.15 \sim 0.3$ MeV, and this effect can be understood now in a fully quantitative way.

4. Summary

Work is now in progress for exploring the origin of PSS and its breaking mechanism by combining the SRG, SUSY quantum mechanics, and perturbation theory. It is shown that while the spin-symmetry-conserving term appears in the single-particle Hamiltonian H, the PSS-conserving term appears naturally in its SUSY partner Hamiltonian $\tilde{H}$. The eigenstates of Hamiltonians H and $\tilde{H}$ are exactly one-to-one identical except for the so-called intruder states. In such a way, the origin of PSS deeply hidden in H can be traced in its SUSY partner Hamiltonian $\tilde{H}$. Furthermore, the perturbative nature of PSS is demonstrated by the perturbation calculations, and the PSS-breaking term can be regarded as a small perturbation on the exact PSS limits, so that the special patterns of the PSO splittings can be interpreted in a quantitative way.

Acknowledgements

This work was partly supported by the RIKEN iTHES Project, the Grant-in-Aid for JSPS Fellows under Grant No. 24-02201; the Major State 973 Program No. 2013CB834400; the NSFC under Grants No. 11105005, No. 11105006, and No. 11175002; the China Postdoctoral Science Foundation under Grant No. 2012M520101; and the Research Fund for the Doctoral Program of Higher Education under Grant No. 20110001110087.

References

[1] Liang H, Shen S, Zhao P and Meng J 2013 *Phys. Rev. C* **87** 014334
[2] Shen S, Liang H, Zhao P, Zhang S and Meng J 2013 *Phys. Rev. C* **88** 024311
[3] Arima A, Harvey M and Shimizu K 1969 *Phys. Lett. B* **30** 517
[4] Hecht K and Adler A 1969 *Nucl. Phys. A* **137** 129
[5] Ginocchio J N 1997 *Phys. Rev. Lett.* **78** 436
[6] Meng J, Sugawara-Tanabe K, Yamaji S, Ring P and Arima A 1998 *Phys. Rev. C* **58** R628
[7] Sugawara-Tanabe K and Arima A 1998 *Phys. Rev. C* **58** R3065
[8] Meng J, Sugawara-Tanabe K, Yamaji S and Arima A 1999 *Phys. Rev. C* **59** 154
[9] Alberto P, Fiolhais M, Malheiro M, Delfino A and Chiapparini M 2001 *Phys. Rev. Lett.* **86** 5015
[10] Lu B N, Zhao E G and Zhou S G 2012 *Phys. Rev. Lett.* **109** 072501; 2013 *Phys. Rev. C* **88** 024323
[11] Jolos R V, Shirikova N Y and Sushkov A V 2012 *Phys. Rev. C* **86** 044320
[12] Alberto P, de Castro A S and Malheiro M 2013 *Phys. Rev. C* **87** 031301(R)
[13] Liang H, Zhao P, Zhang Y, Meng J and Van Giai N 2011 *Phys. Rev. C* **83** 041301(R)
[14] Li F Q, Zhao P W and Liang H Z 2011 *Chin. Phys. C* **35** 825
[15] Typel S 2008 *Nucl. Phys. A* **806** 156
[16] Leviatan A 2004 *Phys. Rev. Lett.* **92** 202501; 2009 *ibid.* **103** 042502
[17] Guo J Y 2012 *Phys. Rev. C* **85** 021302(R)
[18] Li D P, Chen S W and Guo J Y 2013 *Phys. Rev. C* **87** 044311
[19] Cooper F, Khare A and Sukhatme U 1995 *Phys. Rep.* **251** 267
[20] Koepf W and Ring P 1991 *Z. Phys. A* **339** 81

Journal of Physics: Conference Series **533** (2014) 012021 doi:10.1088/1742-6596/533/1/012021

Recent progresses on the pseudospin symmetry in single particle resonant states

Bing-Nan Lu[1], **En-Guang Zhao**[1,2] **and Shan-Gui Zhou**[1,2]

[1] State Key Laboratory of Theoretical Physics, Institute of Theoretical Physics, Chinese Academy of Sciences, Beijing 100190, China

[2] Center of Theoretical Nuclear Physics, National Laboratory of Heavy Ion Accelerator, Lanzhou 730000, China

E-mail: sgzhou@itp.ac.cn (Shan-Gui Zhou)

Abstract. The pseudospin symmetry (PSS) is a relativistic dynamical symmetry directly connected with the small component of the nucleon Dirac wave function. Much effort has been made to study this symmetry in bound states. Recently, a rigorous justification of the PSS in single particle resonant states was achieved by examining the asymptotic behaviors of the radial Dirac wave functions: The PSS in single particle resonant states in nuclei is conserved exactly when the attractive scalar and repulsive vector potentials have the same magnitude but opposite sign. Several issues related to the exact conservation and breaking mechanism of the PSS in single particle resonances were investigated by employing spherical square well potentials in which the PSS breaking part can be well isolated in the Jost function. A threshold effect in the energy splitting and an anomaly in the width splitting of pseudospin partners were found when the depth of the square well potential varies from zero to a finite value.

1. Introduction

In 1969, the pseudospin symmetry (PSS) was found in atomic nuclei: Doublets of single particle levels with quantum numbers $(n_r, l, j = l+1/2)$ and $(n_r - 1, l+2, j = l+3/2)$ in the same major shell are nearly degenerate [1, 2]. Since the PSS was observed, much effort has been devoted to investigate its origin [3, 4, 5, 6, 7]. Finally, it was revealed that the PSS in nuclei is a relativistic symmetry which is exactly conserved when the scalar potential $S(r)$ and the vector potential $V(r)$ have the same size but opposite sign, i.e., $\Sigma(r) \equiv S(r) + V(r) = 0$ [8], or, more generally, when $d\Sigma(r)/dr = 0$ [9, 10]. The PSS is usually viewed as a dynamical symmetry [11, 12] because, in either limit, $\Sigma(r) = 0$ or $d\Sigma(r)/dr = 0$, there are no bound states in realistic nuclei, thus in realistic nuclei the PSS is always broken.

The spin symmetry (SS) and PSS have been investigated extensively. The readers are referred to Ref. [13] for a review and to Refs. [14, 15] for overviews of recent progresses. Here we simply mention that the study of the PSS has been extended along several lines:

(i) From single particle states in the Fermi sea to those in the Dirac sea, i.e., the SS in anti-nucleon spectra [16, 17, 18, 19].

(ii) From normal nuclei to hypernuclei [20, 21, 22].

(iii) From spherical systems to deformed ones [23, 24, 25, 26].

(iv) From bound states to resonant states; next we will focus on this extension.

There has been a great interest in the exploration of continuum and resonant statesin atomic nuclei [27, 28, 29]. The study of symmetries, e.g., the PSS, in resonant states is very interesting. There have been some numerical investigations of the PSS in single particle resonances [30, 31, 32, 33, 34] and the SS in single particle resonant states [35]. Recently, we gave a rigorous justification of the PSS in single particle resonant states and investigated several open problems related to the exact conservation and breaking mechanism of the PSS in single particle resonances [36, 37, 15]. In this contribution, we will present some results concerning the PSS in single particle resonant states.

2. Justification of the PSS in resonant states

In atomic nuclei, the equation of motion for nucleons, the Dirac equation, reads,

$$[\boldsymbol{\alpha} \cdot \boldsymbol{p} + \beta(M + S(\boldsymbol{r})) + V(\boldsymbol{r})]\,\psi(\boldsymbol{r}) = \epsilon\psi(\boldsymbol{r}), \tag{1}$$

where $\boldsymbol{\alpha}$ and β are Dirac matrices, M is the nucleon mass, and $S(\boldsymbol{r})$ and $V(\boldsymbol{r})$ the scalar and vector potentials.

For a spherical nucleus, the upper and lower components of the Dirac spinor are $iF_{n\kappa}(r)/r \cdot Y^l_{jm}(\theta,\phi)$ and $-G_{\tilde{n}\kappa}(r)/r \cdot Y^{\tilde{l}}_{jm}(\theta,\phi)$, respectively. One can derive a Schrödinger-like equation for the lower component,

$$\left[\frac{d^2}{dr^2} - \frac{1}{M_-(r)}\frac{d\Sigma(r)}{dr}\frac{d}{dr} - \frac{\tilde{l}(\tilde{l}+1)}{r^2} + \frac{1}{M_-(r)}\frac{\kappa}{r}\frac{d\Sigma(r)}{dr} - M_+(r)M_-(r)\right]G(r) = 0, \tag{2}$$

where $M_+(r) = M + \epsilon - \Delta(r)$ and $M_-(r) = M - \epsilon + \Sigma(r)$.

For the continuum in the Fermi sea, $\epsilon \geq M$; the regular solution for Eq. (2) vanishes at the origin and at large r,

$$G(r) = \frac{i}{2}\left[\mathcal{J}^G_\kappa(p)h^-_{\tilde{l}}(pr) - \mathcal{J}^G_\kappa(p)^* h^+_{\tilde{l}}(pr)\right], \quad r \to \infty, \tag{3}$$

where $h^{\pm}_{\tilde{l}}(pr)$ are the Ricatti-Hankel functions. $\mathcal{J}_\kappa(p)$ is the Jost function which is an analytic function of p. The zeros of $\mathcal{J}^G_\kappa(p)$ on the lower p plane and near the real axis correspond to resonant states. The resonance energy E and width Γ are determined by the relation $E - i\Gamma/2 = \epsilon - M$.

In the PSS limit, Eq. (2) is reduced as

$$\left[\frac{d^2}{dr^2} - \frac{\tilde{l}(\tilde{l}+1)}{r^2} + (\epsilon - M)\,M_+(r)\right]G(r) = 0. \tag{4}$$

For pseudospin doublets with κ and $\kappa' = -\kappa + 1$, the small components satisfy the same equation because they have the same pseudo orbital angular momentum $\tilde{l}$. For continuum states, $G_\kappa(\epsilon, r) = G_{\kappa'}(\epsilon, r)$ is true for any energy ϵ thus we have $\mathcal{J}^G_{\kappa'}(p) = \mathcal{J}^G_\kappa(p)$. Thus the zeros are the same for $\mathcal{J}^G_{\kappa'}(p)$ and $\mathcal{J}^G_\kappa(p)$ and the PSS in single particle resonant states in nuclei is exactly conserved [36].

3. Numerical Results and Discussions

Spherical square well potentials for $\Sigma(r)$ and $\Delta(r)$ read,

$$\Sigma(r) = \begin{cases} C, & r < R, \\ 0, & r \geq R, \end{cases} \qquad \Delta(r) = \begin{cases} D, & r < R, \\ 0, & r \geq R, \end{cases} \tag{5}$$

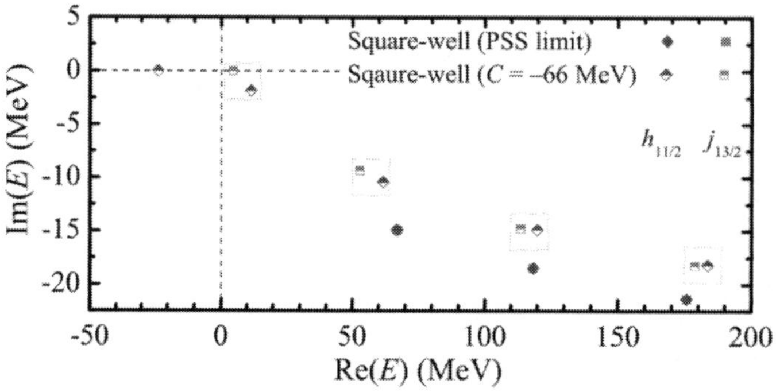

Figure 1. (Color online) The zeros of the Jost function $\mathcal{J}_\kappa^G$ on the complex energy plane in square-well potentials (5) with $C = 0$ (solid symbols) and $C = -66$ MeV (half-filled symbols) for PS partners $h_{11/2}$ (diamond) and $j_{13/2}$ (square).

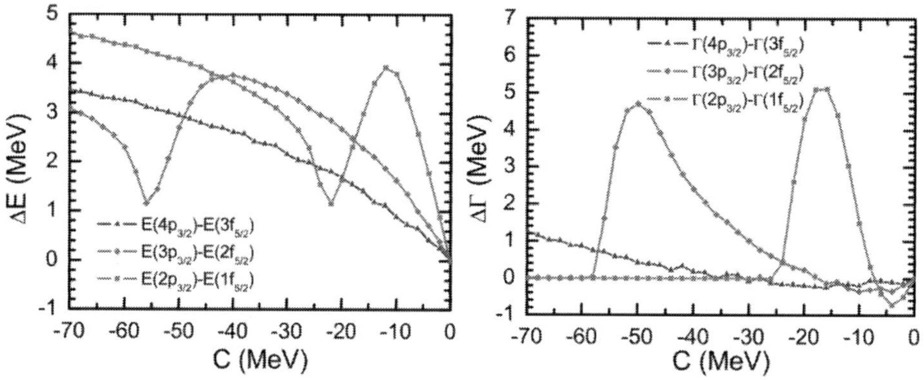

Figure 2. (Color online) The energy (the left panel) and width (the right panel) splitting between PSS partners with the pseudo orbital angular momentum $\tilde{l} = 2$, $p_{3/2}$ and $f_{5/2}$, as a function of the potential depth C. Taken from Ref. [15].

where C and D are depths and R is the width. The Jost function $\mathcal{J}_\kappa^G(p)$ is derived as [36],

$$\mathcal{J}_\kappa^G(p) = -\frac{p^{\tilde{l}}}{2ik^{\tilde{l}+1}}\left\{j_{\tilde{l}}(kR)ph_{\tilde{l}}^{+\prime}(pR) - kj_{\tilde{l}}^{\prime}(kR)h_{\tilde{l}}^{+}(pR) - b\left[kj_{\tilde{l}}^{\prime}(kR) - \frac{\kappa}{R}j_{\tilde{l}}(kR)\right]h_{\tilde{l}}^{+}(pR)\right\}, \quad (6)$$

with $b = C/(\epsilon - M - C)$ and $k = \sqrt{(\epsilon - C - M)(\epsilon - D + M)}$. By examining the zeros of the Jost function (6), one gets single particle bound states and resonant states.

Figure 1 shows solutions for PSS doublets with $\tilde{l} = 6$, i.e., $h_{11/2}$ with $\kappa = -6$ and $j_{13/2}$ with $\kappa' = 7$ for square-well potentials with $D = 650$ MeV and $R = 7$ fm. The conservation of the PSS for single particle resonant states is seen: When $C = 0$, the roots for $h_{11/2}$ and $j_{13/2}$ are the same. When $C = -66$ MeV, the PSS is broken.

By examining the Jost function, one can trace continuously the PSS partners from the PSS limit to the case with a finite potential depth. This has been done in Ref. [15] in which we found a threshold effect in the energy splitting and an anomaly in the width splitting of pseudospin partners when C varies from zero to a finite value. As the depth of the single particle potential becomes deeper, the PSS is broken more and a threshold effect in the energy splitting appears which can be seen in Fig. 2: The energy splitting first increases then decreases until the pseudospin doublets encounter the threshold where one of the levels becomes a bound state and the splitting takes a minimum value; When the potential becomes even deeper, the splitting increases again. When the depth of the single particle potential increases from zero, there also appears an anomaly in the width splitting of PS partners which is shown in Fig. 2: It first decreases from zero to a maximum value with negative sign, then increases and becomes zero again; after the inversion of the width splitting, the splitting increases and reaches a maximum and positive value, then it becomes smaller and finally reaches zero. Note that for the doublet $4p_{3/2}$ and $3f_{5/2}$, the threshold is much below -70 MeV and in Fig. 2 the splittings in energy and width always increase with $|C|$ increasing.

Finally we mention that there are still many open problems concerning the exact and breaking of the PSS in single particle resonant states; the readers are referred to Refs. [15, 37] for further discussions.

Acknowledgments
This work has been supported by MOST of China (973 Program: 2013CB834400), NSF of China (11121403, 11175252, 11120101005, 11211120152, and 11275248), and the Knowledge Innovation Project of CAS (KJCX2-EW-N01).

References

[1] Arima A, Harvey M and Shimizu K 1969 *Phys. Lett. B* **30** 517–522
[2] Hecht K T and Adler A 1969 *Nucl. Phys. A* **137** 129–143
[3] Bohr A, Hamamoto I and Mottelson B R 1982 *Phys. Scr.* **26** 267–272
[4] Dudek J, Nazarewicz W, Szymanski Z and Leander G A 1987 *Phys. Rev. Lett.* **59** 1405–1408
[5] Castanos O, Moshinsky M and Quesne C 1992 *Phys. Lett. B* **277** 238–242
[6] Bahri C, Draayer J P and Moszkowski S A 1992 *Phys. Rev. Lett.* **68** 2133–2136
[7] Blokhin A L, Bahri C and Draayer J P 1995 *Phys. Rev. Lett.* **74** 4149–4152
[8] Ginocchio J N 1997 *Phys. Rev. Lett.* **78** 436–439
[9] Meng J, Sugawara-Tanabe K, Yamaji S, Ring P and Arima A 1998 *Phys. Rev. C* **58** R628–631
[10] Meng J, Sugawara-Tanabe K, Yamaji S and Arima A 1999 *Phys. Rev. C* **59** 154–163
[11] Arima A 1999 Dynamical symmetries and nuclear structure RIKEN-AF-NP-276
[12] Alberto P, Fiolhais M, Malheiro M, Delfino A and Chiapparini M 2001 *Phys. Rev. Lett.* **86** 5015–5018
[13] Ginocchio J N 2005 *Phys. Rep.* **414** 165–261
[14] Liang H, Shen S, Zhao P and Meng J 2013 *Phys. Rev. C* **87** 014334–13
[15] Lu B N, Zhao E G and Zhou S G 2013 *Phys. Rev. C* **88** 024323–11
[16] Ginocchio J N 1999 *Phys. Rep.* **315** 231–240
[17] Zhou S G, Meng J and Ring P 2003 *Phys. Rev. Lett.* **91** 262501–4
[18] He X T, Zhou S G, Meng J, Zhao E G and Scheid W 2006 *Eur. Phys. J. A* **28** 265–269
[19] Liang H, Long W H, Meng J and Van Giai N 2010 *Eur. Phys. J. A* **44** 119–124
[20] Song C Y, Yao J M and Meng J 2009 *Chin. Phys. Lett.* **26** 122102–3
[21] Song C Y, Yao J M and Meng J 2011 *Chin. Phys. Lett.* **28** 092101–4
[22] Song C Y and Yao J M 2010 *Chin. Phys. C* **34** 1425–1427
[23] Lalazissis G A, Gambhir Y K, Maharana J P, Warke C S and Ring P 1998 *Phys. Rev. C* **58** R45–R48
[24] Sugawara-Tanabe K and Arima A 1998 *Phys. Rev. C* **58** R3065–R3068; *ibid*, 1999 **60** 019901(Erratum)
[25] Ginocchio J N, Leviatan A, Meng J and Zhou S G 2004 *Phys. Rev. C* **69** 034303–11
[26] Guo J Y, Chen S W, Niu Z M, Li D P and Liu Q arXiv:1309.1523 [nucl-th]
[27] Yang S C, Meng J and Zhou S G 2001 *Chin. Phys. Lett.* **18** 196–198
[28] Zhang S Q, Meng J and Zhou S G 2003 *Sci. China G* **46** 632–658
[29] Meng J, Toki H, Zhou S G, Zhang S Q, Long W H and Geng L S 2006 *Prog. Part. Nucl. Phys.* **57** 470–563
[30] Guo J Y, Wang R D and Fang X Z 2005 *Phys. Rev. C* **72** 054319–8
[31] Guo J Y and Fang X Z 2006 *Phys. Rev. C* **74** 024320–8
[32] Liu Q, Niu Z M and Guo J Y 2013 *Phys. Rev. A* **87** 052122–7
[33] Zhang S S, Zhang W, Sun B H, Guo J Y and Zhou S G 2006 *High Ener. Phys. Nucl. Phys.* **30(S2)** 97–99
[34] Zhang S S, Sun B H and Zhou S G 2007 *Chin. Phys. Lett.* **24** 1199–1202
[35] Xu Q and Guo J Y 2012 *Int. J. Mod. Phys. E* **21** 1250096–12
[36] Lu B N, Zhao E G and Zhou S G 2012 *Phys. Rev. Lett.* **109** 072501–5
[37] Lu B N, Zhao E G and Zhou S G 2013 *AIP Conf. Proc.* **1533** 63–69

Journal of Physics: Conference Series **533** (2014) 012022 doi:10.1088/1742-6596/533/1/012022

Spectroscopy of exotic nuclei with MINIBALL and FATIMA

Thorsten Kröll

Institut für Kernphysik, TU Darmstadt, Schlossgartenstrasse 9, 64289 Darmstadt, Germany

E-mail: tkroell@ikp.tu-darmstadt.de

Abstract. The investigation of the structure of exotic nuclei is the aim of many experimental and theoretical efforts in nuclear physics. Of particular interest is the evolution of nuclear structure near to shell closures. An important experimental tool for such investigations is prompt γ-ray spectroscopy. Some examples employing state-of-the-art intrumentation will be presented.

1. Introduction

Many efforts in experimental and theoretical nuclear physics investigate the structure of nuclei in regions around shell closures and magic numbers. Their reproduction is a benchmark for any nuclear theory. Going away from the valley of β-stability, traditional magic numbers may disappear or new ones may appear in exotic nuclei. These fingerprints indicate changes of the nuclear interaction relevant for their existence. However, the investigation of unstable nuclei is a challenge for experiments. In this contribution, some examples illustrating different techniques for both the population of exotic nuclei and their investigation by γ-ray spectroscopy are presented.

The mechanisms to populate excited states in the exotic nuclei discussed cover a broad range from gentle Coulomb excitation via nucleon-transfer reactions to more violent nucleon-knockout reactions and neutron-induced fission. The experiments have been performed at REX-ISOLDE (CERN), ILL (Grenoble, France), and GSI (Darmstadt, Germany). Although the experimental method applied is always prompt γ-ray spectroscopy for all presented physics cases, the actual set-ups are quite different and use both large arrays of segmented high-resolution HPGe detectors, e.g. MINIBALL [1], and fast scintillators, e.g FATIMA consisting of $LaBr_3(Ce)$ crystals [2]. The accessible observables to characterise nuclear states are excitation energies, spin-parity assignments and electromagnetic transition probabilities as well as spectroscopic factors.

2. Physics examples

2.1. Island of inversion

The "island of inversion" is a region of the nuclear chart where the traditional magic number $N = 20$ disappears. The structure of neutron-rich Mg isotopes is characterised by a competition of nearly spherical sd configurations and deformed fp intruder configurations. In ^{30}Mg, a nearly spherical ground state and a deformed excited 0^+ state have been found, for results from ISOLDE see Refs. [3, 4, 5], locating this isotope just outside of the island. Going to ^{32}Mg, the inversion occurred and the deformed 0^+ state is now the ground state. Mg isotopes at the border of the "island of inversion" have been populated in one- and two-neutron transfer reactions in inverse kinematics [6, 7] at REX-ISOLDE. The spectroscopy has been done by a combination of the MINIBALL γ-ray spectrometer and the T-REX Si detector array for particle spectroscopy [8].

Journal of Physics: Conference Series **533** (2014) 012022 doi:10.1088/1742-6596/533/1/012022

The most prominent result was the identification of the long-sought excited 0^+ state in ^{32}Mg [6], the analogue of the ground state in ^{30}Mg, which is expected to have a nearly spherical shape too. This state was populated in a (t,p) reaction in inverse kinematics employing a radioactive tritium target. The 0^+ spin-parity assigment is based on the characteristic angular distribution of the outgoing protons. From the observed γ-ray yield a lower limit for its lifetime of > 10 ns could be estimated, similar to the excited 0^+ state in ^{30}Mg [3].

Indeed, the cross section for the population of the excited 0^+ state in ^{32}Mg can be explained assuming a *sd* configuration with a small $p_{3/2}$ admixture in its wave function which is consistent with what is observed in the one-neutron knockout from the ground state of ^{30}Mg. However, it remains a challenge to nuclear theory to reproduce consistently both the low excitation energy of 1058 keV and the large cross section in the two-neutron transfer reaction.

2.2. *New shell closure at $N = 32, 34$*

The shell model reproduced until recently the traditional magic number at $N = 28$ in Ca only if empirical interactions are used. Newer developments with realistic interactions derived from NN potentials applying chiral EFT achieved with the inclusion of NNN forces this goal too [9]. Extrapolated to more neutron-rich isotopes both the shell model with empirical and realistic interactions, but also beyond-mean-field calculations [10], predict new shell closures at $N = 32$ and/or $N = 34$. The changes in the level ordering and the single-particle configurations contributing to the wave functions of the states have been investigated by one-neutron knockout reactions on isotopes north-east of ^{48}Ca at relativistic beam energies. The experiment was performed at the fragment separator FRS of GSI. The MINIBALL array, positioned at the intermediate focal point S2, allowed to obtain exclusive momentum distributions of the outgoing reaction products. These were measured using the second half of the FRS as high-resolution spectrometer.

In ^{55}Ti, for the ground state as well as for a newly found excited state at 955 keV a p configuration was found. This is in agreement with predictions obtained with the shell model [11]. For the Sc isotopes, it was observed that the cross sections observed cannot be reproduced using occupation numbers obtained by the shell model even if empirical interactions are applied. The observed $\Delta L = 1$ contribution indicating the knockout of a p neutron is larger whereas the $\Delta L = 3$ indicating the knockout of a f neutron is smaller than theory currently predicts [12].

2.3. *Nuclei around ^{132}Sn*

Many experiments at REX-ISOLDE with MINIBALL employ γ-ray spectroscopy following "safe" Coulomb excitation as a tool to study collective properties of exotic nuclei (see also [13]). Here, the beam energy is chosen such that projectile and target interact only via the well-known electromagnetic interaction. The radioactive isotope of interest impinges as beam on a target and either of them (or both) can be excited. From the double-differential cross sections, extracted from the measured γ-ray–particle yields, the electromagnetic matrix elements are determined. As the cross section depends on both transitional but also diagonal matrix elements (reorientation), reduced transition probabilities $B(E\lambda)$ as well as spectroscopic electric moments Q_λ can be extracted. In the analysis, the experimental data are compared with results of calculations and the matrix elements are obtained applying the maximum likelihood method, see e.g. [14]. Additional information, like lifetimes from independent experiments (DSAM, RDS or fast timing) can be considered too and helps in the analysis, in particular in cases where more than one state has been excited and many matrix elements have to be considered.

The region around the isotope ^{132}Sn is in the focus of many experimental and theoretical studies as it is one out of only two heavy doubly-magic nuclei on the neutron-rich side of the nuclear chart, ^{78}Ni is the other, which can be reached experimentally. Although near two shell closures, several isotopes exhibit a suprisingly irregular behaviour. One of the most prominent examples is ^{136}Te whose $B(E2; 0^+_{gs} \rightarrow 2^+_1)$ value is considerably smaller than expected by theory [15]. However, a new analysis of the data [16] as well as a direct lifetime measurement using fast timing [17] corrected the value to about 20% larger values. At REX-ISOLDE, neutron-rich Cd and Xe isotopes have been studied in "safe" Coulomb excitation.

The excitation energy of the first 2^+ state of ^{128}Cd exibits a strange irregularity not observed anywhere else. Approaching the shell closure at $N = 82$ the energy does not increase from ^{126}Cd to ^{128}Cd, it even slightly decreases! This trend is not reproduced by the shell model, but only by beyond-mean field calculations resulting in the prediction of a considerable deformation of this nucleus. However, our results find only for the lighter isotopes ^{122,124}Cd $B(E2)$ values above the shell model predictions whereas for the heavier isotopes ^{126,128}Cd the values roughly agree [18, 19]. For the 2^+ state in ^{126}Cd the additional result of a DSAM measurement has been included in the analysis. Indeed, also the Q_2 values, although having large errors, are more consistent with spherical shapes. Hence, the excitation energy of ^{128}Cd remains an open challenge for theory.

For the Xe isotopes beyond $N = 82$, previously only lifetimes of the first 2^+ and 4^+ states in ^{140}Xe were known, however for the 2^+ state two contradicting values can be found in literature. The longer lifetime corresponding to a smaller $B(E2)$ value would indicate a continuation of small $B(E2)$ values above $N = 82$ observed in ^{136}Te. Assuming vanishing quadrupole moments, $Q_2 = 0$, the $B(E2)$ values for the even isotopes $^{138-144}$Xe obtained from the Coulomb excitation data follow a smooth trend given by a simple Grodzins-type systematics along the Xe isotopic chain and no irregular behaviour is observed.

A different approach to direct lifetime measurements is the fast-timing technique involving fast scintillator crystals like $LaBr_3$. The set-up comprised the EXILL (EXOGAM@ILL) HPGe array and FATIMA, consisting of 16 $LaBr_3$ crystals. A detailled description of the method and the set-up is presented in a further contribution to this school [2]. Lifetimes in neutron-rich Xe isotopes populated in neutron-induced fission of ^{235}U and ^{241}Pu have been measured at ILL (Grenoble) in March 2013.

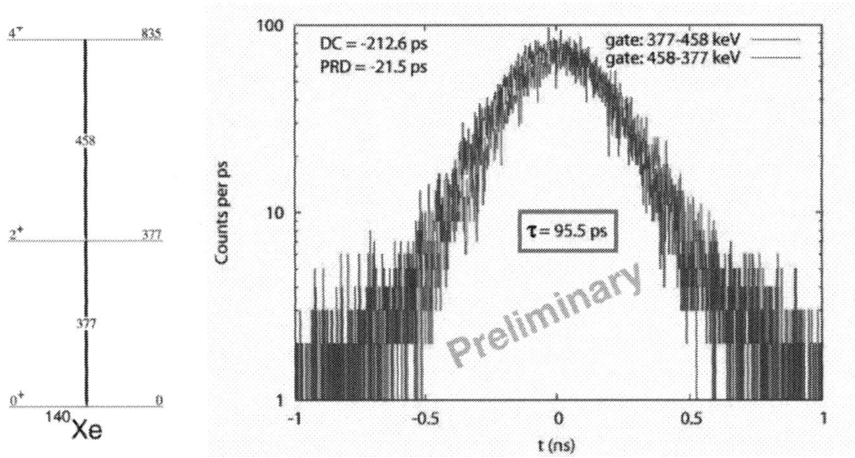

Figure 1. Partial level scheme of ^{140}Xe (left) and preliminary time difference spectrum analysing $LaBr_3$ doubles only (right). Plot by courtesy of S. Ilieva (TU Darmstadt).

In Fig. 1, one preliminary result of the analysis of $LaBr_3$ only coincidence data for a small part of the data set is shown. The obtained lifetime of $\tau = 95.5$ ps for the 2^+_1 state in ^{140}Xe is

Journal of Physics: Conference Series **533** (2014) 012022 doi:10.1088/1742-6596/533/1/012022

already near to the previously measured value of $\tau = 101.7(32)$ ps [20]. In conclusion, in Xe no irregular behaviour can be evidenced.

However, since the energy resolution of $LaBr_3$ is inferior compared to HPGe detectors, in many cases, in particular for weakly populated channels or the odd Xe isotopes, a further coincidence with Ge detectors may be required in the ongoing analysis to clean the spectra.

3. Future prospects

The Coulomb excitation programme at ISOLDE with MINIBALL will be continued at the upgraded facility HIE-ISOLDE which will offer a higher beam energy, an increased beam intensity and an improved beam quality. In particular, the higher beam energy will allow for multiple Coulomb excitation and the extension of the transfer programme towards heavier nuclei. The study of nucleon knockout reactions and quasi-free scattering is an important programme of R^3B at FAIR employing the spectrometer and calorimeter CALIFA. Large set-ups of $LaBr_3$ detectors like FATIMA will be used in future at several places like ILL, RIKEN, ISOLDE, JYFL or DESPEC at FAIR for prompt and delayed γ-ray spectroscopy.

Acknowledgments

This contribution represents the work of several groups. This work is supported by BMBF (06DA9036I, 05P12RDNUP, and 05P12RDCIA), HIC for FAIR, EU through EURONS (No. 506065) and ENSAR (No. 262010), ILL, and the REX-ISOLDE, MINIBALL, and FATIMA/EXILL collaborations.

References

[1] Warr N *et al.* 2013 *Eur. Phys. J. A* **49** 40
[2] Jolie J *et al.* contribution to this school
[3] Mach H *et al.* 2005 *Eur. Phys. J. A* **25** 105
[4] Niedermaier O *et al.* 2005 *Phys. Rev. Lett.* **94** 172501
[5] Schwerdtfeger W *et al.* 2009 *Phys. Rev. Lett.* **103** 012501
[6] Wimmer K *et al.* 2010 *Phys. Rev. Lett.* **105** 252501
[7] Bildstein V *et al.* to be published
[8] Bildstein V *et al.* 2012 *Eur. Phys. J. A* **48** 85
[9] Holt JD *et al.* 2012 *J. Phys. G* **39** 085111
[10] Rodriguez TR and Egido JL 2007 *Phys. Rev. Lett.* **99** 062501
[11] Maierbeck P *et al.* 2009 *Phys. Lett. B* **675** 22
[12] Schwertel S *et al.* 2012 *Eur. Phys. J. A* **48** 191
[13] Scheck M *et al.* contribution to this school
[14] Review of Particle Physics 2008 *Phys. Lett. B* **667** 13
[15] Radford D *et al.* 2002 *Phys. Rev. Lett.* **88** 222501
[16] Danchev M *et al.* 2011 *Phys. Rev. Lett.* **84** 061306(R)
[17] Fraile LM *et al.* 2008 *INPC2007 proceedings, Nucl. Phys. A* **805** 218
[18] Ilieva S *et al.* to be published
[19] Bönig S *et al.* to be published
[20] Lindroth A *et al.* 1999 *Phys. Rev. Lett.* **82** 4783

Journal of Physics: Conference Series **533** (2014) 012023 doi:10.1088/1742-6596/533/1/012023

Magnetic moment measurements - extending isotopic chains beyond the stable elements.

G J Kumbartzki

Department of Physics and Astronomy, Rutgers University, New Brunswick, NJ, USA

E-mail: kum@physics.rutgers.edu

Abstract. The magnetic moments in many isotopic chains have been systematically measured using the transient field technique on beams of separated isotopes excited in inverse kinematics. Such experiments have provided insight into how the structure of nuclei evolves by successively adding nucleons. Since naturally occurring isotopic chains are relatively short, efforts are underway to make unstable isotopes available. In limited cases the use of an α-particle transfer to beam projectiles has been successfully employed in measurements on unstable nuclei. In this investigation beams of ^{78}Kr and ^{86}Kr were used to measure magnetic moments of excited states in the unstable ^{82}Sr and ^{90}Sr nuclei utilizing the transfer of an α particle from ^{12}C nuclei in the target.

The systematics of nuclei with isotopic chains around the closed neutron shell $N = 50$ are of special interest. A compilation of g factors of the first 2^+ states from Zn ($Z = 30$) to Cd ($Z = 48$) is shown in Figure 1.

At shell closure $N = 50$ the positive large g factors indicate single-particle proton states, while away from this closed neutron shell both protons and neutrons contribute to the wave

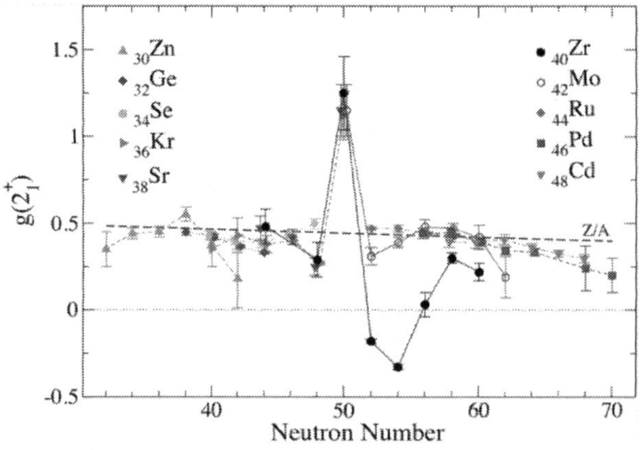

Figure 1. The $g(2_1^+)$ factors of isotopic chains on both sides of the closed neutron shell $N = 50$.

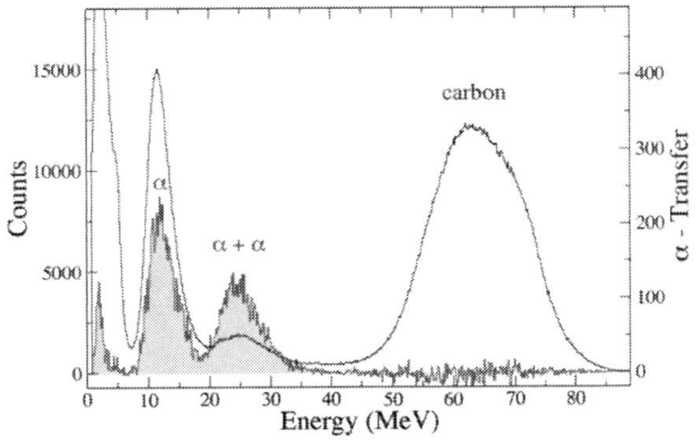

Figure 2. Particle spectrum and insert (shaded area) showing particles in coincidence with γ transitions in ^{90}Sr.

functions. As a matter of fact, most g factors are close to the collective value of Z/A. But Zr with 40 protons clearly stands out. Adding neutrons beyond $N = 50$ results in negative g factors. Obviously, neutron excitation dominates in the structure of these excited Zr states. If this effect were simply attributed to a $2p_{1/2}$ subshell closure for protons at $Z = 40$, a similar behavior may be expected for the neighboring Sr isotopes. At $Z = 38$ the $1f_{5/2}$ shell is complete and for neutron numbers above $N = 50$ negative g factors could be expected.

Experiment

By far the best way to measure the g factors is the Coulomb excitation of projectiles in inverse kinematic conditions followed by the observation of the perturbed angular correlation of the de-exciting γ transitions.

The isotope ^{90}Sr is unstable with a half–life of 28.79 yr. Rare isotope beams are rare and when available have notoriously low intensities. For this experiment ^{90}Sr was produced from a ^{86}Kr beam by pickup of α particles from a ^{12}C target.

The isotopically pure ^{86}Kr beam with 3.2 MeV/A was delivered by the K500 Texas A&M cyclotron. The beam struck a multilayered target composed of carbon (0.9 mg/cm^2), gadolinium (5.0), tantalum (1.1) and copper (5.0). Coulomb excitation of beam projectiles and the α transfer both occur in the carbon layer. The reaction products go forward and traverse the gadolinium and tantalum layers. While the heavier reaction products are stopped in the copper layer of the target, the lighter products – carbon nuclei and α particles from the breakup of the ^{8}Be – go through and are detected in a conventional particle detector placed 20 mm downstream of the target at zero degrees to the beam. The magnetic interaction happens while the ions traverse the ferromagnetic gadolinium layer.

Experimental details of the transient field technique can be found in References. [1] and [2]. An example of an α–transfer experiment is found in Reference [3]. Four clover HP Ge detectors were used to detect the γ rays. The preamplifier output signals of the particle and γ detectors were digitized. The energies and times were recorded as singles events, from which offline particle – γ coincidence spectra were selected.

In Figure 2 particle spectra associated with both the Coulomb excitation of the beam and the α transfer are shown. Appropriate gates on time and on γ energies were set to produce the α–particle spectrum.

The particle spectrum associated with the α transfer shows two peaks, depending upon whether both α particles from the ^{8}Be breakup or only one hit the detector. The particle detector is 100 μm thick. The α particles (~40 MeV) and other light particles do not stop in

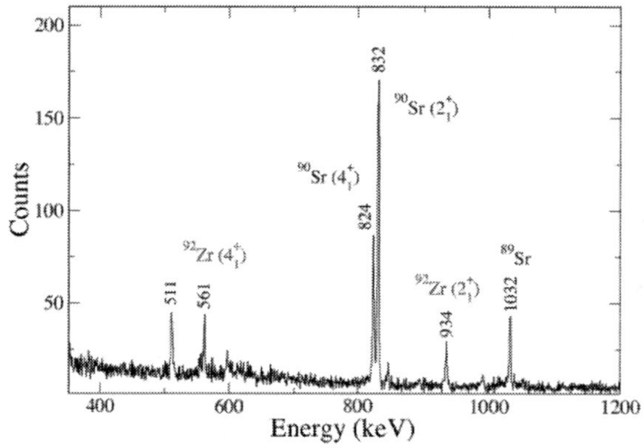

Figure 3. Coincident γ–particle spectrum showing the close proximity of the 4_1^+ and 2_1^+ transitions in ^{90}Sr.

Table 1. Remeasured g factors of low-lying states in ^{86}Kr and ^{78}Kr and previous results from Reference [4].

^{86}Kr	Reference[4]	this work
2_1^+	+1.12(14)	+1.07(6)
4_1^+		+1.07(15)
^{78}Kr		
2_1^+	+0.43(3)	+0.48(2)
4_1^+	+0.46(7)	+0.49(5)
2_2^+	+0.54(10)	+0.53(7)

the detector. They deposit only a small part of their total energy in the depletion layer of the silicon detector.

A coincidence double–α – γ spectrum is shown in Figure 3. The γ spectra gated on the single–α peak have a higher background and were not used in the data analysis.

The spectrum in Figure 3 also contains transitions in ^{92}Zr which is produced by a ^{86}Kr(^{12}C,α2n)^{92}Zr reaction. This reaction channel is unusually strong for ^{86}Kr. The ^{86}Kr also favors single proton pickup, all of which may be related to its neutron number $N = 50$. Both channels are virtually absent with the ^{78}Kr beam.

The α–transfer reaction is not well studied. No detailed excitation function measurements exist. The onset of the reaction is observed already at slightly below the Coulomb barrier. Above the barrier other reaction channels start to dominate. The intensity ratio of the single–α – to the double–α – particle peak increases with beam energy.

The α–transfer reaction excites the nuclei at higher energies and therefore in higher excited states. Their precession in the transient field and subsequent feeding to the states of interest has to be considered. In addition the α–transfer nuclei have a lower spin alignment than that achieved by Coulomb excitation, which reduces the sensitivity of the precession measurement.

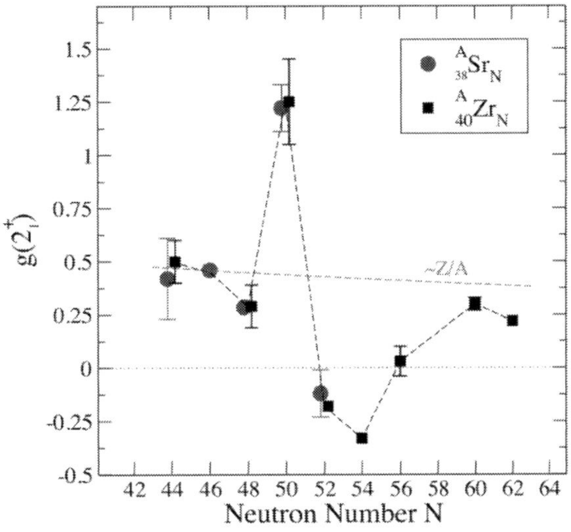

Figure 4. The g factors of the newly extended Sr chain.

Results

The Coulomb excitation of ^{86}Kr and ^{78}Kr were measured simultaneously with the ^{86}Kr(^{12}C,2α)^{90}Sr and ^{78}Kr(^{12}C,2α)^{82}Sr reactions. The Kr results are in excellent agreement with an earlier measurement [4] (Table 1). The $g(4_1^+)$ factor of ^{86}Kr is measured for the first time.

The $g(2_1^+)$ of ^{90}Sr (Figure 4 at N = 52) is negative or close to zero suggesting that, like in Zr, neutrons dominate in the wave function of the first excited state and the protons tend to stay in the filled $f_{5/2}$ (or $2p_{1/2}$ for Zr) shell. The $g(2_1^+)$ result for ^{82}Sr (N = 44) is also shown.

The results presented in this report should be considered preliminary. More measurements were performed and are being analyzed.

Acknowledgments

Special thanks to my collaborators Benczer-Koller N, Torres D.A, Ratkiewicz A, Gürdal G and Sharon Y.Y from Rutgers University, Speidel K-H from University of Bonn and McCleskey M from Texas A&M University. The experimental work was done at the Texas A&M cyclotron accelerator; the help of local students, postdoctoral fellows and staff is greatly appreciated. The work was supported in part by the U.S. National Science Foundation.

References

[1] Kumbartzki G J XIX International School on Nuclear Physics, Neutron Physics and Applications (VARNA 2011) *J. Phys.: Conf. Ser.* **366** 012028 (2012)

[2] Benczer-Koller N and Kumbartzki G J *J.Phys.G:Nucl.Part.Phys.* **34**, 321 (2007).

[3] Torres D A, Kumbartzki G J, Sharon Y Y, Zamick L, Manning B, Benczer-Koller N *et al.* 2011 *Phys. Rev.* C **84** 044327

[4] Mertzimekis T J, Benczer-Koller N, Holden J, Jakob G, Kumbartzki G, *et al.* 2001 *Phys. Rev.* C **64** 024314

Journal of Physics: Conference Series **533** (2014) 012024 doi:10.1088/1742-6596/533/1/012024

Fission at intermediate neutron energies

S Lo Meo[1,4], D Mancusi[2], C Massimi[3,4], G Vannini[3,4] and A Ventura[4]

[1] ENEA, Centro Ricerche Clementel, 40129 Bologna, Italy
[2] CEA, Centre de Saclay, Irfu/SPhN, F91191 Gif-sur-Yvette, France
[3] Dipartimento di Fisica e Astronomia dell' Università di Bologna, 40126 Bologna, Italy
[4] INFN, Sezione di Bologna, 40127 Bologna, Italy

E-mail: alventura@bo.infn.it

Abstract. In the present work, as a theoretical support to the campaign of neutron cross section measurements at the n_TOF facility at CERN[1], Monte Carlo calculations of fission induced by neutrons in the energy range from 100 MeV to 1 GeV are carried out by means of a recent version of the Liège Intranuclear Cascade Model, INCL++[6], coupled with different evaporation-fission codes, such as Gemini++[7] and ABLA07[8]. Theoretical cross sections are compared with experimental data obtained by the n_TOF collaboration and perspectives for future theoretical work are outlined.

1. Introduction

High-accuracy data of neutron-induced fission cross sections are essential to the design of Generation IV reactors and, as far as neutron intermediate energies are concerned, of accelerator-driven subcritical systems (ADS). This is the reason why fission cross sections in the unprecedented range from thermal energies up to about 1 GeV are measured at the n_TOF facility at CERN[1], exploiting the neutron beam produced by 20 GeV/c protons from the PS accelerator impinging on a lead spallation target.

Fission cross sections for natural lead and bismuth[2], forming the eutectic system that should act as a spallation target and a coolant in an ADS, as well as for actinides of the $U - Th$ cycle and minor actinides[3], have been measured relative to the $^{235}U(n,f)$ cross section, which is considered as a standard up to a neutron energy of 200 MeV, the maximum energy at which reliable data of the absolute cross section are available.

In order to obtain absolute fission cross sections of actinides and pre-actinides in the whole energy range measured at n_TOF, experimental ratios have been normalized to an evaluated $^{235}U(n,f)$ cross section, taken from the ENDF/B-VII library[4] up to 30 MeV and from the JENDL/HE-2007 library[4] from 30 MeV to 1 GeV. It is then natural to investigate whether the choice made for the normalization in the intermediate energy range, from about 100 MeV to 1 GeV, is compatible with recent absolute data of the (p,f) reaction, since it is expected that at energies well above the fission barrier neutrons and protons have a similar behaviour and the corresponding cross sections a similar energy trend. In Ref.[5], (p,f) cross sections for several pre-actinides and actinides, including, of course, ^{235}U, have been measured at nine proton energies in the range from 200 MeV to 1 GeV and are used to calibrate the fission model parameters for our (n,f) calculations.

Journal of Physics: Conference Series **533** (2014) 012024 doi:10.1088/1742-6596/533/1/012024

2. Cross Section Calculations

In the energy range of interest, fission induced by nucleons can be seen as a two-stage process: a fast cascade stage, initiated by the high energy projectile, and representable as a succession of two-body collisions, with emission of fast nucleons, light clusters, pions, etc, leaving an excited residual nucleus, and the slow decay stage, where the remnant decays by evaporation, fission, or other mechanisms. In our system of codes, the intranuclear cascade is described by a recent C++ version of the Liège Intranuclear Cascade Model, INCL[6], the evaporation-fission model by a C++ version of GEMINI[7] or a Fortran version of ABLA[8].

In keeping with the philosophy of the authors[6], no parameters have been modified in the INCL model, which was already optimized by reproducing a large amount of experimental data connected with the cascade stage, such as total reaction cross sections, double-differential spectra of emitted nucleons and light clusters, residue production, etc, while we have taken the freedom to adjust fission parameters in both GEMINI and ABLA. The basic parameters are essentially two: the height of the fission barrier and the level density parameter in the fission channel, a_f, or its ratio to the level density parameter in the neutron channel, a_n. In both models, the fission barrier is calculated by means of the finite-range liquid-drop model of Sierk[9], with ground-state shell and pairing corrections taken from Ref.[10], while the level density formalisms are different and cannot be discussed in this short note. The barrier height is expected to play a role only at low energy, the level density is very important in the whole energy range. Therefore, the present preliminary analysis has been carried out by adjusting only a_f on experimental (p, f) or (n, f) data.

Reproducing the experimental (p, f) cross sections of Ref.[5] already allows us to outline a a systematic trend of level density parameters, which increase with increasing fissility parameter, proportional to Z^2/A, from ^{209}Bi to ^{233}U, and decrease beyond that nucleus: for instance, GEMINI calculations suggest $a_f/a_n = 1.045$ for ^{209}Bi, 1.060 for ^{235}U and 1.020 for ^{239}Pu. The (n, f) data in the same energy range are not yet rich enough to confirm the trend.

In the case of ^{235}U we have first determined the value of a_f that permits the best reproduction of the experimental (p, f) cross section in the energy range 200 MeV -1 GeV[5] and then used it as a first guess to reproduce the experimental (n, f) cross section from 100 to 200 MeV[11][12]. Of course, for the procedure to make sense, the final value of a_f should be close to the first guess, so that (p, f) and (n, f) data are reproduced with similar model parameters.

The results of our (n, f) calculations with the INCL + GEMINI and INCL + ABLA07 chains are compared in Fig. 1 to the experimental data up to 200 MeV and to the JENDL/HE-2007 evaluation, used in Refs.[2][3] to obtain absolute (n, f) cross sections.

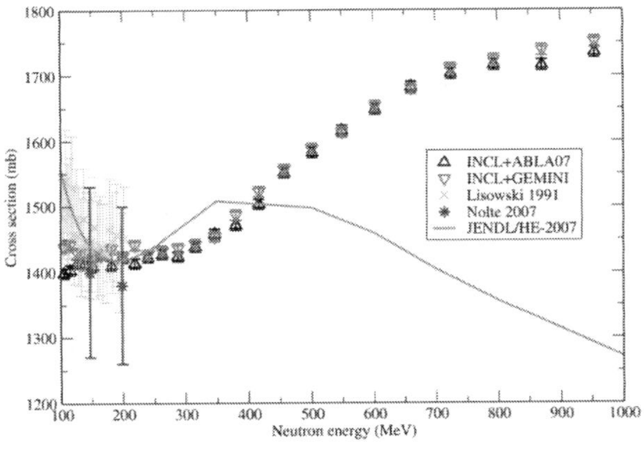

Figure 1. Calculated $^{235}U(n, f)$ cross section compared with the JENDL/HE-2007 evaluation[4] and experimental data[11][12].

As shown in the figure, the INCL + GEMINI and INCL + ABLA07 chains yield cross sections in close agreement in the whole energy range, which reproduce (n, f) data up to 200 MeV and are larger than the experimental (p, f) cross section at high energy, since σ_{pf} (1 GeV) = 1591 ± 113 mb[5]. At the same energy, the calculated (p, f) cross sections are 1670 mb (GEMINI) and 1700 mb (ABLA). The a_f value used in (n, f) calculations is higher by 2% than the one used in (p, f) calculations. The discrepancy might be reduced by modifying the height of the fission barrier. While our calculated cross sections increase up to 1 GeV (and start decreasing beyond that energy), the JENDL/HE-2007 evaluation steadily decreases in the same energy range.

As far as the n_TOF measurements are concerned, ^{234}U and ^{237}Np are published in Ref.[3], natural lead and ^{209}Bi in Ref.[2], while preliminary data are available for ^{232}Th and $^{233,238}U$. In the case of ^{234}U, since no (p, f) data exist in Ref.[5], we have assumed the same a_f value as for ^{235}U. Fig. 2 shows that theoretical and experimental $\sigma_{nf}(^{234}U)/\sigma_{nf}(^{235}U)$ ratios are in excellent agreement. Fig. 3 compares absolute (n, f) cross sections, which diverge at high energy, owing to the normalization to the JENDL/HE-2007 cross section of ^{235}U, adopted in Ref.[3] .

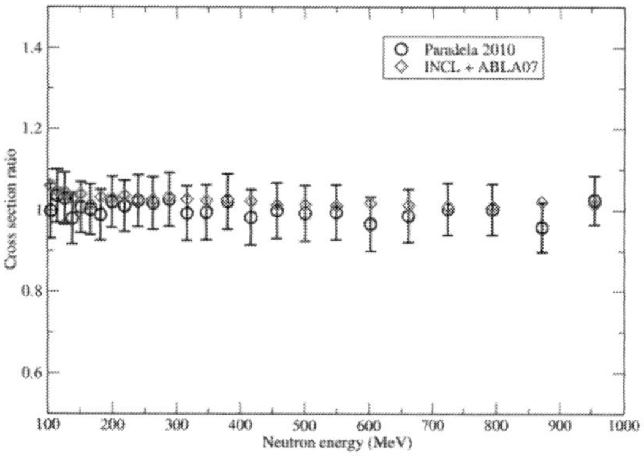

Figure 2. Calculated *vs.* experimental $\sigma_{nf}(^{234}U)/\sigma_{nf}(^{235}U)$ ratio. Data are taken from Ref.[3].

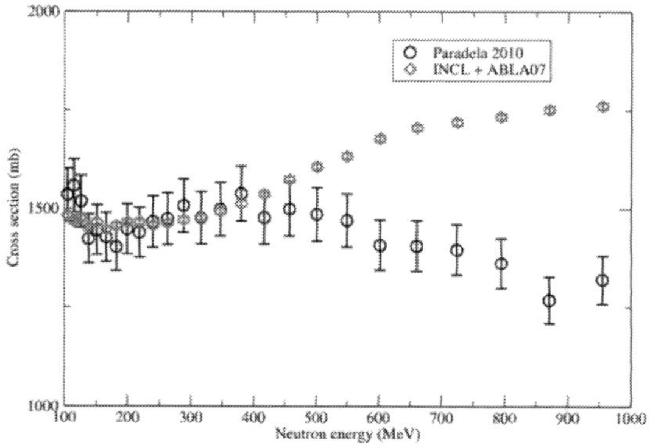

Figure 3. Calculated $\sigma_{nf}(^{234}U)$ *vs.* the experimental ratio of Fig. 2 normalized to JENDL/HE-2007[4].

In the case of ^{209}Bi, (p, f) and (n, f) data can be reproduced with the same value of a_f. Theoretical and experimental[2] ratios to $^{235}U(n, f)$ are again in excellent agreement, as shown in Fig. 4, while absolute cross sections are compared in Fig. 5; again, the normalization to the Japanese evaluation explains the disagreement at high energy.

Summing up, experimental (n, f) ratios measured at n_TOF for pre-actinides and actinides are satisfactorily reproduced by using fission model parameters that are similar, or rather close, to the ones adopted in reproducing experimental (p, f) cross sections. Absolute cross sections are larger than those obtained by the n_TOF Collaboration in the energy range from 500 MeV to 1 GeV because of different normalization.

3. Conclusions and Outlook

The preliminary calculations presented in this short note indicate that the Liège Intranuclear Cascade Model, INCL, coupled with an appropriate evaporation-fission model like GEMINI or ABLA, can reproduce the experimental data obtained at n_TOF on fission induced by intermediate energy neutrons. The fits can be improved by acting on additional fission model

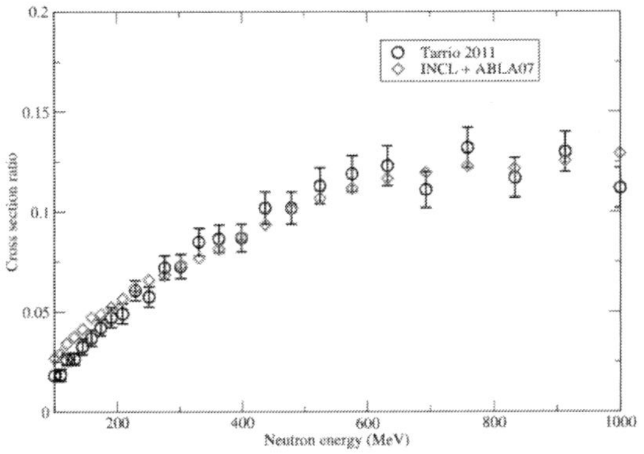

Figure 4. Calculated *vs.* experimental $\sigma_{nf}(^{209}Bi)/\sigma_{nf}(^{235}U)$ ratio. Data are taken from Ref.[2].

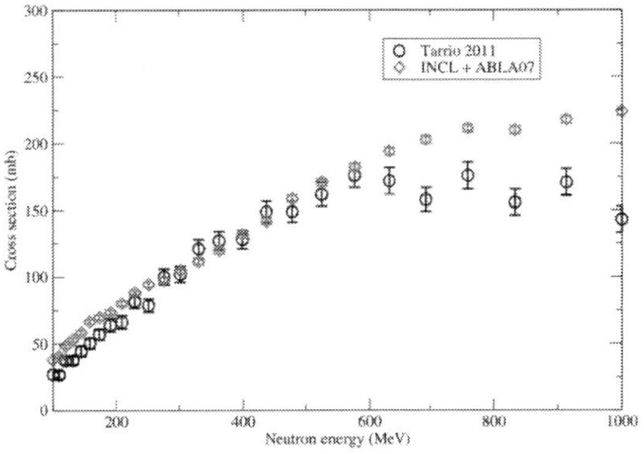

Figure 5. Calculated $\sigma_{nf}(^{209}Bi)$ *vs.* the experimental ratio of Fig. 4 normalized to JENDL/HE-2007[4].

parameters, with a view to obtaining a consistent set for the simultaneous description of (n, f) and (p, f) reactions in the same energy range. This is necessary to the derivation of absolute (n, f) cross sections from experimental ratios: on the basis of the present study, in particular, the JENDL evaluation of the $^{235}U(n, f)$ cross section adopted in Refs.[2][3] does not seem to be consistent with the experimental (p, f) cross section in the same energy range.

As a by-product of such a systematic study, it will be of interest to examine the dependence of model parameters that reproduce (n, f) data on characteristics of target nuclei, such as the fissility parameter, and check whether the trend already appearing in (p, f) data is confirmed.

Finally, it is our intention to make predictions for other isotopes whose measurements are already planned at CERN, not only of fission cross sections, but also of angular distributions of fission fragments, which are also measurable at the n_TOF facility.

Acknowledgments
We are grateful to Igor A. Pshenichnov for valuable comments.

References

[1] Guerrero C *et al.* (the n_TOF Collaboration) 2013 *Eur. Phys. J. A* **49** 27.
[2] Tarrío D *et al.* (the n_TOF Collaboration) 2011 *Phys. Rev. C* **83** 044620.
[3] Paradela C *et al.* (the n_TOF Collaboration) 2010 *Phys. Rev. C* **82** 034601.
[4] See, for instance, *www-nds.iaea.org*
[5] Kotov AA *et al.* 2006 *Phys. Rev. C* **74** 034605.
[6] Boudard A, Cugnon J, David J-C, Leray S and Mancusi D, 2013 *Phys. Rev. C* **87** 014606.
[7] Charity RJ, *Proceedings of the Joint ICTP-IAEA Advanced Workshop on Model Codes for Spallation Reactions, ICTP Trieste, Italy, 4-8 February 2008, D. Filges et al., Eds.*, IAEA report INDC(NDS)-530, Vienna, 2008, p. 139; Mancusi D, Charity RJ and Cugnon J, 2010 *Phys. Rev. C* **82** 044610.
[8] Kelić A, Ricciardi MV and Schmidt K-H, *Proceedings of the Joint ICTP-IAEA Advanced Workshop on Model Codes for Spallation Reactions, ICTP Trieste, Italy, 4-8 February 2008, D. Filges et al., Eds.*, IAEA report INDC(NDS)-530, Vienna, 2008, p. 181.
[9] Sierk AJ 1986 *Phys. Rev. C* **33** 2039.
[10] Möller P *et al.* 1995 Atomic Data Nucl. Data Tables **59** 185.
[11] Lisowski PW *et al.* 1988 *Proceedings of the International Conference on Nuclear Data for Science and Technology, Mito, Japan*, p. 97; 1991 *Proceedings of the Specialists' Meeting on Neutron Cross-Section Standards for the Energy Region above 20 MeV, Uppsala, Sweden*, NEA report NEANDC-305/U, p. 177.
[12] Nolte R *et al.* 2007 *Nucl. Sci. Eng.* **156** 197.

Journal of Physics: Conference Series **533** (2014) 012025 doi:10.1088/1742-6596/533/1/012025

Multi-nucleon transfer reactions for production and study of heavy neutron rich nuclei

V I Zagrebaev[1], S G Zemlyanoy[1], E M Kozulin[1], Yu Kudryavtsev[2], V Fedosseev[3], R Bark[4], Z Janas[5] and H A Othman[6]

[1]FLNR, Joint Institute for Nuclear Research, 141980 Dubna, Russia
[2]Instituut voor Kern- en Stralingsfysica, K.U. Leuven, B-3001 Leuven, Belgium
[3]CERN, CH-1211, Geneva23, Switzerland
[4]iThemba LABS, Nat. Research Foundation, Somerset West 7129, South Africa
[5]Faculty of Physics, University of Warsaw, Hoza 69, PL-00681 Warsaw, Poland
[6]Phys. Dep., Faculty of Science, Menoufiya University, Egypt

E-mail: zemlya@jinr.ru

Abstract. The present limits of the upper part of the nuclear map are very close to stability while the unexplored area of heavy neutron-rich nuclides along the neutron closed shell N = 126 is extremely important for nuclear astrophysics investigations and, in particular, for the understanding of the r-process of astrophysical nucleosynthesis. A new way was recently proposed for the production of these nuclei via low-energy multi-nucleon transfer reactions. The estimated yields of neutron-rich nuclei are found to be rather high in such reactions and several tens of new nuclides can be produced, for example, in the near-barrier collision of 136Xe with 208Pb. A new setup is proposed to produce and investigate heavy neutron-rich nuclei located along the neutron closed shell N=126.

1. Introduction

Properties of light and medium mass nuclei located far from the stability line have been studied already for many years. As a rule, the light exotic nuclei are produced in fragmentation processes and the medium mass neutron rich nuclei with A~100 – in fision reactions. Production and study of heavier neutron rich nuclei with A>160 encounters both physical and technical problems. Due to the "curvature" of the stability line (its increasing bending to the neutron axis), fusion reactions of stable nuclei produce only proton rich isotopes of heavy elements. For example, in the fusion of rather neutron rich ^{18}O and ^{186}W isotopes one may get only the neutron deficient ^{204}Pb excited compound nucleus, which after evaporation of several neutrons shifts even more to the proton rich side. As a result, on the nuclear map there are, for example, 19 known neutron rich isotopes of cesium (Z = 55) and only 4 of platinum (Z = 78). For elements with Z > 100 only neutron deficient isotopes (located to the left of the stability line) have been synthesized so far. That is the main reason also for the impossibility to reach the center of the "island of stability" (Z~114 and N~184) in the superheavy mass region, and neutron deficient isotopes of elements with Z>120 (being synthesized in fusion reactions) should have very short half-lives (less than one microsecond), insufficient for their separation and identification.

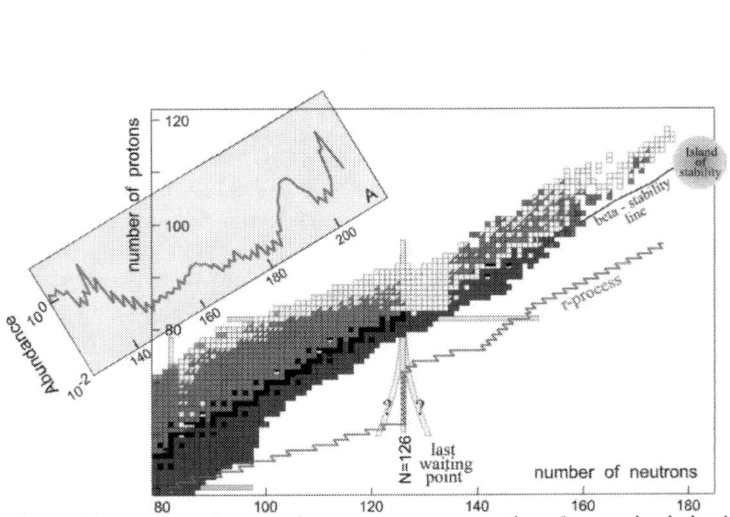

Fig. 1. Upper part of the nuclear map. r-process of nucleosynthesis is shown schematically.

At the same time, the unexplored area of heavy neutron rich nuclei is extremely important for nuclear astrophysics investigations and, in particular, for the understanding of the r-process of astrophysical nucleogenesis (a sequence of neutron-capture and β^-–decay processes). Just in this region the closed neutron shell N=126 is located which is the last "waiting point" in the r-process (see Fig. 1). The half-lives and other characteristics of these nuclei feed directly into the r-process scenario of supernovae explosions. Study of the structural properties of nuclei along the neutron shell N = 126 could also contribute to the present discussion of the quenching of shell gaps in nuclei with large neutron excess.

2. Physics objectives

Three possibilities for the production of heavy neutron rich nuclei are known. These are the multi-nucleon transfer reactions [1], fusion reactions with extremely neutron rich radioactive nuclei and rapid neutron capture processes. Today the two last methods look unrealizable because of low intensity of radioactive beams and low neutron fluxes in existing nuclear reactors. On the contrary, the low-energy multi-nucleon transfer reactions, as well as the quasi-fission processes [2] (which are similar) are quite practical. They can be used for the production of new neutron rich isotopes not only in the region of Z~80 but also in the superheavy mass area [3]. Theoretical estimations show that several tens of new nuclides in the region of N=126 and Z~75 can be produced, for example, in the near-barrier collision of ^{136}Xe with ^{208}Pb (see Fig. 2).

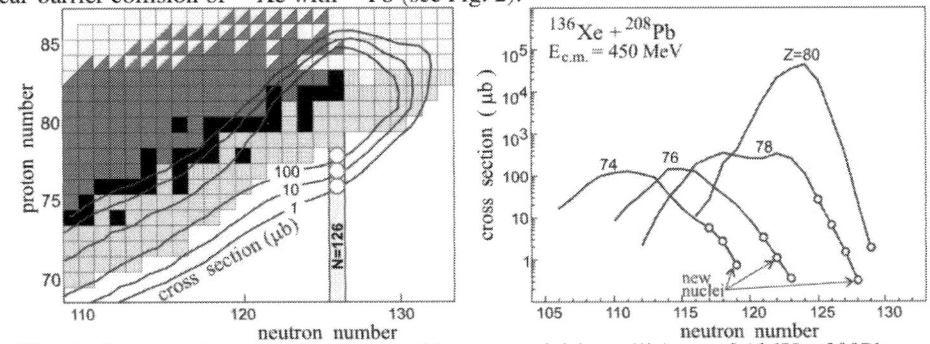

Fig. 2. Cross sections for formation of heavy nuclei in collisions of 136Xe+208Pb at center-of-mass energy of 450 MeV [1]. Open circles in the right panel correspond to unknown isotopes.

Neutron rich isotopes of transfermium elements can also be produced in the multi-nucleon transfer reactions at low-energy collisions of actinide nuclei. Production and study of properties of neutron rich

Journal of Physics: Conference Series **533** (2014) 012025 doi:10.1088/1742-6596/533/1/012025

fermium isotopes (A > 260) are extremely interesting for several reasons. Firstly, as mentioned above, all known isotopes of fermium (and of more heavy elements) are located to the left side of the beta-stability line (see Fig. 1). Secondly, the well known "fermium gap" (isotopes $^{258-260}$Fm with very short half-lives for spontaneous fission) impedes formation of nuclei with Z>100 by the weak neutron fluxes realized in existing nuclear reactors. In this connection, it is extremely interesting to know what is the first β^-–decaying fermium isotope and how long is its half-life. This is important not only for reactor-based physics but also for explosive nucleosynthesis in which this fermium gap might be bypassed.

Unfortunately, the neutron rich heavy nuclei with Z>70 formed in the multi-nucleon transfer reactions cannot be separated and studied at available setups created quite recently just for studying the products of deep inelastic scattering (such as VAMOS, PRISMA and others). These fragment separators (as well as other setups) cannot separate heavy nuclei with Z>70 by their atomic number.

3. Scheme of setup

During the last several years a combined method of Z and A separation has been intensively studied and developed based on stopping nuclei in gas and subsequent resonance laser ionization [5-6]. This method was used up to now for separation and study of light exotic nuclei and fission fragments. Such techniques allows one to extract nuclei with a given atomic number, while a separation of the single-ionized isotopes by their mass number can be done rather easily by a magnetic field. Half-lives of heavy neutron rich nuclei, which we are interested in (as a rule, β^-–decaying), are much longer than the extraction time of ions at such a setup.

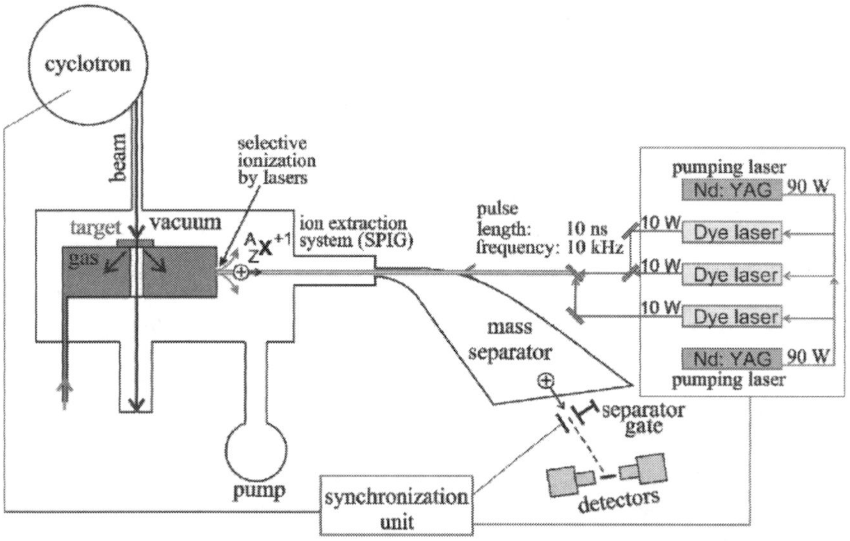

Fig. 3. Schematic view of setup for resonance laser ionization of nuclear reaction products stopped in gas with subsequent mass separation of them and transportation to detecting system.

The scheme of this setup is shown in Fig. 3. Neutron rich isotopes of heavy elements are produced in multi-nucleon transfer reactions with heavy ions accelerated up to 5–10 MeV/nucleon (depending on projectile-target combination). The target, a foil of about 300 μg/cm^2 thickness (or larger), is placed at the window of gas cell (or inside it). Nuclear reaction products recoiling from the target as multi-charged ions, are decelerated and neutralized due to collisions with the electrons and atoms of buffer gas in the gas cell filled with pure argon or helium. Then the desired atoms (with a given Z number) are ionized by means of two or three-step resonance laser irradiation and extracted through the exit hole or supersonic nozzle into the vacuum volume as positive charged ions (Q=+1) with slow energies of about 0.2 eV. In vacuum the ions are transported through the sextupole or quadrupole ion–guide system, which helps to pump away the residual buffer gas and keep the ions.

Then the ions are accelerated up to 30–60 keV and separated by the magnet of the mass–separator. A detailed description of the technique of stopping and cooling the ion beams in buffer gas, laser ionization and separation can be found in [4-14]. In this way a low-energy beam of singly-ionized ions is produced with an extremely small emittance. Atomic nuclei of this beam have a definite value of charge number and a given (chosen) mass value, there are neither isobars nor isotopes. This allows one to perform subsequent high sensitive analysis of spectroscopic and decay properties of these nuclei, as well as a measurement of spins, magnetic dipole and electric quadrupole moments and charge radii of these nuclei by means of laser spectroscopy.

4. Conclusions

The use of the designed setup for separation and study of heavy neutron rich nuclei does not require any specific beams of accelerated ions. The ion beams available at FLNR fully satisfy our requirements which look as follows: the ions - from $^{16,18}O$ to ^{238}U (i.e., quite different depending on the problem to be solved), beam energies 4.5 – 9 MeV/nucleon (slightly above the Coulomb barrier), beam intensity up to 10^{13} pps with beam size at the target 3–10 mm and beam emittance 20π mm mrad. The different heavy targets, including actinides it is expected to use.

The proposed method allows one to extract not only heavy transfer reaction products but also any other nuclei with half-lives longer than a few tens of milliseconds including neutron rich fission fragments, fusion reaction products and light exotic nuclei. Such studies are already performed at the similar setups, for example, in Finland and Belgium, and at some different with a hot cavity facility in CERN. Estimated efficiency of such setups is about 10% and depends on the half-life or the extracted ion. This is quite sufficient for the study of nuclear reaction products formed with cross sections of about 1 microbarn at a beam intensity of 0.1 pμA. This method allows one to study the structure and decay properties of new exotic nuclei as well as spectroscopic properties of the corresponding atoms. At higher beam intensity this setup could be used also for separation and study of neutron rich long living superheavy nuclei produced in multi-nucleon transfer reactions with actinide targets (see above). This setup could definitely open a new opportunity in the studies at heavy-ion facilities and will have significant impact on future experiments.

References

[1] Zagrebaev V, Greiner W, Phys. Rev. Lett. 101 (2008) 122701.
[2] Itkis M G et al., Nucl. Phys. A734 (2004) 136.
[3] Zagrebaev V I, Oganessian Yu Ts, Itkis M G, and Greiner W, Phys. Rev. C73 (2006) 031602(R).
[4] Köster U et al., Spectrochim. Acta B58 (2003) 1047.
[5] Kudryavstev Yu et al., Nucl. Instr. and Meth. Phys. Res. B204 (2003) 336.
[6] Moore I D et al., J. Phys. G: Nucl. Part. Phys. 31 (2005) S1499.
[7] Ärje J, Äystö J, Hyvönen H, Taskinen P, Koponen V, Honkanen J, Hautojärvi A, Vierinen K, Phys. Rev. Lett. 54 (1985) 99.
[8] Savard G et al., Nucl. Instr. and Meth. Phys. Res. B204 (2003) 582.
[9] Äystö J, Nucl. Phys. A693 (2001) 477.
[10] Vermeeren L, Bijnens N, Huyse M, Kudryavtsev Yu A, Van Duppen P, Wauters J, Qamhieh ZN, Silverans RE, Thoen P, Vandeweert E, Phys. Rev. Lett.73 (1994) 1935.
[11] Kudryavtsev Yu et al., Nucl. Instr. and Meth. Phys. Res. B114 (1996) 350.
[12] Kessler T, Moore I., Kudryavtsev Yu, Peräjärvi K, Popov A, Ronkanen P, Sonoda T, Tordoff B, Wendt K, Äystö J, Nucl. Instr. and Meth. Phys. Res. B266 (2008) 681.
[13] Kudryavtsev Yu et al., Nucl. Instr. and Meth. Phys. Res. B267 (2009) 2908.
[14] Ravn H L and Allardyce B W "On-Line Mass Separators". In: Treatise on Heavy-Ion Science, Vol. 8 ed. by D. Allan Bromley (Plenum Publishing Corporation, 1989) 363

Journal of Physics: Conference Series **533** (2014) 012026 doi:10.1088/1742-6596/533/1/012026

First results of the (n,γ) EXILL campaigns at the Institut Laue Langevin using EXOGAM and FATIMA

J Jolie[1a], J.-M. Régis[1], D. Wilmsen[1], S. Ahmed[1b], M. Pfeiffer[1], N. Saed-Samii[1], N. Warr [1], A. Blanc[2], M. Jentschel[2], U. Köster[2], P. Mutti[2], T Soldner[2] G. Simpson[3], G. de France[4], W. Urban[5], A.M. Bruce[6], O.J. Roberts[6], L.M. Fraile[7], V. Paziy [7], A. Ignatov[8], S. Ilieva[8], Th. Kröll[8], M. Scheck[8], M. Thürauf[8], D. Ivanova[9], S. Kisyov[9], S. Lalkovski[9], Zs. Podolyak[10], P.H. Regan[10], W. Korten[11], D. Habs[12], P. G. Thirolf[12], C. A. Ur[13]

[1] IKP, University of Cologne, Zülpicher Str. 77, D-50937 Köln, Germany,
[2] ILL, 6 rue Jules Horowitz, 38042 Grenoble Cedex 9, France
[3] LPSC, 53 rue des Martyrs, 38026 Grenoble Cedex, France
[4] GANIL, BP 55027, 14076 Caen Cedex 5, France
[5] Institute of Experimental Physics, University of Warsaw, ul. Hoza 69,00681 Warsaw, Poland
[6] SCEM, University of Brighton, Lewes Road,Brighton BN2 4GJ, UK
[7] Departamento de Fisica Atómica y Nuclear, Universidad Complutense, 28040 Madrid Spain
[8] Institut für Kernphysik, TU Darmstadt, Germany
[9] Faculty of Physics, University of Sofia, Bulgaria
[10] Dept. of Physics, University of Surrey, Guildford GU2 7XH, and National Physical Laboratory, Teddington, Middlesex, UK TW11 0LW
[11] CEA, Centre de Saclay, IRFU, F-91191 Gif-sur-Yvette, France
[12] Fakultät für Physik, Ludwig Maximilian Universität, 85748 Garching, Germany
[13] INFN Sezione di Padova, 35131 Padova, Italy

E-mail: j.jolie@ikp.uni-koeln.de

Abstract. At the PF1B cold neutron beam line at the Institut Laue Langevin the EXILL array consisting of EXOGAM, GASP and LOHENGRIN detectors was used to perform (n,γ) measurements under very high coincidence rates. About ten different reactions were then measured in autumn 2012. In spring 2013 the EXOGAM array was combined with 16 $LaBr_3(Ce)$ scintillators in the FATIMA@EXILL campaign for the measurement of lifetimes using the generalised centroid difference method. We report on the properties of both set-ups and present first results on Pt isotopes from both campaigns.

1. Introduction

During autumn 2012 and spring 2013 the EXOGAM spectrometer[1] and additional detectors were installed on the high intensity cold neutron guide PF1B of the Institut Laue Langevin (ILL) for the EXILL campaign. EXILL is partially a follow-up of a previous campaign using 8 EUROBALL capsule Ge

[a] To whom any correspondence should be addressed.

detectors and a highly collimated cold neutron beam for (n,γ) measurements [2]. The EXILL campaigns, which took two reactor cycles of 49 days, used beside the (n,γ) also the (n,fission) reaction on ^{235}U and ^{241}Pu targets. Here we will only report on very first results from the (n,γ) reaction on ^{194}Pt and ^{195}Pt.

2. The used set-ups

The same collimator used in Ref [2] provided, at the target position, a 12-mm diameter cold neutron beam of 10^8n/(s.cm^2). Perpendicular to the beam 8 EXOGAM Clover detectors with their BGO shields were mounted in a ring while the additional detectors were mounted in the forwards and backwards directions under 45°. During the first reactor cycle and part of the second one, 6 shielded GASP Ge detectors and 2 unshielded LOHENGRIN Clovers were mounted in addition to EXOGAM. Digital data acquisition allowed to handle event rates up to 0.84 MHz. With this set-up we have measured gamma-rays from the ^{194}Pt(n,γ)^{195}Pt reactions. During 30 days of the second cycle 16 LaBr$_3$(Ce) scintillators from the fast-timing array (FATIMA) collaboration were combined with the EXOGAM detectors for fast-timing measurements. From this campaign we report on results with the ^{195}Pt(n,γ)^{196}Pt reaction.

3. First results for the ^{194}Pt(n,γ)^{195}Pt reaction

Three 96% enriched targets of masses 200,140 and 70 mg were irradiated for a total of 3 days. Of interest was the population of the 4 days 13/2$^+$ isomer, which is populated with a thermal capture cross section of less than 0.1 barn. Gamma-ray cascades leading to this isomer after thermal neutron capture were first observed by D. D. Warner *et al.* in 1982 [3] and a 3/2$^+$,5/2$^+$ → 5/2$^+$ → 9/2$^+$ → 13/2$^+$ cascade was identified populating the isomer from the capture state. Using only the EXOGAM data, the population of the isomer could be corrected and extended as shown in Figure 1 [4]. The full analysis of these data is still ongoing.

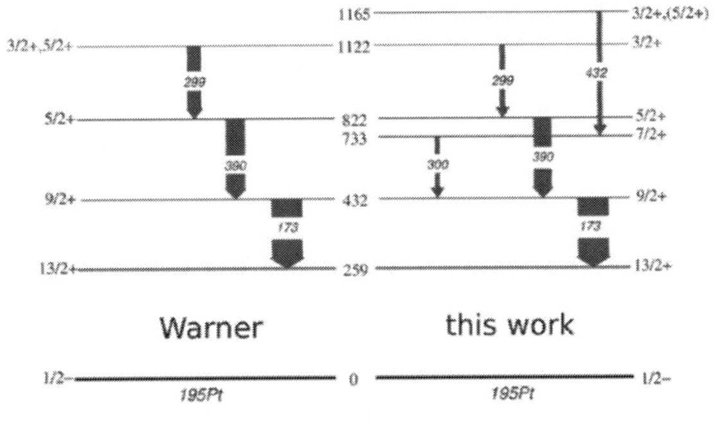

Figure1: Comparison between the Warner cascade and the extended cascade.

3. First results for the ^{195}Pt(n,γ)^{196}Pt reaction

Many years ago the even-even nucleus ^{196}Pt was proposed as an example of the SO(6)-limit of the Interacting Boson Model (IBM) [5]. When the Hamiltonian for N s,d bosons is composed by the linear combination of second order Casimir operators of the group chain: $SO(3) \subset SO(5) \subset SO(6) \subset U(6)$, the quantum numbers classifying the irreducible representation (irrep) of these groups: L, v, σ, N are valid and the Hamiltonian is analytically solvable. The lowest states have $\sigma=N$. States at higher energies can have $\sigma=N\text{-}2, N\text{-}4,\ldots$. Recently, the need to test the SO(6) symmetry and not only the SO(5) properties in ^{196}Pt was stressed [6]. When testing the goodness of the description of an SO(6) nucleus one thus should look for properties involving states with different σ. One of these properties concerns the electric quadrupole transition operator. In the SO(6) limit, the E2 transition operator is a generator of SO(6). Now generators cannot act outside their irrep and the strict selection rule $\Delta\sigma= 0$ follows. In order to test this selection rule, absolute B(E2) values between $\sigma=N\text{-}2$ and $\sigma=N$ states need to be measured. This is not an easy task as the first $\sigma=N\text{-}2$ states in ^{196}Pt are at high energy and have low spins. They are not populated by standard in-beam reactions with ions, but well by neutron capture. It is, therefore, not an accident that Ref [5] relied on ILL data. In 1990 we made an attempt to measure the lifetime of the lowest $\sigma=N\text{-}2$ state at 1402.7 keV using the GRID method [6] at GAMS4 [7]. For this 0^+_3 state a lower lifetime limit of $\tau> 1.8$ ps could be established showing that the B(E2) towards the 2^+_1 was smaller than $0.034e^2b^2$ and, as such, at least an order of magnitude smaller than allowed transitions between states with $\sigma=N$. Using the FATIMA set-up at EXILL triple coincidences can be used in which a Ge gate selects the cascade of gamma-rays to be measured by the $LaBr_3(Ce)$ scintillators. These fast signals can then be used for the generalised mirror symmetric centroid difference method [9,10] to measure lifetimes. The prompt response difference curve (PRD) was obtained using an ^{152}Eu source and transitions in ^{49}Ti after neutron capture. The method can be used to measure lifetimes down to about 10 ps. Figure 2 shows the time spectra of the 333-356 keV cascade gated by the 1978 keV transition in ^{196}Pt. The observed centroid difference between both spectra corresponds to 108.4(48) ps. Using the PRD curve a lifetime of $\tau = 50(5)$ ps is deduced which agrees very well with the literature value of 49.2(2) ps. Being confident that the method works, the lifetime of the 1402 keV 0^+ state was then measured. Again the 356 keV transition provided the decay signal for the fast timing. The feeding signal was obtained from the 566 keV transition feeding the 1402 keV state. The Ge gate was set on the 1048 keV transition of the decay towards the first excited state. The effective lifetime deduced from the centroid shift $\tau_{eff} = \tau_2 + \tau_0$ was 54(7) ps leading to an upper limit for the lifetime of $\tau_0 < 12$ ps. Using also the lower limit from [8] we find that 0.56 W.u $< B(E2;0^+_3 \rightarrow 2^+_1) < 5$ W.u and 0.05 W.u $< B(E2;0^+_3 \rightarrow 2^+_2) < 0.41$ W.u. Clearly no collective B(E2) values are found between the states, confirming the validity of the SO(6) symmetry.

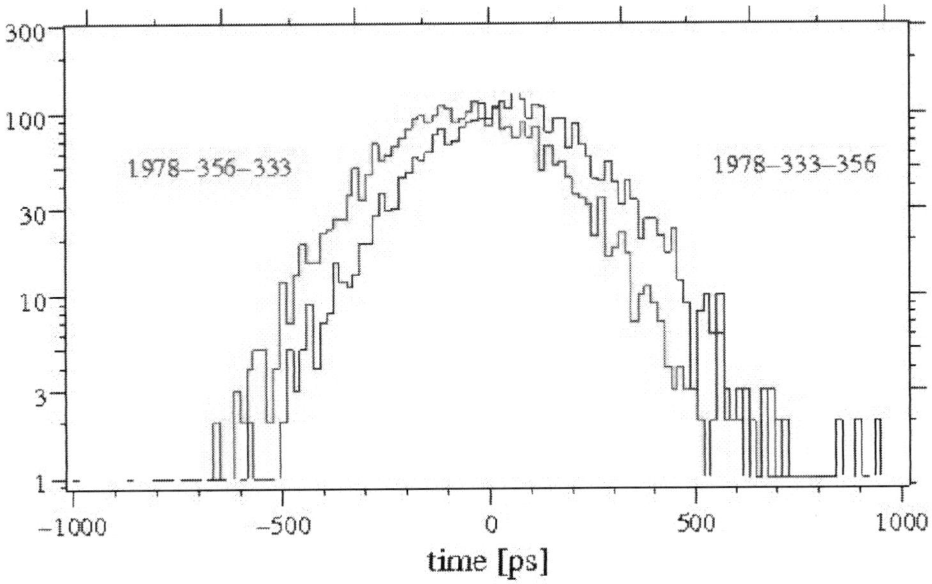

Figure 2: The binned time spectra using the 356 keV ground state transition as start for the TAC (left curve) and once as stop (right curve) (color online).

4. Acknowledgements

This work was supported by NUPNET by contract 05P12PKNUF (BMBF) and DNC7RP01/4, by the Science and Technology Facilities Councils (UK), by the DFG cluster of Excellence 'Origin and Structure of the Universe. The EXILL campaign would not have been possible without the support of several services at the ILL and the LPSC. We are grateful to the EXOGAM collaboration for the loan of the detectors, to GANIL for assistance during installation and dismantling, and to the INFN Legnaro laboratory for the loan of the GASP detectors.

References

[1] J.Simpson et al., 2000, Acta Physica Hungarica, New Series, Heavy Ion Physics **11**, 159 .
[2] W. Urban et al, 2013, JINST **8** 3014.
[3] D. D. Warner, et al. 1982, Phys. Rev. C **26** 1921.
[4] D. Wilmsen 2013, Diploma thesis, University of Cologne.
[5] J.A. Cizewski et al. 1978, Phys. Rev. Lett. 40 167.
[6] G. Rainovski *et al.* 2010, Phys. Lett. B683 11.
[7] H.G. Börner, J. Jolie1993, Journ. Phys. G19 217.
[9] H.G. Börner, et al.1990, Phys. Rev. C42 R2271.
[10] J.M Régis, G. Pascovici, J. Jolie, M. Rudigier 2010, Nucl. Instr. Meth. A622 83.
[11] J.-M. Régis et al. Nucl. Instr. Meth. Phys. Res. A 726 (2013) 191.

XX International School on Nuclear Physics, Neutron Physics and Applications (Varna2013) IOP Publishing
Journal of Physics: Conference Series **533** (2014) 012027 doi:10.1088/1742-6596/533/1/012027

A microscopic model beyond mean-field: from giant resonance properties to the fit of new effective interactions

M. Brenna[1], G. Colò[1], X. Roca-Maza[1], P. F. Bortignon[1], K. Moghrabi[2] and M. Grasso[2]

[1] Dipartimento di Fisica, Università degli Studi di Milano, and INFN, Sezione di Milano, Via Celoria 16, I-20133, Milano, Italy

[2] Institut de Physique Nucléaire, Université Paris-Sud, IN2P3-CNRS, F-91406 Orsay Cedex, France

E-mail: marco.brenna@mi.infn.it

Abstract. Self-consistent mean-field models are able to reproduce well the overall properties of nuclei for a wide range of masses. Nevertheless, they are intrinsically unsuitable for the description of some important observables like the single-particle strength distribution or, in connection with collective states, their damping width and their gamma decay to the ground state or to low lying states. For this reason, a completely microscopic approach beyond mean-field has been implemented recently, based on the Skyrme functional.

When beyond mean-field theories are handled, the mean-field-fitted effective interaction should be refitted at the desired level of approximation. If zero-range interactions are used, divergences arise. We present some steps towards the refitting of Skyrme interactions, for its application in finite nuclei.

1. Introduction

Self-consistent mean-field (SCMF) models [1] have become increasingly reliable in the descriptions of ground state (masses, radii, deformations) and of high-energy collective states, as giant resonances (GRs). Nonetheless, the SCMF approaches present well-known limitations. For instance, they do not reproduce, as a rule, the level density around the Fermi energy. Moreover, the fragmentation of single particle and GR strength functions, and the associated decay properties are outside the framework of these models.

A possible solution which can be used to overcome these drawbacks is the inclusion of beyond mean-field (BMF) correlations produced by the interweaving between single particle states and collective phonons. The basic ideas of the so-called particle-vibration coupling (PVC) models have been discussed in Ref. [2]. When these coupling are included, the standard shell model acquires a dynamical content: the average potential becomes non-local in time or, which is the same, energy dependent [3]. Recently, some of us has developed a fully microscopic self-consistent model, based on Skyrme functionals, to treat properly single particle states [4]. In this contribution we report on the application of this model to the calculation of inclusive (strength function) as well as exclusive (γ-decay width) GRs properties.

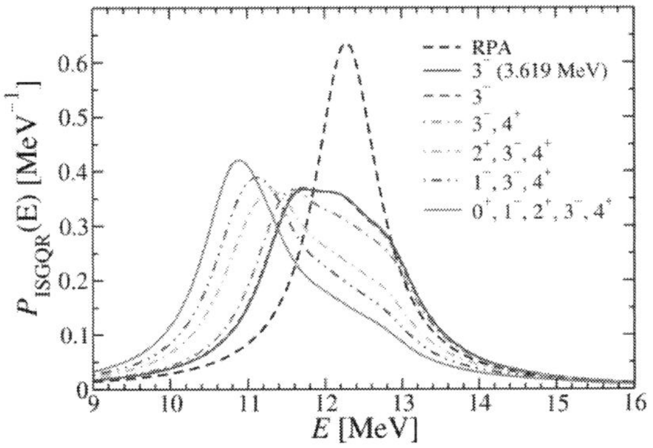

Figure 1. (Color online) Probability P to find the ISGQR state at an energy E. Different curves are obtained when the phonons listed in the legend are used as intermediate states. The label RPA [black-dashed line] refers to the curve calculated in the RPA with a Lorentzian width of 1 MeV.

Table 1. γ-decay widths Γ_γ. The first four rows are from this work, with no adjustable parameters. The following three are previous theoretical calculations. The last one refers to the experiment. See text and Ref. [5].

Interaction	E_{ISGQR} [MeV]	Γ_γ^{GS} (eV)	E_{3^-} [MeV]	$\Gamma_\gamma^{3^-}$ (eV)
SLy5 [7]	12.28	231.54	3.62	3.39
SkP [9]	10.28	119.18	3.29	8.34
Ref. [10]	10.60	130 ± 40	2.61	5±5

A problem of overcounting may arise in the BMF realm, since, so far, effective forces have been fitted at mean-field, thus including in an implicit way a class of higher order correlations. Therefore, we should expect to be obliged to re-fit the interactions at the required level of approximation. Moreover, if zero-range forces are used, like the Skyrme one, divergences arise when they are employed at BMF level. In the second part of this contribution, we present some preliminary results on the way to reabsorb these divergences in a simplified Skyrme interaction applied to finite nuclei.

2. The giant resonances widths within the PVC approach

In this section we present the results obtained for two observables: the strength function of the isoscalar giant quadrupole resonance (ISGQR) in ^{208}Pb, and the γ-decay width of the ISGQR in ^{208}Pb to the ground state and to the first collective octupole state. The main feature of these calculations is the consistent use of the Skyrme force in both the HF+RPA solution and in the PVC vertex, added on top of the former. Details about the formalism used and a more accurate discussion of the results can be found in [5].

2.1. The strength function

The present calculation follows closely Ref. [6], in which the GR strength functions were computed in the PVC model using a separable phenomenological force. In Fig. 1 the probability per unit energy of finding the ISGQR state is plotted, with the SLy5 [7] parametrization of the Skyrme force. Phonons with multipolarity $L = 0, 1, 2, 3, 4$ and natural parity are included in the model space. The RPA peak is shifted downwards to an energy of $E = 10.9$ MeV, in good agreement with the experiment ($E^{exp} = 10.9 \pm 0.3$ MeV [8]). The main contribution to the width comes from the first 3^- state, producing a spreading width of the order of 2 MeV ($\Gamma^{exp} = 3.0 \pm 0.3$).

2.2. *The γ-decay width*

In the model described in Ref. [5], the γ decay of the GRs to the ground state is evaluated at the RPA level, while the decay to low-lying collective states is accounted for at the first contributing order beyond the mean-field. It should be noted that only the direct decay, not the statistical one coming from the compund nucleus, can be computed in this model [11]. In this contribution we focus on the ISGQR in ^{208}Pb and its decay to the ground state and to the first 3^- state, using different Skyrme interactions. In table 1, some results for the decay widths (Γ_γ) are listed, together with the experimental outcome.

For the decay to the ground state the main issue is the overestimation of the energy of the resonance: the decay width is proportional to the energy raised to the power $2\lambda+1$, being λ the multipolarity of the transition. If we rescale the energy to the experimental value, all the interactions can reproduce the experimental decay width within the experimental error. On the other hand, concerning the decay to the 3^- state, only two interactions (the ones in table 1) are in agreement with the experiment. Actually, it should be recognized that it is just remarkable that Skyrme interactions can reproduce the order of magnitude (few eV) of this exclusive observable, given the fact that this functional form are fitted to reproduce basically macroscopic properties of nuclei at the scale of hundreds of keV.

3. The problem of the divergences beyond mean-field

If zero-range interactions are used, divergences arise in beyond mean-field quantities. Until now, the problem is circumvented by imposing a truncation to the model space with a somewhat arbitrary recipe, and this is clearly unsatisfactory.

In Refs. [12, 13], the problem of the renormalization of the whole Skyrme interaction was faced in nuclear matter with different degrees of neutron-proton asymmetry (from uniform to pure neutron matter): a new parameter, namely the maximum value of the transferred momentum, was introduced among the Skyrme ones. In a finite system, like the nucleus, because of the absence of translational invariance, the transferred momentum is not well defined. However, we can write the velocity-independent part of the Skyrme interaction as

$$V(\boldsymbol{r}_1',\boldsymbol{r}_2',\boldsymbol{r}_1,\boldsymbol{r}_2)=\frac{1}{2\sqrt{2}}g\left(\frac{\boldsymbol{R}}{\sqrt{2}}\right)\delta_3(\boldsymbol{r}')\delta_3(\boldsymbol{r})\delta_3(\boldsymbol{R}-\boldsymbol{R}')=\frac{1}{2\sqrt{2}}g\left(\frac{\boldsymbol{R}}{\sqrt{2}}\right)v(\boldsymbol{r}',\boldsymbol{r})\delta_3(\boldsymbol{R}-\boldsymbol{R}'),$$

where $\boldsymbol{r}^{(\prime)}=\frac{\boldsymbol{r}_1^{(\prime)}-\boldsymbol{r}_2^{(\prime)}}{\sqrt{2}}$, $\boldsymbol{R}^{(\prime)}=\frac{\boldsymbol{r}_1^{(\prime)}+\boldsymbol{r}_2^{(\prime)}}{\sqrt{2}}$ and $g(\boldsymbol{R})=t_0+\frac{t_3}{6}\left[\rho(\boldsymbol{R})\right]^\alpha$. We introduce a cutoff on relative momenta [14], modifying the term $v(\boldsymbol{r}',\boldsymbol{r})$, by including a cutoff λ and λ',

$$v^{\lambda\lambda'}(\boldsymbol{r}',\boldsymbol{r})=\frac{1}{\Omega}\int \mathrm{d}_3k\mathrm{d}_3k'\,\mathrm{e}^{i\boldsymbol{k}'\cdot\boldsymbol{r}'}v\left(\boldsymbol{k}',\boldsymbol{k}\right)\theta(\lambda-k)\theta(\lambda'-k')\mathrm{e}^{-i\boldsymbol{k}\boldsymbol{r}}=\frac{1}{4\pi^4}\frac{\lambda^2\lambda'^2}{rr'}j_1(r\lambda)j_1(r'\lambda').$$

This choice of the cutoff can be motivated by the fact that the Skyrme interaction can be seen as a low-relative-momentum expansion of the effective nuclear potential. By using the Goldstone theorem [15], the second order contribution to the total energy reads

$$\Delta E=\sum_{pp'hh'}\sum_{J}\frac{(2J+1)|\langle(pp')J|v^{\lambda\lambda'}|(hh')J\rangle|^2}{\epsilon_h+\epsilon_{h'}-\epsilon_p-\epsilon_{p'}},\tag{1}$$

where $\langle(pp')J|v^{\lambda\lambda'}|(hh')J\rangle$ is the J-coupled matrix element of the truncated interaction and $\epsilon_{h(p)}$ are the Hartree-Fock energies of the holes (particles). The presence of a diverence is caused by the fact that the sum on the particles p and p' is unbound.

In figure 2, the second order contribution to the total energy in ^{16}O is shown as a function of the maximum energy $\epsilon_p^{\max}$ of the particles included in the model space. The existence of a divergence is clear and it is linear with $\epsilon_p^{\max}$.

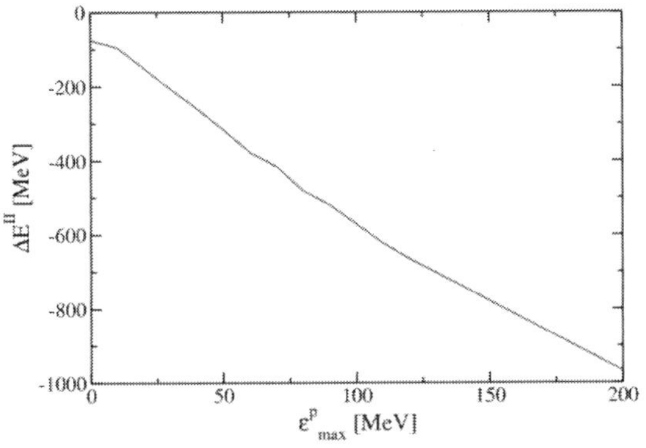

Figure 2. (Color online) Second order contribution to the total energy of ^{16}O as a function of the maximum energy of the particles included in the model space. The $t_0 - t_3$ part of the SkP [16] Skyrme interaction is used.

The analysis of the results are currently undergoing, in order to verify how the interactions fitted in nuclear matter can cure the divergence in nuclei.

4. Conclusion

Beyond mean-field correlations are important to overcome some intrinsic limitations of the SCMF approach in nuclei. A completely microscopic model, based on the particle-vibration coupling idea, was recently developed. This model has been so far applied to single particle observables, like the self-energy, and, as reported here, to inclusive (strength and energy) and exclusive (γ decay) properties of the GRs. The results obtained for the strength function of the ISGQR and its γ decay in ^{208}Pb are in fairly good agreement with the experimental findings.

Anyhow, the employment of interactions fitted at mean-field level in a higher order framework is the main limitation of these works. Moreover, zero-range forces cause the divergence of beyond mean-field quantities. We presented here a first attempt to treat this problem in finite systems.

Acknowledgments

M.B. thanks IPN Orsay for the warm hospitality during the time when part of this work was carried out.

References

[1] Bender M, Heenen P H and Reinhard P G 2003 *Rev. Mod. Phys.* **75** 121
[2] Bohr A and Mottelson B R 1975 *Nuclear Stucture* vol II (W. A.Benjamin Inc.)
[3] Mahaux C, Bortignon P F, Broglia R A and Dasso C H 1985 *Phys. Rep.* **4** 1
[4] Colò G, Sagawa H and Bortignon P F 2010 *Phys. Rev. C* **82** 064307
[5] Brenna M, Colò G and Bortignon P F 2012 *Phys. Rev. C* **85** 014305
[6] Bortignon P F and Broglia R A 1981 *Nucl. Phys. A* **371** 405
[7] Chabanat E, Bonche P, Haensel P, Meyer J and Schaeffer R 1998 *Nucl. Phys. A* **635** 231
[8] Youngblood D H, Lui Y W, Clark H L, John B, Tokimoto Y and Chen X 2004 *Phys. Rev. C* **69** 1
[9] Dobaczewski J, Flocard H and Treiner J 1984 *Nucl. Phys. A* **422** 103
[10] Beene J R, Bertrand F E, Halbert M L, Auble R L, Hensley D C, Hören D J, Robinson R L, Sayer R O and Sjoreen T P 1989 *Phys. Rev. C* **39** 1307
[11] Beene J R, Bertsch G F, Bortignon P F and Broglia R A 1985 *Phys. Lett. B* **164** 19
[12] Moghrabi K, Grasso M, Colò G and Van Giai N 2010 *Phys. Rev. Lett.* **105** 262501
[13] Moghrabi K, Grasso M, Roca-Maza X and Colò G 2012 *Phys. Rev. C* **85** 044323
[14] Carlsson B G, Toivanen J and von Barth U 2013 *Phys. Rev. C* **87**(5) 054303
[15] Fetter A L and Walecka J D 2003 *Quantum theory of many-particle systems* (Dover publications inc.)
[16] Van Giai N and Sagawa H 1981 *Nucl. Phys. A* **371** 1

Journal of Physics: Conference Series **533** (2014) 012028 doi:10.1088/1742-6596/533/1/012028

Pygmy dipole resonance

Vladimir Ponomarev

Institut für Kernphysik, TU Darmstadt, Schlossgartenstrasse 9, 64289 Darmstadt, Germany

E-mail: ponomare@theorie.ikp.physik.tu-darmstadt.de

Abstract. The pygmy dipole resonance (PDR) is a low-energy debris of the $1\hbar\omega$ $E1$-strength which is pushed by an isovector residual interaction to higher energies to form the giant dipole resonance. It exhausts about 1% of the EWSR below the particle threshold. High energy resolution experiments performed during the last decade reveal fine structure of the PDR in many nuclei. We report on the studies of the PDR fine structure performed within the Quasiparticle-Phonon model. Excited states are described by a wave function which includes one-, two-, and three-phonon configurations, i.e. the configuration space in calculations below the threshold is almost complete. We discuss also some particular features of the PDR excitation in different nuclear reactions.

A low-energy dipole strength below the threshold has been observed for the first time in experiments with tagged photons (see, e.g., [1]) in the late 70-ies. It was observed as a kind of a bump on top of the tail of the giant dipole resonance and received the name of the "pygmy dipole resonance" (PDR). Theoretical calculations of those days also reported substructures at low energies [2]. With a new generation of germanium detectors it became possible in the late 90-ies to resolve the fine structure of the PDR establishing that many dozens of 1^- states are excited in the nuclear resonance fluorescence (NRF) experiments to form the PDR [3, 4]. During the last decade the properties of the PDR have been studied within many different theoretical approaches. In my talk I limit the presentation by only the results obtained within the quasiparticle phonon model (QPM) [5].

From the theoretical point of view, the PDR may be considered as a low-energy debris of the $1\hbar\omega$ $E1$-strength which is pushed to higher energies by an isovector residual interaction forming the giant dipole resonance (GDR) (see figure 1). As figure 2 demonstrates, while many two-quasiparticle configurations contribute coherently to the E1 transition matrix element of a state belonging to the GDR, we deal with a destructive interference in the case of a PDR state.

The process of the strength fragmentation in the PDR energy region is demonstrated in figure 3 for ^{136}Xe. The major part of the E1 strength below the threshold is carried by a few QRPA (or one-phonon) states (figure 3a). These states are coupled to more complex configurations. Theoretically, to describe this process we write the wave function of excited states as a composition of one-, two-, etc. phonon configurations and diagonalize the model Hamiltonian to obtain the eigenstates. Calculation performed on the basis of one- and two-phonon configurations is presented in figure 3b. In this calculation each 1^- state carries only a small fraction of one-phonon configurations and accordingly, its B(E1) value is much smaller than in figure 3a. At the same time, the total B(E1) strength does not change much for the PDR energy region. The fragmentation process continues when three-phonon configurations are added to the wave function (figure 3c). One should keep in mind that the density of four-, five-,

Journal of Physics: Conference Series **533** (2014) 012028 doi:10.1088/1742-6596/533/1/012028

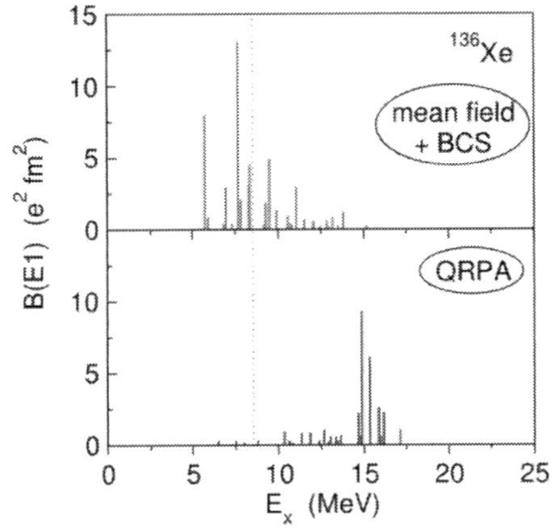

Figure 1. E1-strength distribution in ^{136}Xe calculated in the mean field and QRPA approximations. Dashed line shows $1\hbar\omega$ energy.

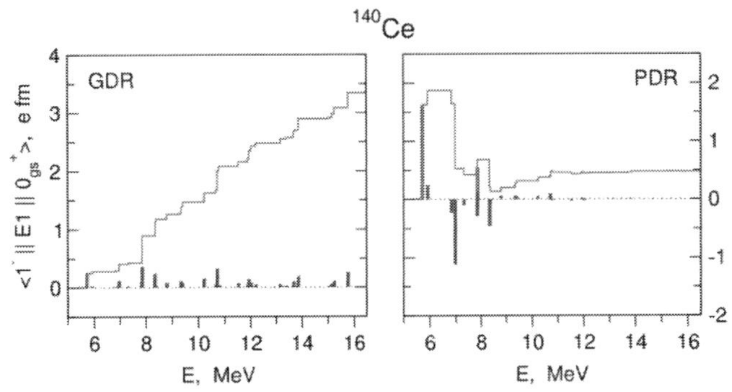

Figure 2. Running sum of the E1 transition matrix element for a state belonging to the GDR and PDR. Vertical lines present contribution of different two-quasiparticle configurations to the matrix element.

etc. phonon configurations is still very low below the threshold. It means that calculation in figure 3c is performed on almost complete basis of states for this energy region.

In addition to 1^- states, 1^+ states are also excited by an electromagnetic probe below the threshold. The later are either isoscalar 1^+ states or the ones belonging to the tail of the M1 resonance (see figure 4).

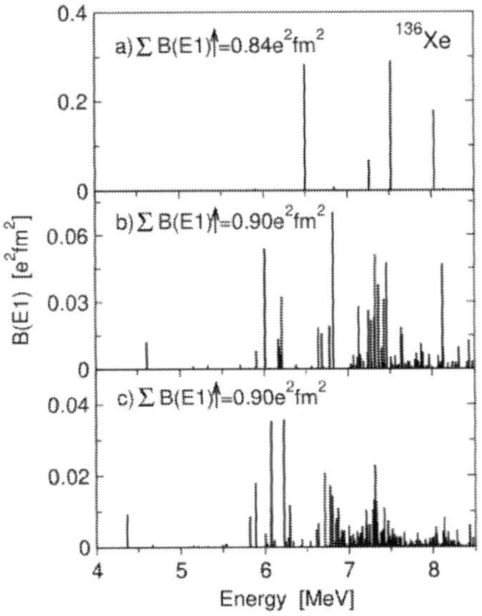

Figure 3. Fragmentation of the PDR in ^{136}Xe. Calculations are performed: a) within the QRPA, and with coupling to b) two- and c) two- and three-phonon configurations.

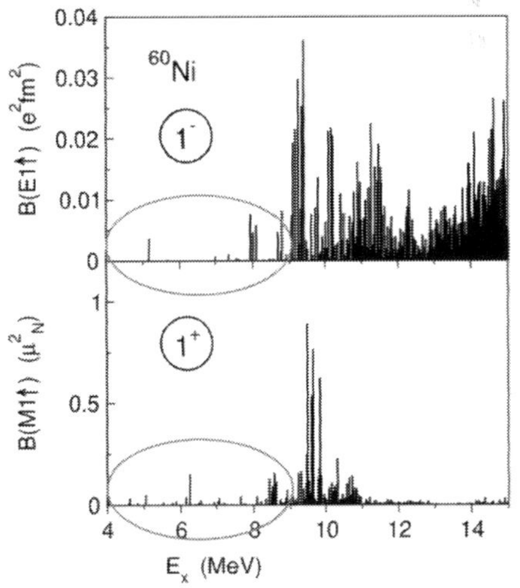

Figure 4. Fragmentation of the E1 and M1 strength in ^{60}Ni. Oval marks the energy interval which is typically covered in NRF experiments.

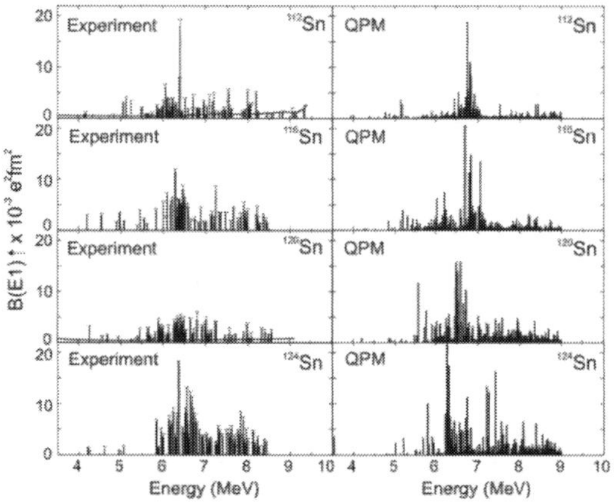

Figure 5. Fragmentation of the PDR in Sn isotopes.

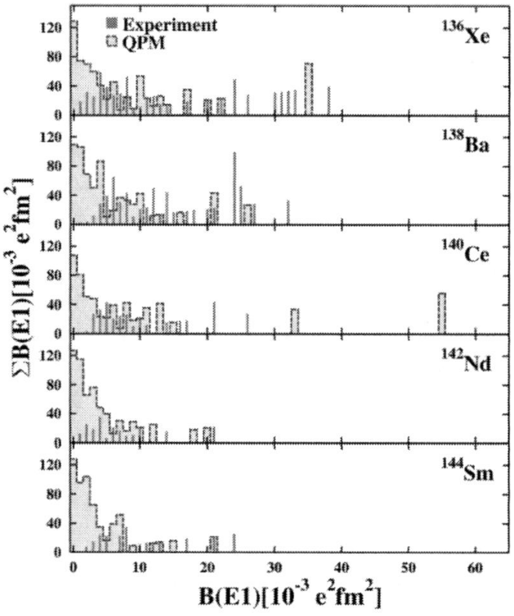

Figure 6. Fragmentation in the experimentally observed distribution and within the QPM for the N = 82 isotones.

The QPM calculations of the PDR have been performed for nuclei belonging to different parts of the nuclei chart: Cr-Fe-Ni [6, 7, 8], Zr-Mo [9] regions, Sn isotopes [4, 10, 11], N=82 isotones [12], and Pb isotopes [13]. The model describes rather well fragmentation of the E1 strength in the PDR region (see, e.g., figures 5 and 6) when almost complete basis of complex configurations is employed. In addition to many observed 1^- states it predicts many more weak 1^- states below detection limit (see left part of figure 6).

The PDR has been recently studied in $(\alpha, \alpha'\gamma)$ reaction [11, 14]. A good correspondence between the levels observed in this reaction with the (γ, γ') data has been established for the low energy part of the PDR. But no 1^- levels have been detected in the $(\alpha, \alpha'\gamma)$ reaction at higher excitation energy. Calculation in figure 7 also predicts that excitation of the low energy part of the PDR by the isoscalar r^3Y_1 external field which mimics (α, α') reaction, is enhanced in comparison to the excitation by the electromagnetic field. Analysis of transition densities indicate that their tails out of nucleus are responsible for this enhancement.

The PDR has been also observed in the (p, p') reaction at small scattering angles [15]. To extract information on 1^- states in the energy region of overlapping levels the multipole decomposition analysis (MDA) has been performed. The spectrum has been split into energy bins of about 100 keV and the cross section behaviour in each of them has been fit assuming contribution from 1^-, 1^+, and 2^+ levels (see figure 8 top). For 1^- levels cross sections have been calculated with the PDR and GDR wave functions. It has been found that the best χ^2 fit for the low energy part of the spectra is obtained with the PDR wave function while at higher energies the GDR wave function gives better results (see figure 8 bottom).

A new direction in the PDR studies concerns its decay properties. In analysis of the NRF spectra branching ratios to low-lying states are often neglected. Nevertheless, decays into them have been observed (see, e.g., [7]).This type of studies allows not only to correct extracted from

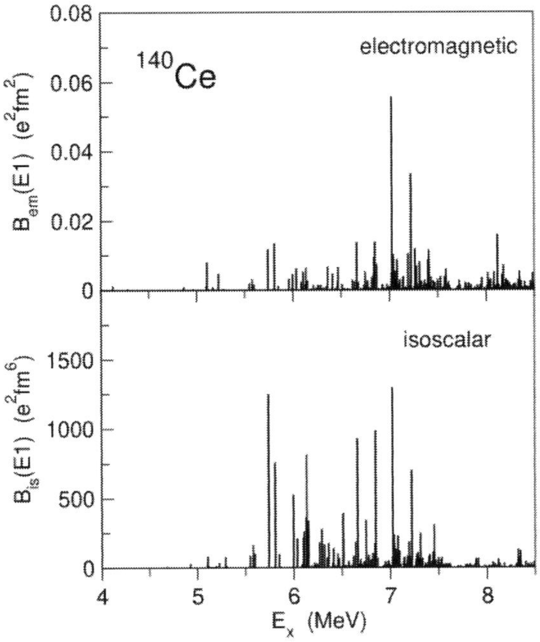

Figure 7. Electromagnetic and isoscalar response for 1^- states in ^{140}Ce.

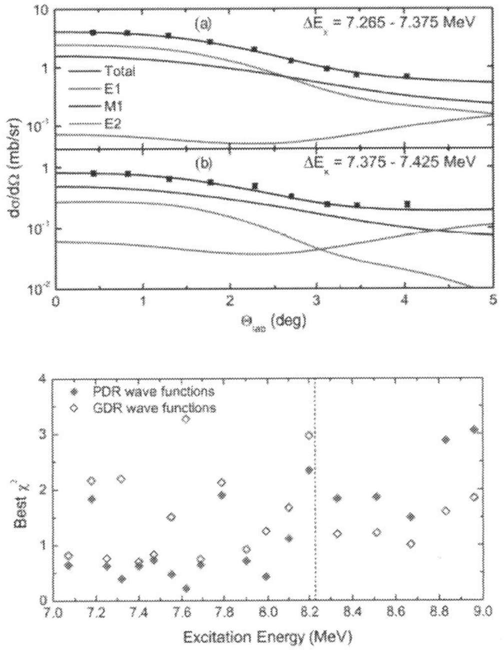

Figure 8. Examples of the MDA fits for two adjacent energy bins (top) and best χ^2 values in the MDA using either PDR- or GDR-type angular distributions for excitation energies $E_x = 7$ to 9 MeV (bottom).

NRF data B(E1) value but also investigate coupling of doorway 1^- states to some particular complex configurations.

Acknowledgments

I thank my colleagues experimentalists for fruitful many years collaboration on this topic. This work has been supported by the DFG through the grant SFB 634.

References

[1] Laszewski R M 1986 *Phys. Rev.* C **34** 1114; Laszewski R M and Axel P 1979 *Phys. Rev.* C **19** 342
[2] Soloviev V G, Stoyanov Ch and Voronov V V 1978 *Nucl. Phys.* A **304** 503
[3] Herzberg R-D *et al* 1997 *Phys. Lett.* B **390** 49
[4] Govaert K *et al* 1998 *Phys. Rev.* C **57** 2229
[5] Soloviev V G 1992 *Theory of Atomic Nuclei, Quasiparticles, and Phonons* (IOP, London)
[6] Bauwens F *et al* 2000 *Phys. Rev.* C **62** 024302
[7] Scheck M *et al* 2013 *Phys. Rev.* C **87** 051304; Scheck M *et al* 2013 *Phys. Rev.* C **88** 044304
[8] Pai H *et al* 2013 *Phys. Rev.* C **88** 054316
[9] Romig C *et al* 2013 *Phys. Rev.* C **88** 044331
[10] Özel B *et al* 2009 *arXiv:0901.2443*
[11] Endres J *et al* 2010 *Phys. Rev. Lett.* **105** 212503; Endres J *et al* 2012 *Phys. Rev.* C **85** 064331
[12] Savran D *et al* 2008 *Phys. Rev. Lett.* **100** 232501; Savran D *et al* 2011 *Phys. Rev.* C **84** 024326
[13] Enders J *et al* 2003 *Nucl. Phys.* A **724** 243
[14] Savran D *et al* 2006 *Phys. Rev. Lett.* **97** 172502
[15] Tamii A *et al* 2011 *Phys. Rev. Lett.* **107** 062502; Poltoratska I *et al* 2012 *Phys. Rev.* C **85** 041304

Fusion, breakup and scattering of weakly bound nuclei

P R S Gomes and J Lubian

Instituto de Física, Universidade Federal Fluminense, Av. Litorânea s/n, Niterói, R.J., 24210-340, Brazil

E-mail: paulogom@if.uff.br

Abstract. We discuss the systematic results reached in the investigation of the breakup of weakly bound nuclei, at near barrier energies, and its effects on fusion cross sections, elastic scattering and reaction cross sections. Particularly we show that the complete fusion of weakly bound systems are suppressed by dynamic breakup effects at energies above the Coulomb barrier and enhanced at sub-barrier energies. We also show that the energy dependence of the optical potential obtained in the elastic scattering of weakly bound projectiles corresponds to the so-called Breakup Threshold Anomaly (BTA).

The investigation of fusion cross sections between heavy ions has been extensively performed in the last 40 years. It is already well understood that low lying collective inelastic excitations of the colliding nuclei and transfer channels may lead to huge sub-barrier fusion enhancement when compared to predictions of unidimensional barrier penetration models. Particularly important is this kind of enhancement when highly deformed nuclei fuse, such as ^{154}Sm [1-4].

The elastic scattering between heavy ions at near barrier energies has also been extensively investigated and systematically presents a behaviour of the energy dependence of the interacting optical potential known as Threshold Anomaly (TA) [5-7]. This anomaly is rapid variation of the real and imaginary parts of this potential when the energy decreases towards the Coulomb barrier energy. The real part of the potential shows a localized peak, whereas there is a sharp decrease of the imaginary part of this potential at energies close to the barrier. The decrease of the imaginary potential is due to the closure of the non-elastic channels at energies below the Coulomb barrier. Causality considerations connect the real and imaginary parts of the optical potential through a dispersion relation [8,9]. The real potential can be written as $V_{eff} = V_0 + \Delta V$, where V_0 is the real potential at higher energies and ΔV is called polarization potential, which is attractive.

In the last years, the understanding of scattering and reaction mechanisms involving weakly bound nuclei, particularly halo nuclei, has been a subject of great interest, both theoretically and experimentally [10]. Some of the main questions on this field are (i) how the breakup of those weakly bound nuclei affect the complete fusion cross section; (ii) how is the behaviour of the optical potential energy dependence of such systems. (iii) how is the effect of breakup on elastic, quasi-elastic scattering and its barrier distribution [11-13]; (iv) how is the effect of breakup on the total reaction cross sections [14,15].

Concerning the effect of the breakup on the fusion cross section, there are two kinds of effects to be investigated [16, 17]. The first is static effects, caused by the longer tail of the optical potential of the weakly bound and specially halo nuclei. This effect gives rise to lower and thicker barriers when compared with tightly bound systems, and consequently enhances fusion cross section at sub-barrier energies. The second kind of effect is the dynamic, which is due to the strong coupling between the elastic channel and the continuum states representing the breakup channel.

In order to investigate the dynamic effects on fusion, one has to disentangle these effects from the static ones, and so one has to eliminate the static effects. Canto et al. [16, 17] have shown that in order to reach this goal, one has to use the dimensionless function F(x), called Fusion Function, and energy variable x, defined as $F(x) = (2E\sigma_f)/(\hbar\omega\ R_B^2)$ and $x = (E-V_B)/(\hbar\omega))$, where V_B, R_B and $\hbar\omega$ are the barrier height, radius and the curvature of the Coulomb barrier. For situations where channel coupling effects can be neglected and the fusion cross section can be approximated by Wong's formula [18], F(x) becomes system independent and can be written as an Universal Fusion Function (UFF) written as $F_0(x) = \ln[1+\exp(2\pi x)]$, which becomes a benchmark curve. However, this procedure may lead to wrong conclusions, since it has two shortcomings. The first is the fact that the Wong model is not valid at sub-barrier energies for light systems, as it is the situation of some those involving the above mentioned weakly bound projectiles. The second is that if one compares experimental fusion functions with the UFF, where no coupled channels are taken into account, the differences would be due to the effects of all channels not considered in the calculations, such as all important low lying collective couplings, and not just effect of the breakup which is the subject under investigation. To overcome those difficulties, Canto et al. [16] suggested to compare the UFF curve with a renormalized experimental fusion functions which takes into account the possible failure of the Wong model and the effects of inelastic couplings.

In the following, we mention the results of systematic studies for the fusion cross sections of tens weakly bound systems, both stable and radioactive, neutron-halo, proton-halo and no-halo, using the fusion function method. The bare potential used in the calculations was the double folding São Paulo potential (SPP) [18, 19]. This potential has been widely and successfully used in the last years, for a large variety of systems and energy ranges [20, 21], including fusion barrier distributions for weakly bound nuclei and fusion of halo nuclei [22, 23]. Coupled channels calculations were performed using the code FRESCO [24]. The first systematic results were presented in Ref. [16, 17, 25]. They show that complete fusion of stable weakly bound nuclei ^{6,7}Li, ^{9}Be and neutron-halo nuclei ^{6,8}He are suppressed at energies above the Coulomb barrier by a factor around 20 to 30% in relation with the UFF, and show some small enhancement at sub-barrier energies, when compared with the UFF. Those effects are interpreted as due to dynamic breakup couplings plus transfer couplings, since transfer channels were not included in the coupled channel calculations. Figure 1 shows this behaviour for a selected number of systems. The linear scale is better to observe the behaviour at energies above the barrier, whereas the log scale is more suitable for sub-barrier energies. For the total fusion (when the whole or part of the projectile fuses) of stable weakly bound system, there is neither suppression nor enhancement above the barrier. It is interesting to mention that for proton-halo fusion, there are only two fusion excitation functions available in literature, for ^{8}B + ^{28}Si, ^{58}Ni [26, 27]. For ^{28}Si [26], the results are compatible with the systematics mentioned above, whereas for the ^{58}Ni the fusion cross section shows an anomalous behaviour of enhancement above the barrier when compared with UFF. A recent work [28] shows that there are some signatures that there might be some problems with the ^{8}B + ^{58}Ni fusion data, what shows that the method of the UFF is very powerful to detect such problems.

Concerning the elastic scattering, the energy dependence of the optical potential when weakly bound nuclei are involved has been shown, in several works, to have a different behaviour as that for tightly bound systems, namely, the TA and the attractive polarization potential owing to the couplings to inelastic and transfer channels. This is because the breakup channel may be important even below the barrier and produces repulsive polarization potential. So, the imaginary potential does not decrease at the barrier energy. Indeed, it may increase. Consequently, the real potential decreases at this energy region and fusion is suppressed owing to this effect. This behaviour is called breakup threshold anomaly (BTA) [29, 30].

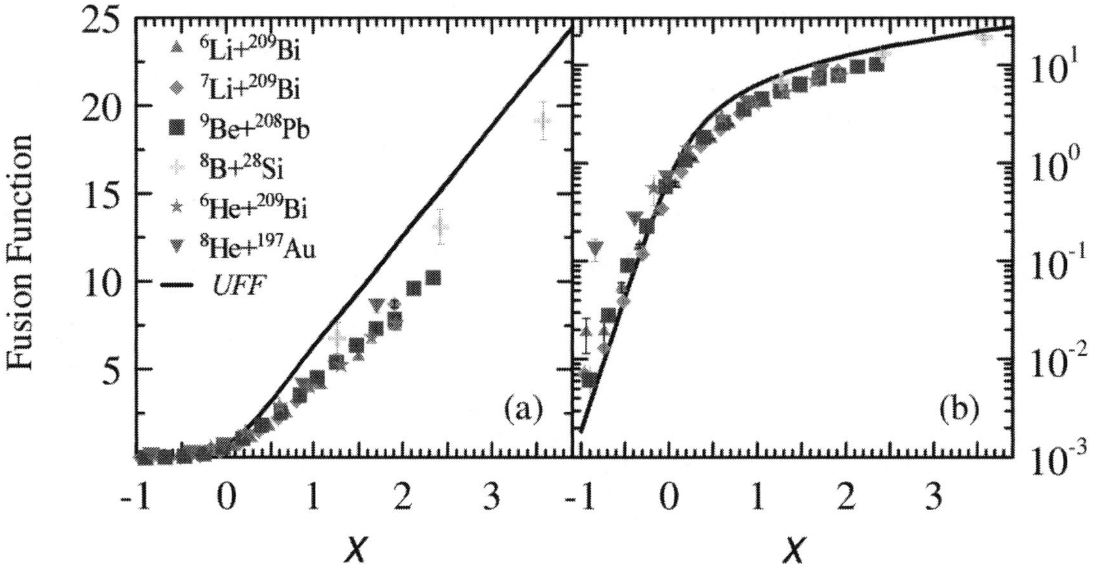

Figure 1: Renormalized experimental fusion function for some selected weakly bound systems. The curve is the UFF (see text for details)

Acknowledgements
The authors would like to thank CNPq, CAPES, FAPERJ and PRONEX for the partial financial support.

References

[1] Stokstad R G et al. 1978 *Phys. Rev. Lett.* **41** 465
[2] Stokstad R G et al. 1980 *Phys. Rev. C* **39** 2427
[3] Gil S et al. 1990 *Phys. Rev. Lett.* **65** 3100
[4] Gomes P R S et al. 1994 *Phys. Rev. C* **49** 245
[5] Nagarajan M A, Mahaux C C and Satchler G R 1985 *Phys. Rev. Lett.* **54** 1136
[6] Satchler G R 1991 *Phys. Rep.* **199** 147
[7] Brandan M E and Satchler G R 1997 *Phys. Rep.* **285** 143
[8] Mahaux C, Ngo H and Satchler G R 1986 *Nucl. Phys. A* **449** 354
[9] Satchler G R and Love W 1979 *Phys. Rep.* **55** 183
[10] Canto L F , Gomes P R S. Gomes, Donangelo R. and Hussein M S 2006 *Phys. Rep.* **424** 1
[11] Monteiro D S et al. 2009 *Phys. Rev. C* **79** 014601
[12] Lubian J, Correa T, Gomes P R S, Canto L F, 2008 *Phys. Rev. C* **78**, 064614
[13] Lubian J et al. 2009 *Phys. Rev. C* **79**, 064605
[14] Gomes P R S, Lubian J, Padron I , Anjos R M 2005 *Phys. Rev. C* **71** 017601
[15] Shorto J M B et al. 2009 *Phys. Lett. B* **678** 77
[16] Canto L F et al. 2009 *J. Phys. G* **36** 015109
[17] Canto L F et al. 2009 *Nucl. Phys A* **821** 51
[18] Chamon L C et al. 1997 *Phys. Rev. Lett.* **79** 5218

[19] Chamon L C et al. 2002 *Phys. Rev. C* **66** 014610
[20] Alves J J S et al. 2005 *Nucl. Phys. A* **748** 59
[21] Gasques L R, Chamon L C, Gomes P R S, Lubian J 2006 *Nucl. Phys. A* **764** 135
[22] Crema E , Chamon L C , Gomes P R S, 2005 *Phys. Rev. C* **72** 034610
[23] Crema E, Gomes P R S, Chamon L C 2007 *Phys. Rev. C* **75** 037601
[24] Tompson I J, 1988 Comp. *Phys. Rep.* **7,** 167
[25] Gomes P R S, Lubian J , Canto L F 2009 *Phys. Rev. C* **79** 027606
[26] Pakou A et al. 2013 *Phys. Rev. C* **87** 014619
[27] Aguilera E F et al 2011 *Phys. Rev. Lett.* **107** 092701
[28] Rangel J et al. 2013 *Eur. Phys. J. A* **49**, 57
[29] Gomes P R S et al. 2005 J. *Phys. G* **31** S1669
[30] Hussein M S, Gomes P R S, Lubian J, Chamon L C 2006 *Phys. Rev. C* **73** 044610

XX International School on Nuclear Physics, Neutron Physics and Applications (Varna2013) IOP Publishing
Journal of Physics: Conference Series **533** (2014) 012030 doi:10.1088/1742-6596/533/1/012030

Systematic spectroscopic study of neutron rich nuclei within a new shell model context

D Bianco[1,2], **N Lo Iudice**[1,2], **F Andreozzi**[1,2], **A Porrino**[1,2] **and F Knapp**[3]

[1]Dipartimento di Fisica, Università di Napoli Federico II, Monte S. Angelo, via Cintia, I-80126 Napoli, Italy
[2]Istituto Nazionale di Fisica Nucleare, Sezione di Napoli, Monte S. Angelo, via Cintia, I-80126 Napoli, Italy
[3]Faculty of Mathematics and Physics, Charles University, Prague, Czech Republic

E-mail: bianco@na.infn.it

Abstract. An iterative approach for diagonalizing nuclear Shell Model (SM) Hamiltonian matrices, developed few years ago, has been used for performing large scale shell model calculations around the doubly-magic ^{132}Sn. Calculated spectra and transition strengths correctly reproduces the available experimental data, and have been used for probing the prediction of collective algebraic models in this region, in particular for isotopes on which few measures are available. The calculation predicts an increasing neutron weight in the isotopes departing from the doubly-magic ^{132}Sn and moving toward the neutron drip line.

1. Introduction

In recent years, the study of low-energy collective states in the region around the doubly magic ^{132}Sn has attracted considerable interest, from both an experimental and a theoretical point of view. For these isotopes, the proton-neutron interacting boson model (IBM-2) [1] predicts the occurrence of states with F-spin $F = F_{max} - 1$ in the lower part of the spectrum.

These mixed-symmetry states, originally observed in ^{94}Mo [2], have been systematically studied in experiments devoted to nuclei with neutron number below the $N = 82$ shell closure [3, 4, 5, 6, 7, 8, 9]. Theoretically, this region has been explored from a microscopic point of view in the quasi-particle random-phase-approximation (QRPA) [10], the QPM [11], and in large scale SM approaches [12, 13, 14, 15, 16].

This work presents the results of a large-scale SM study conducted on neutron-rich isotopes above the $N = 82$ shell closure. The calculations which have been performed exploits a new iterative matrix diagonalization algorithm [17, 18, 19], endowed with an importance sampling which selects the basis states relevant to the eigenstates to be generated, allowing a considerable reduction in the computational power needed.

After a brief outline of the algorithm and an analysis of the convergence properties of the iterative diagonalization process, we perform an investigation of the spectra, $E2$ and $M1$ transitions, focusing on the different contribution from the proton and the neutron part of the wave-function and their evolution moving towards the neutron drip-line.

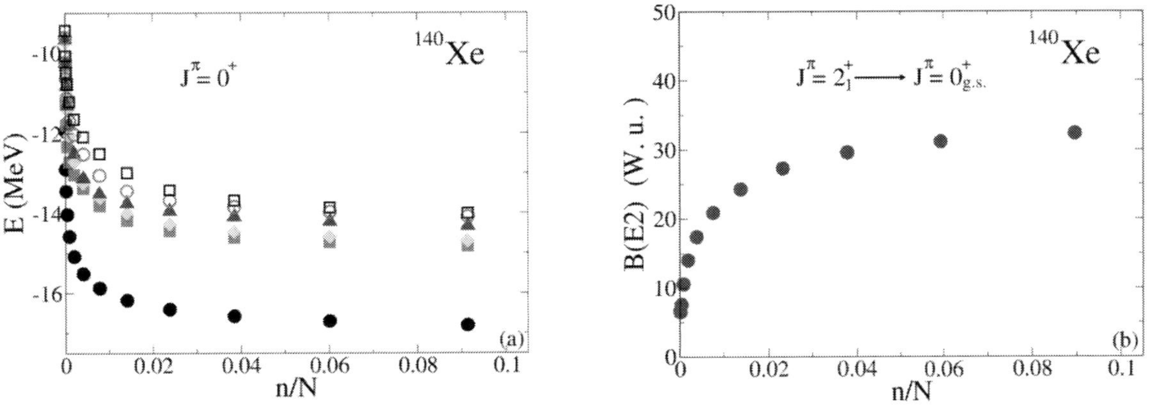

Figure 1. (Color online) Convergence rate of the lowest $J^\pi = 0^+$ energy levels (a) and of the $2_1^+ \to 0_1^+$ $E2$ transition strengths (b) in ^{140}Xe. N and n respectively indicate the total number of basis states and the number of configurations included in the corresponding calculation.

2. SM calculation and algorithm outline

Let the matrix $A = \{a_{ij}\}$ represent a self-adjoint operator $\hat{A}$ in an orthonormal basis $\{| 1\rangle, \dots |i >, \dots |N\rangle\}$. The iterative algorithm starts with an m-dimensional approximation to the v lowest eigenvectors, calculated with the corresponding eigenvalues diagonalizing the principal square submatrix with linear dimension $m(\geq v)$. The components of the eigenvectors on the remaining $d - m$ basis configurations are then recovered iteratively diagonalizing the matrix

$$A_k^{(1)} = \begin{pmatrix} \Lambda_{k-1}^{(1)} & B_k^{(1)} \\ \tilde{B}_k^{(1)} & A_k^{(1)}(p) \end{pmatrix}, \tag{1}$$

with linear dimension $v + p$, where p indicates the number of basis states which are added to the current approximation in the k-th step. In Eq. 1, $\Lambda_k^{(1)}$ is a v-dimensional diagonal matrix with the current values of the eigenvalues on its diagonal entries and $A_k^{(1)}(p)$ the p-dimensional submatrix corresponding to the new p basis states. The two subspaces spanned by the current approximation for the eigenvectors, $|\psi_i^{(1)}(k-1)\langle$, and the p basis configurations, are coupled by the two matrices $B_k^{(1)}$ and its transpose $\tilde{B}_k^{(1)}$, whose matrix elements are

$$b_{ij}^{(k)} = \langle \psi_i^{(1)}(k-1) \mid \hat{A} \mid j \rangle. \tag{2}$$

The v approximate eigenvectors resulting from the first loop, $\psi_i^{(1)}(N) = \psi_i^{(2)}(0)$, are the entries for a new iteration loop. Since this basis is no longer orthonormal and may be even redundant, we have to solve a generalized eigenvalue problem by resorting to the Cholesky decomposition method. With this modification, the iteration loops proceed as the first one and generate a sequence of v current eigenvalues and corresponding eigenvectors, converging to the exact ones. The algorithm has been applied to the modified Hamiltonian

$$H_J = H + c[\mathbf{J}^2 - J(J+1)]^2, \tag{3}$$

where $\mathbf{J}$ is the total spin operator, which allows an effective projection on a good angular momentum J by a proper choice of the positive constant c. The basis states are classified according to their total magnetic quantum number $M = M_\pi + M_\nu$, where $M_\tau = m_{1_\tau} + \dots m_{i_\tau} + \dots m_{n_\tau}$.

Table 1. Proton (π) and neutron (ν) single particle energies (in MeV)

$(nlj)_\pi$	$1g_{7/2}$	$2d_{5/2}$	$2d_{3/2}$	$1h_{11/2}$	$3s_{1/2}$	
ϵ_π	0.00	0.96	2.71	2.80	3.5	
$(nlj)_\nu$	$2f_{7/2}$	$3p_{3/2}$	$1h_{9/2}$	$3p_{1/2}$	$2f_{5/2}$	$1i_{13/2}$
ϵ_ν	0.00	0.85	1.56	1.66	2.00	2.11

The eigenvalue problem has been solved in a subspace spanned by those many-particles configurations fulfilling the condition

$$\frac{|<j|H_J|\psi_i^{(l-1)}(N)>|^2}{a_{jj}-E_i^{(l-1)}(N)} > \epsilon, \tag{4}$$

where ϵ indicates an arbitrary small number, and l is an index corresponding to the current iteration loop. Fixing a sequence of positive small numbers of decreasing values $\epsilon_1 > \dots \epsilon_k > \dots > \epsilon_F$, a set of subspaces of increasing dimension has been selected, in which solving the corresponding eigenvalue problem. An illustration of this sampling procedure and of its rate of convergence is given in Fig. 1. In the limit $\epsilon_F \to 0$, the exact solutions are recovered

$$\lim_{\epsilon_F\to 0}\{E_i(\epsilon_F),\, \psi_i(\epsilon_F)\} = \{E_i,\, \psi_i\}, \tag{5}$$

The two-body potential was chosen to be a renormalized G matrix [20] derived from the CD-Bonn potential [21]. The shell model space includes the proton $\{1g_{7/2}, 2d_{5/2}, 2d_{3/2}, 3s_{1/2}, 1h_{11/2}\}$ and neutron $\{2f_{7/2}, 3p_{3/2}, 1h_{9/2}, 3p_{1/2}, 2f_{5/2}, 1i_{13/2}\}$ shells. The adopted neutron and proton single particle energies are shown in Table 1. They are very close to the experimental energies of ^{133}Sb [22] and ^{133}Sn [23], respectively. The two-body part of the interaction has been optimized for the current calculation scaling by factors 1.1 and 1.2, respectively, its $J^\pi = 0^+$ proton-proton and neutron-neutron pairing-like components.

The $E2$ transition strengths were computed with the proton and neutron effective charges $e_\pi = 1.7$, $e_\nu = 0.7$ for Te and $e_\pi = 1.6$, $e_\nu = 0.7$ for Xe. These are the same charges used respectively for Te [14] and Xe [13]isotopes with $N < 82$. Also the $M1$ transition probabilities were evaluated with the same spin gyromagnetic quenching factor $g_s = 0.5$ used for $N < 82$ [13, 14].

3. Results and case analysis

An extensive review of the theoretical results obtained with the proposed algorithm for the neutron-rich Tellurium and Xenon isotopes lying above the $N = 82$ shell closure has been given in [16]. In the present paper, those results are discussed and analyzed.

The agreement of the SM spectra with the experiments is good apart from some discrepancies in ^{140}Xe, where the SM yrast levels differ from the measured energies by $\sim$ 100 to $\sim$ 200 keV [16, 24], which is the isotope of the chain with the lowest degree of convergence. On the other hand, the information on the $E2$ transition strengths is not rich, but the calculations reproduce fairly well the available data.

The SM 2_1^+ energies decrease and the corresponding $E2$ decay transition strengths increase departing from the neutron shell closure in both regions below and above $N = 82$. Their evolution, however, is not symmetric with respect to the neutron magic number.

The weakness of the $E2$ transition above the shell closure suggests an exceedingly strong neutron dominance in the 2_1^+ wavefunction [12, 8]. In order to investigate this point, the ratio $R_{\nu/\pi}(E2) = \frac{B_\nu(E2;2_1^+\to 0_1^+)}{B_\pi(E2;2_1^+\to 0_1^+)}$ has been evaluated, where the neutron $B_\nu(E2;2_1^+ \to 0_1^+)$ and the

Journal of Physics: Conference Series **533** (2014) 012030 doi:10.1088/1742-6596/533/1/012030

proton $B_\pi(E2; 2_1^+ \to 0_1^+)$ strengths are computed using the charges ($e_\nu = 1$ and $e_\pi = 0$) and ($e_\nu = 0$ and $e_\pi = 1$), respectively.

The neutron weight in the 2_1^+ states is predicted to get more and more enhanced in both Te and Xe isotopes as they move toward the drip line [16]. It grows quite fast in Te isotopes, where the neutron to proton ratio goes from $R_{\nu/\pi}(E2) = 2.6$ in ^{136}Te to $R_{\nu/\pi}(E2) = 7.4$ in ^{140}Te.

On the other hand, it increases more slowly in Xe isotopes. Protons and neutrons give a comparable contribution to the $E2$ strength in ^{138}Xe, consistently with experiments, while the neutrons prevail in ^{140}Xe.

An effect of the neutron dominance is that the neutron-proton exchange symmetry is strongly broken departing from the doubly magic ^{132}Sn and approach the drip line. This symmetry breaking seems to find support in the dramatic drop of the $E2$ and $M1$ $2_i^+ \to 2_1^+$ transition strengths, for all low-lying 2_i^+ states, in going from ^{136}Te to the heavier ^{138,140}Te isotopes and from ^{138}Xe to ^{140}Xe.

The strong suppression of the $2_i^+ \to 2_1^+$ $E2$ and $M1$ transitions are to be ascribed to the increasing prevalence of neutrons in the 2_1^+ states which hinders the action of the proton component of the electric quadrupole operator on the 2_1^+ state and inhibits the transitions induced by the proton orbital motion.

These unexpectedly dramatic drops imply that none of the low energy 2_i^+ ($i \geq 2$), in the explored neutron rich Te and Xe isotopes with $N > 84$, has a definite neutron-proton mixed symmetry and/or a quadrupole collective one- or two-phonon nature.

References

[1] A. Arima, T. Otsuka, F. Iachello, and I. Talmi, Phys. Lett. **B 66**, 205 (1977).
[2] N. Pietralla, C. Fransen, D. Belic, P. von Brentano, C. Friessner, U. Kneissl, A. Linnemann, A. Nord, H. H. Pitz, T. Otsuka, I. Schneider, V. Werner, and I. Wiedenhöver, Phys. Rev. Lett. **83**, 1303 (1999).
[3] N. Pietralla *et al.*, Phys. Rev. C **58**, 796 (1998).
[4] G. Rainovski, N. Pietralla, T. Ahn, C. J. Lister, R. V. F. Janssens, M. P. Carpenter, S. Zhu, and C. J. Barton III, Phys. Rev. Lett. **96**, 122501 (2006).
[5] T. Ahn, N. Pietralla, G. Rainovski, A. Costin, K. Dusling, T. C. Li, A. Linnemann, and S. Pontillo, Phys. Rev. C **75**, 014313 (2007).
[6] R. O. Hughes, N. V. Zamfir, D. C. Radford, C. J. Gross, C. J. Barton, C. Baktash, M. A. Caprio, R. F. Casten, A. Galindo-Uribarri, P. A. Hausladen, E. A. McCutchan, J. J. Ressler, D. Shapira, D. W. Stracener, and C.-H. Yu, Phys. Rev. C **71**, 044311 (2005).
[7] S. F. Hicks, J. R. Vanhoy, and S. W. Yates , Phys. Rev. C **78**, 054320 (2008).
[8] M. Danchev *et al.*, Phys. Rev. C **84**, 061306(R) (2011).
[9] T. Alharbi *et al.*, Phys. Rev. C **87**, 014323 (2013).
[10] J. Terasaki, J. Engel, W. Nazarewicz, and M. Stoitsov , Phys. Rev. C **66**, 054313 (2002).
[11] N. Lo Iudice, Ch. Stoyanov, and D. Tarpanov, Phys. Rev. C **77**, 044310 (2008).
[12] N. Shimizu, T. Otsuka, T. Mizusaki, M. Honma, Phys. Rev. C **70**, 054313 (2004); Phys. Rev. C **74**, 059903(E) (2006).
[13] D. Bianco, F. Andreozzi, N. Lo Iudice, A. Porrino, and F. Knapp Phys. Rev. C **84**, 024310 (2011).
[14] D. Bianco, N. Lo Iudice, F. Andreozzi, A. Porrino, and F. Knapp Phys. Rev. C **86**, 044325 (2012).
[15] D. Bianco, F. Andreozzi, N. Lo Iudice, A. Porrino, and F. Knapp Phys. Rev. C **85**, 034332 (2012).
[16] D. Bianco, F. Andreozzi, N. Lo Iudice, A. Porrino, and F. Knapp Phys. Rev. C **88**, 024303 (2013).
[17] F. Andreozzi, A. Porrino, and N. Lo Iudice, J. Phys. A: Math. Gen. **35**, L61 (2002).
[18] F. Andreozzi, N. Lo Iudice, and A. Porrino, J. Phys. : Nucl. Part. Phys. **29**, 2319 (2003).
[19] D. Bianco, F. Andreozzi, N. Lo Iudice, A. Porrino, and F. Knapp, J. Phys. G: Nucl. Part. Phys. **38**, 025103 (2011).
[20] M. Hjorth-Jensen, T. T. Kuo, E. Osnes, Physics Reports **261**, 125 (1995).
[21] R. Machleidt, Phys. Rev. C **63**, 024001 (2001).
[22] L. A. Stone, J. D. Robertson, S. H. Faller, P. F. Mantica, B. E. Zimmerman, C. Chung, and W. B. Walters, Phys. Scr. **T 56**, 316 (1995).
[23] P. Hoff et al., Phys. Rev. Lett. **77**, 1020 (1996).
[24] N. Nica, Nuclear Data Sheets **108**, 1287 (2007).

Journal of Physics: Conference Series **533** (2014) 012031 doi:10.1088/1742-6596/533/1/012031

Microscopic analysis of ^{11}Li elastic scattering on protons and breakup processes within ^{9}Li+$2n$ cluster model

K Spasova[1,2], V K Lukyanov[3], D N Kadrev[1], A N Antonov[1], E V Zemlyanaya[3], K V Lukyanov[3] and M K Gaidarov[1]

[1] Institute for Nuclear Research and Nuclear Energy, Bulgarian Academy of Sciences, Sofia 1784, Bulgaria
[2] University "Ep. K. Preslavski", Shumen 9712, Bulgaria
[3] Joint Institute for Nuclear Research, Dubna 141980, Russia

E-mail: kyspasova@yahoo.co.uk

Abstract. Theoretical analysis of the elastic scattering and breakup in interactions of the ^{11}Li nucleus with protons are presented. The hybrid model of the microscopic optical potential (OP) is applied. The OP includes the single-folding real part, while its imaginary part is derived within the high-energy approximation (HEA) theory. The spin-orbit contribution to the OP is also included. The differential cross sections of ^{11}Li+p elastic scattering and the total reaction cross sections are calculated at energies of 62, 68.4, and 75 MeV/nucleon and are compared with the available experimental data. The breakup cross sections at 62 MeV/nucleon and the momentum distributions of the fragments using a two cluster model of the ^{11}Li nucleus are obtained. An analysis of the single-particle density of ^{11}Li is performed.

The experiments for the elastic scattering of ^{11}Li+p make it possible to study the mechanism of the reactions and give tests for various theoretical methods for analysis. ^{11}Li is a typical example of a halo nucleus with a large radius of its matter distribution, with a very large interaction radius and a small separation energy. The idea of the existence of a two-neutron halo in ^{11}Li was experimentally verified in measurements and studies of the differential cross sections of the ^{11}Li+p elastic scattering in the energy range 60–75 MeV/nucleon [1-3].

In the present work and in [4] we study the elastic scattering cross section for ^{11}Li+p at three incident energies ($E < 100$ MeV/nucleon) using microscopically calculated OP's within the hybrid model [5]. Following our previous works [5-8] we apply this model to the elastic scattering of ^{11}Li+p.

The main expressions for the OP are

$$U_{opt} = V^F(r) + iW(r), \quad V^F(r) = V^D(r) + V^{EX}(r), \tag{1}$$

where the real part of the nucleon-nucleus OP is a result of a folding of the nuclear density and of the effective NN potential and involves the direct $V^D(r)$ and exchange parts $V^{EX}(r)$. The imaginary OP (ImOP) is based on the HEA. The whole OP is constructed in the form

$$U_{opt}(r) = N_R V^F(r) + iN_I W(r) + 2\lambda_\pi^2 \left\{ N_R^{SO} V_0^F \frac{1}{r}\frac{df_R(r)}{dr} + iN_I^{SO} W_0^H \frac{1}{r}\frac{df_I(r)}{dr} \right\} (\mathbf{l}.\mathbf{s}), \tag{2}$$

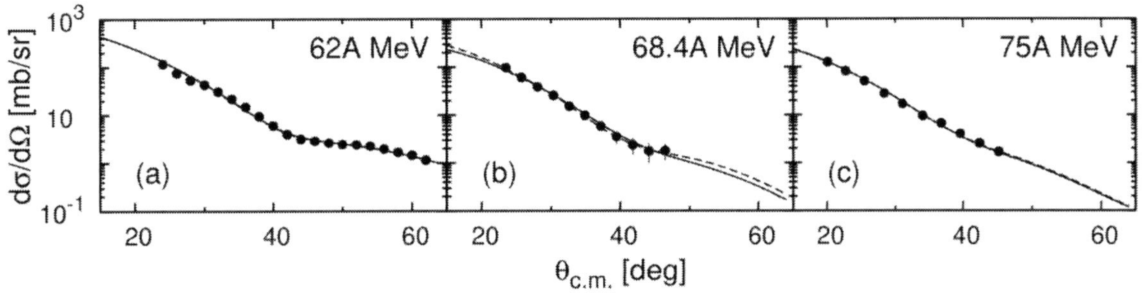

Figure 1. The ^{11}Li+p elastic scattering cross section at E = 62, 68.4, and 75 MeV/nucleon. Solid line: without SO term; dashed line: with SO term. The experimental data are taken from [1], [2] and [3].

Table 1. Values of the N's parameters, volume integrals J_V and J_W (in MeV fm^3), χ^2 and total reaction cross section σ_R (in mb) for the results at three energies E (in MeV/nucleon) shown in Fig. 1.

E	N_R	N_I	N_R^{SO}	N_I^{SO}	J_V	J_W	χ^2	σ_R
62	0.871	0.953			342.474	332.015	1.415	456.97
	0.851	0.974	0.028	0.000	334.610	339.332	1.468	461.21
68.4	0.625	0.186			232.210	60.489	1.328	153.44
	0.543	0.140	0.201	0.000	201.744	45.530	0.316	122.25
75	0.679	0.370			238.048	112.913		232.62
	0.660	0.369	0.045	0.000	231.387	112.607		232.62

where λ_π^2=2 fm^2 is the squared pion Compton wavelength. The ReOP and the ImOP of the SO optical potential in (2) are approximated by WS forms. The parameters of the latter are obtained by a fitting procedure to the respective calculated microscopic potentials $V^F(r)$ and $W^H(r)$. The ImOP is taken in two forms, the microscopically obtained W^H within HEA ($W = W^H$) or the form of the folded real potential V^F ($W = V^F$). The OP $U_{opt}(r)$ (2) is applied to calculate the elastic scattering differential cross sections using the program DWUCK4 [9]. For the protons and neutrons of ^{11}Li we used the Large Scale Shell Model (LSSM) densities [10], that have an exponential asymptotics which is the correct one. A set of N coefficients as parameters is introduced. It can be found by fitting the experimental differential cross sections [4]. The problem of the ambiguity of the N's parameters arises. We use a physical constraint on their choice, namely the behavior of the volume integrals J_V and J_W as functions of the energy [11]. The parameters N give a satisfactory agreement of our results with the experimental data. The values of the total cross sections of the elastic scattering and reaction can serve as another criterion. However the corresponding data are missing, so they are highly desirable.

The results of ^{11}Li+p elastic scattering cross sections are presented in the Fig.1. The corresponding values of the N's parameters together with those of J_V, J_W, χ^2 and σ_R are given in Table 1.

In addition to the analysis of ^{11}Li+p elastic scattering cross section, we study other characteristics of the reaction mechanism, such as the ^{11}Li total reaction cross section, the breakup cross section and related quantities. A simple two-cluster model in which two clusters are suggested, namely the ^{9}Li core (c) and the correlated pair of neutrons $h = 2n$ is considered

[12]. The wave function of the relative motion of the two clusters $\phi_{00}^{(n)}(\mathbf{s})$ ($\mathbf{s}$ is the distance between the clusters) is obtained as a solution of the Schrödinger equation using WS potential for $0s$ or $1s$ states for a particle with a reduced mass of both clusters. The WS parameters are obtained by fitting the energy of a given state to the empirical separation energy value of the h-cluster $\varepsilon = 0.247$ MeV and the rms radius of the cluster function. For the latter we choose the value of 4.93 fm.

In the framework of the ^{9}Li+$2n$ model of ^{11}Li one can estimate the ^{11}Li+p OP as a sum of two OP's of interactions of the clusters with protons $c+p$ (U_c) and $h+p$ (U_h) folded with the density $\rho_0^{(n)}(s)$ (n=0, 1 and ($\rho_0^{(n)}(s) = |\phi_{00}^{(n)}(\mathbf{s})|^2$):

$$U^{(b,n)}(r) = V^{(b,n)} + iW^{(b,n)} = \int d\mathbf{s}\rho_0^{(n)}(s)\left[U_c^{(n)}(\mathbf{r}+(2/11)\mathbf{s}) + U_h^{(n)}(\mathbf{r}-(9/11)\mathbf{s})\right]. \tag{3}$$

This cluster model is used to calculate the characteristics of the breakup reactions of ^{11}Li with the proton target. The eikonal formalism can be applied to obtain the total breakup and absorbtion cross sections and their sum, the total reaction cross section. In this work the cross sections of the diffractive and stripping (when $h = 2n$ cluster leaves the elastic channel) ^{11}Li+p reactions at $E = 62$ MeV/nucleon are obtained. The differential and the total reaction cross sections (for elastic scattering, as well as for diffractive breakup and absorption) require calculations of the probability functions $d^3P(\mathbf{b},\mathbf{k})/d\mathbf{k}$ that depend on the impact parameter $\mathbf{b}$. The general expression for the probability functions can be written as [13]:

$$\frac{d^3P_\Omega(\mathbf{b},\mathbf{k})}{d\mathbf{k}} = \frac{1}{(2\pi)^3}\left|\int d\mathbf{s}\phi_{\mathbf{k}}^*(\mathbf{s})\Omega(\mathbf{b},\mathbf{r}_\perp)\phi_{00}^{(n)}(\mathbf{s})\right|^2, \tag{4}$$

where $\Omega(\mathbf{b},\mathbf{r}_\perp)$ depends on the two profile functions $S_i(b_i)$, ($i = c, h$) (given in [4]), $\phi_{\mathbf{k}}$ is the continuum wave functions, $\mathbf{k}$ is the relative momentum of both clusters in their center-of-mass frame. One can integrate over the transverse angle of the momenta φ_k to get the double-differential probability $d^2P_\Omega(\mathbf{b},\mathbf{k})/dk_L dk_\perp$. The diffraction breakup cross section has the form:

$$\left(\frac{d\sigma}{dk_L}\right)_{diff} = \int_0^\infty b_h\, db_h \int_0^{2\pi} d\varphi_h \int_0^\infty dk_\perp \frac{d^2P(\mathbf{k},\mathbf{b})}{dk_L dk_\perp}. \tag{5}$$

In the case of the stripping reaction when the h - cluster leaves the elastic channel, following [13], the stripping cross section has the form:

$$\left(\frac{d\sigma}{dk_L}\right)_s = \frac{1}{2\pi^2}\int b_h db_h d\varphi_h\left[1-|S_h(b_h)|^2\right]\int \rho d\rho d\varphi_\rho |S_c(b_c)|^2\left[\int_0^\infty dz\cos(k_L z)\phi_0\left(\sqrt{\rho^2+z^2}\right)\right]^2. \tag{6}$$

The results for the diffraction breakup and stripping ^{11}Li+p scattering are presented in Figs. 2 and 3. They give predictions because of missing experimental data for such processes accompanying the ^{11}Li+p scattering at $E \leq 100$ MeV/nucleon. For the diffractive scattering we obtain values of the widths 98 MeV/c (for n=0) and 85 MeV/c (for n=1) and for the stripping reaction 79 MeV/c (for n=0) and 72 MeV/c (for n=1), respectively.

The analysis of the breakup reaction makes it possible to understand its significant role in the formation of the OP responsible for the ^{11}Li+p elastic scattering. It turns out that the breakup channel gives the total breakup cross section σ_{bu}^{tot} that exceeds 80% from the total reaction cross section σ_R^{tot}, while it is around a half of σ_R^{tot} in the case of ^{6}He+^{12}C (as obtained in [12]).

The single-particle density distribution of ^{11}Li can be derived in the form $\rho(r) = \int d\mathbf{s}\rho_0^{(n)}(\mathbf{s})\left[\rho_c(\mathbf{r}_c) + \rho_h(\mathbf{r}_h)\right]$, where $\rho_c(\mathbf{r}_c)$ is the density for the ^{9}Li cluster from the LSSM

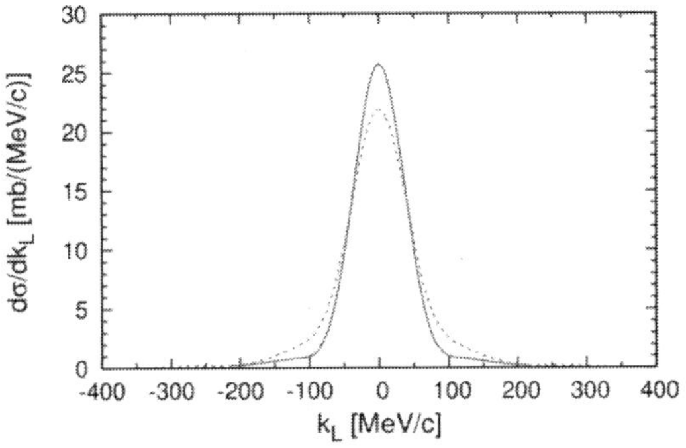

Figure 2. Cross section of diffraction breakup in ^{11}Li+p scattering at E = 62 MeV/nucleon for the cases of n=0 (green dashed line) and n=1 (blue solid line).

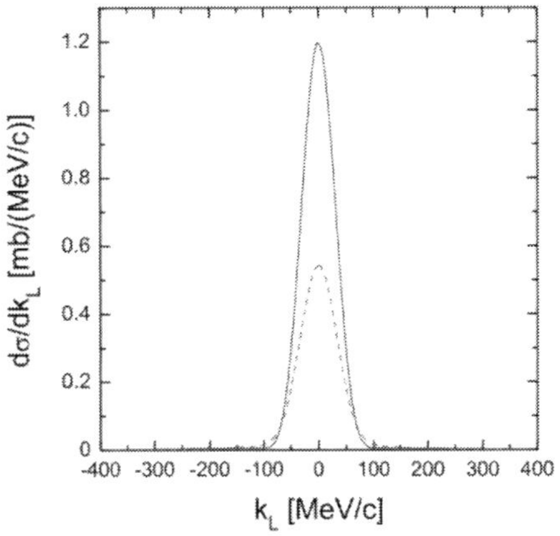

Figure 3. The same as in Fig. 2 but for the stripping reaction.

calculations with rms radius R_c=2.31 fm. The density of the h-halo $\rho_h(\mathbf{r}_h)$ is taken in two forms: Gaussian distribution (G density) and a symmetrized Fermi distribution (SF density). The values of the parameters of G and SF densities and the results of our calculations are obtained in [4] and they are close to the phenomenological density related to the experimental data. We should mention that at the energy of the range 60-70 MeV/nucleon a distortion due to the nuclear and Coulomb forces could affect the cross sections. We have in mind also that our simplified two-cluster model could not give the correct answer and that it can be found in a more complicated three-body approach. This problem remains open and requires further analysis. Future measurements of the cross sections and momentum distribution of the ^{9}Li fragments in the ^{11}Li breakup reactions might provide a supplemental information on the internal spatial structure of the ^{11}Li nucleus.

The work is partly supported by the Project between the INRNE-BAS and JINR. K.S. acknowledges the support of the Project BG-051P0001-3306-003. D.N.K. thanks Bulgarian NSF for the support under Contract No. DID–02/16–17.12.2009.

References

[1] Moon C-B *et al* 1992 *Phys. Lett.* B **297** 39
[2] Korsheninnikov A A *et al* 1997 *Phys. Rev. Lett.* **78** 2317
[3] Korsheninnikov A A *et al* 1996 *Phys. Rev.* C **53** R537
[4] Lukyanov V K, Kadrev D N, Zemlyanaya E V, Antonov A N, Lukyanov K V, Gaidarov M K and Spasova K 2013 *Phys. Rev.* C **88** 034612
[5] Lukyanov V K, Zemlyanaya E V and Lukyanov K V 2006 *Phys. At. Nucl.* **69** 240
[6] Lukyanov V K *et al* 2007 *Eur. Phys. J.* A **33** 389
[7] Lukyanov V K, Zemlyanaya E V, Lukyanov K V, Kadrev D N, Antonov A N, Gaidarov M K and Massen S E 2009 *Phys. Rev.* C **80** 024609
[8] Lukyanov V K, Kadrev D N, Zemlyanaya E V, Antonov A N, Lukyanov K V and Gaidarov M K 2010 *Phys. Rev.* C **82** 024604
[9] Kunz P D and Rost E 1993 *Computational Nuclear Physics* (Springer-Verlag, New York) Vol.2 p.88.
[10] Karataglidis S *et al* 1997 *Phys. Rev. Lett.* **79** 1447; Karataglidis S *et al* 2000 *Phys. Rev.* C **61** 024319
[11] Romanovsky E A *et al* 1998 *Bulletin of the Russian Acad. of Sciences: Physics* **62** No.1 150
[12] Lukyanov V K, Lukyanov K V and Zemlyanaya E V 2011 *Int. J. Mod. Phys.* E **20** 2039
[13] Hencken K, Bertsch G and Esbensen H 1996 *Phys. Rev.* C **54** 3043

On the ^{11}Li two-cluster model for calculations of the ^{11}Li+p elastic scattering at low and high energies

E.V. Zemlyanaya, V.K. Lukyanov, K.V. Lukyanov
Joint Institute for Nuclear Research, 141980 Dubna, Russia

Abstract. The two-cluster model is utilized to the ^{11}Li nucleus consisting of the ^{9}Li and $h = 2$n clusters for calculations of the microscopic optical potentials and the ^{11}Li+p scattering cross sections. The obtained density of ^{11}Li is compared to densities known from the p+^{11}Li scattering data at $T_{lab} = 700$ MeV fitted using the Glauber theory. Effect of $2n$-halo and the ^{9}Li core contribution to the cross sections is analysed.

1. Motivation

In [1], cross sections of the ^{11}Li+p elastic scattering at *low* energies were successfully explained with a help of the microscopic OP [2] where the Large scale shell model (LSSM) [3] density of ^{11}Li was employed (see dashed line in the left panel of Fig. 1). Also, the two-cluster model developed in [4] to the ^{6}He nucleus, was utilized in [1] to estimate the ^{11}Li breakup and the momentum distributions of fragments.

On the other hand, the phenomenological density of ^{11}Li (denoted 0.2GG+0.8GO in [5], see solid line in the left panel of Fig. 1) was obtained in [5] by fitting the Glauber theory to experimental data on the ^{11}Li+p elastic scattering at *high* energy about 700 MeV. It seems interesting to check (i) if this density can explain experimental data at low energies; (ii) if the two-cluster model density is applicable in both cases of low and high energies.

Below we construct the one-particle ^{11}Li density inherent to the dynamic two-cluster model (D2C) of ^{11}Li consisting of the ^{9}Li core and $h = 2n$ halo bounded with energy about 0.3 MeV. The parameters of clusters are varied to get the ^{11}Li-density being close to the 0.2GG+0.8GO-form, and the effect is studied of the halo and core contributions to the ^{11}Li structure. Then we calculate the respective microscopic OPs to reproduce experimental data on the ^{11}Li+p elastic scattering at both *low* and *high* energies.

2. One-particle density in two-cluster model of ^{11}Li

Relative motion of two clusters in ^{11}Li is described by the two-node s-state wave function $\Phi(\mathbf{R})$ which is a solution of the Schrödinger equation with WS-potential, that reproduces the respective binding energy and rms-radius of the (^{9}Li+h) system (this fit was done in [1]). Then the ^{11}Li one-particle density can be calculated as folding integral of the sum of ^{9}Li- and h-densities ρ_c and ρ_h with the density probability $\rho_{cl} = |\Phi(\mathbf{R})|^2$ obtained in [1]:

$$\rho_m(r) = \int d^3R \left[\rho_h(\mathbf{r} + \frac{9}{11}\mathbf{R}) + \rho_c(\mathbf{r} - \frac{2}{11}\mathbf{R})\right] \cdot |\Phi(\mathbf{R})|^2. \quad (1)$$

The halo-density we take in the form of the symmetrized Fermi function (SF):

$$\rho_h(r) = \rho_0 \cdot \sinh(R/a) \cdot [\cosh(R/a) + \cosh(r/a)]^{-1}, \quad R^2 = [5\bar{R}_h^2 - 7(\pi a)^2]/3. \quad (2)$$

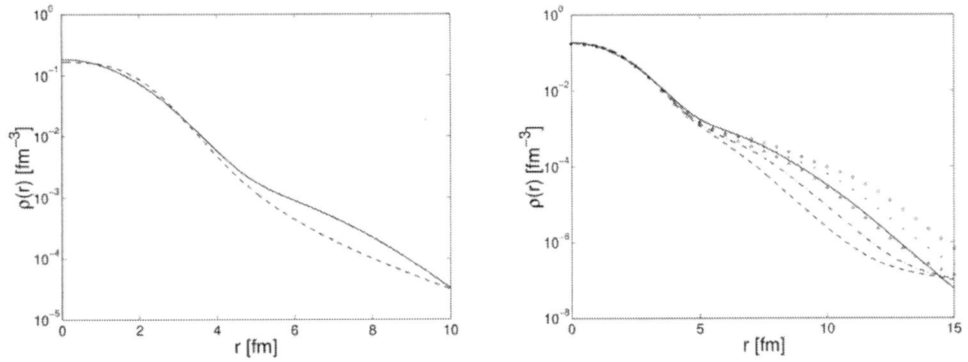

Figure 1. Left panel: the 0.2GG+0.8GO-density of ^{11}Li with the rms-radius $\bar{R}_m$=3.42 fm (solid) and LSSM density with $\bar{R}_m$=2.94 fm (dashed). Right panel: the D2C density with the halo parameters from Table 1.

The halo effect on the ^{11}Li density is shown in Fig. 1, right panel. For the halo density we put $a = 0.5$ and vary rms-radius $\bar{R}_h = 2 \div 6$ fm while the ^{9}Li density is taken in the LSSM form. Respective rms-radii of halo $\bar{R}_h$ and of whole ^{11}Li nucleus $\bar{R}_m$ are given in the Table 1. It is seen that when $\bar{R}_h$ is varied from 2 to 6 fm then the $\bar{R}_m$ radius changes only 30%. The last line in the Table 1 shows the "bestfit" parameters of our D2C density, and the respective curve is given in the left panel of Fig. 2 (dashed) in comparison with the 0.2GG+0.8GO-density (solid).

Table 1. Parameters of $2n$-halo in D2C-density

$\bar{R}_h$	R	$\bar{R}_m$	curve	$\bar{R}_h$	R	$\bar{R}_m$	curve
6	7.365	3.77	circles	3	3.04	3.02	dash-points
5	6	3.49	points	2	0.95	2.79	dashes
4	4.573	3.24	triangles	4.22	4.94	3.29	$a = 0.48$ best

3. Cross sections at low energies

Here we apply the D2C densities (1) to calculate the ^{11}Li+p microscopic OP [2]. Then the DWUCK4 code [6] is utilized to compute the respective elastic scattering cross sections. Right panel of Fig. 2 demonstrates a weak effect on them of the halo contribution in the case of low energies. Also, one sees that in spite of the "bestfit" D2C-based density is slightly different to the 0.2GG+0.8GO-form, both cross sections are in good agreement with experiment as well as the curve with $\bar{R}_h = 5$.

In order to analyze the core contribution effect we fix parameters of halo and consider three forms of ^{9}Li densities, see Fig. 3, left panel. Respective cross sections are presented on the right panel of Fig. 3. When we increase the core radius about 30% then the cross section curves change significantly. So, one can conclude that effect of the ^{9}Li dimension on the cross sections is strong.

4. Cross sections at high energy

At energy 697 MeV we need to account for relativization. Relativistic momentum p in c.m. system is as follows:

$$p = p_{lab} \cdot m_2 / \sqrt{2Em_2 + m_1^2 + m_2^2} \quad p_{lab} = \sqrt{T_{lab}(T_{lab} + 2m_1)} \tag{3}$$

Journal of Physics: Conference Series **533** (2014) 012032 doi:10.1088/1742-6596/533/1/012032

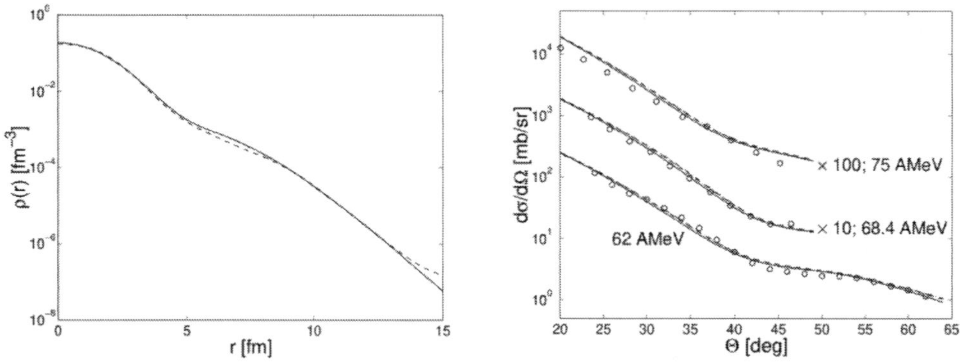

Figure 2. Left: D2C density with "bestfit" parameters (dashed) in comparison with 0.2GG+0.8GO-form (solid). Right: p+^{11}LI cross sections at energies 62, 68.4, 75 MeV calculated with 0.2GG+0.8GO-density (solid), "bestfit" D2C density (dashed) and D2C density with $\bar{R}_h = 5$ (dash-points)

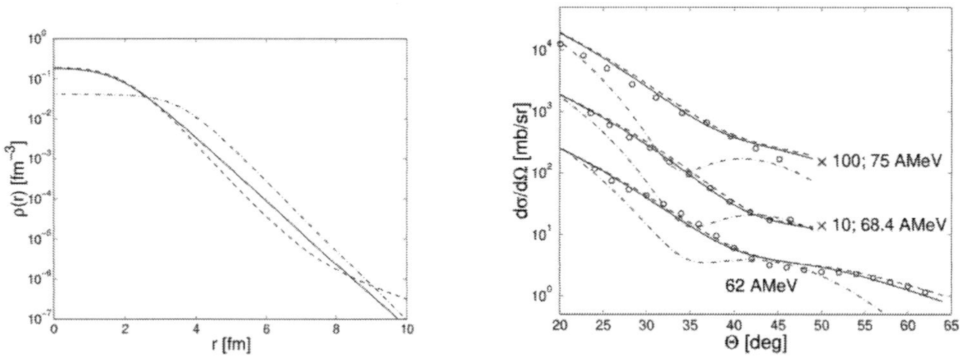

Figure 3. Effect of the ^{9}Li-cluster structure. Left: the ^{9}Li-density: LSSM-form with rms-radius of ^{9}Li $\bar{R}_c$=2.31 and rms-radius of whole ^{11}Li $\bar{R}_m$=3.3 (dashed); SF-density with a=0.55, R=1.77, $\bar{R}_c$=2.46, $\bar{R}_m$=3.4 (solid); SF-density with a=0.5, R=3.5, $\bar{R}_c$=3.35, $\bar{R}_m$=3.94 (dash-points). Right: respective p+^{11}LI cross sections at energies 62, 68.4, 75 MeV.

Here $E = T_{lab} + m_1$ is the total energy, T_{lab} – kinetic energy in laboratory system, m_1 and m_2 are the proton mass and the ^{11}Li mass. The Klein–Gordon–Fock equation at $E \gg |U|$ is reduced to the form of Schrödinger equation with relativistic momentum p and relativistic reduced mass $\bar{\mu} = E \cdot M_A/(E + M_A)$:

$$(\Delta + p^2)\psi(\mathbf{r}) = 2\bar{\mu}\, V(r)\psi(\mathbf{r}), \quad V(r) = U(r) + U_C(r). \tag{4}$$

The potential consists of the Coulomb potential $U_C(r)$ and the microscopic OP $U(r)$.

In Fig. 4 one sees a good agreement of our D2C-based calculations with the high energy data. Fig. 5 shows cross sections calculated using different forms of halo density (left panel) and different forms of the ^{9}Li density (right panel). It is seen that at high energy, the effect of halo contribution remains to be weak while effect of the core-density dimension is strong.

5. Summary

The one-particle ^{11}Li density is constructed within the frame of the dynamic two-cluster model of ^{11}Li consisting of bounded ^{9}Li-core and $h = 2n$ halo clusters. Microscopic OPs obtained with

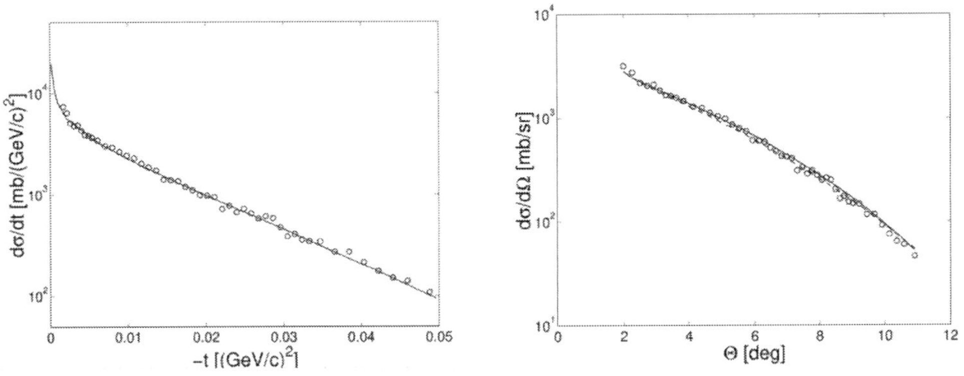

Figure 4. p+^{11}Li elastic scattering at 697 MeV. Left: experimental data and calculations with 0.2GG+0.8GO-density as they done in [5]. Right: our calculations with D2C-density using HEA OP (solid) and folding real and HEA imaginary OP (dashed) compared with the same data but in the other presentation of axes.

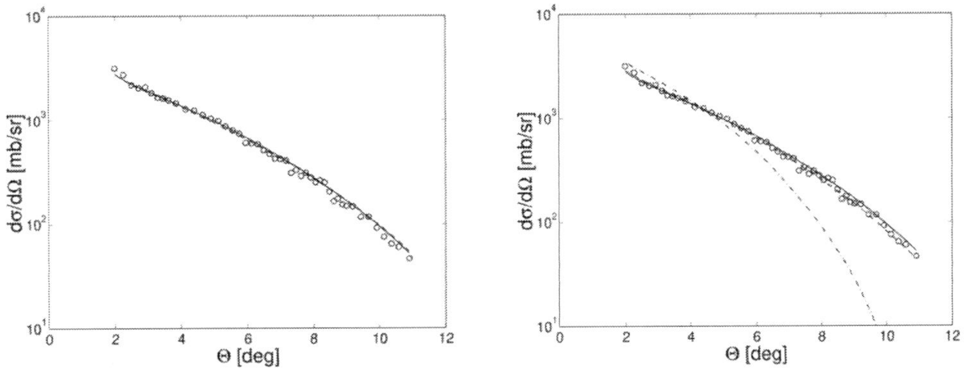

Figure 5. Effect of halo and core structure at high energies. Left: the same as in Fig. 2(right) but for 697 MeV. Right: the same as in Fig. 3(right) but for 697 MeV.

such a density can reproduce experimental data on the ^{11}Li+p elastic scattering at both *low* and *high* energies. Effect of the halo-density is shown to be weak while effect of the core-density dimension is strong at both *low* and *high* energies.

The D2C model makes possible to calculate not only the one-particle density of ^{11}Li and respective OPs but also the $(p+h)$ and $(p+c)$ optical potentials and therefore one can estimate the breakup cross sections and momentum distributions of the ^{11}Li fragments.

The work was partly supported by RFBR (grant 13-01-00060a).

References

[1] Lukyanov V, e.a. 2013, *Phys. Rev. C* **88** 034612
[2] Lukyanov V, Zemlyanaya E, Lukyanov K. 2006 *Phys. At. Nucl.* **69** 240
[3] Karataglidis S, e.a. 2000 *Phys. Rev. C* **61** 024319
[4] Lukyanov V, Zemlyanaya E, Lukyanov K. 2011 *Int. J. Mod. Phys. E* **20** 2039
[5] Dobrovolsky A V, e.a. 2006 *Nucl. Phys. A* **766** 1
[6] Kunz P D, Rost E, *The Distorted-Wave Born Approximation* (Computational Nuclear Physics V.2, Eds: Langanke K. et al. N.Y.: Springer-Verlag, 1993. P. 88)

Journal of Physics: Conference Series **533** (2014) 012033 doi:10.1088/1742-6596/533/1/012033

Investigation of exotic nuclei with absolute transition probabilities

A. Dewald
Institute of Nuclear Physics, University of Cologne

Abstract. Transition probabilities are crucial for the understanding of nuclear structure. Deep inelastic reactions, knockout reactions and projectile Coulomb excitation with fast radioactive beams are suited to populate excited states in exotic nuclei. Examples are presented which demonstrate that recoil Doppler shift lifetime measurements can be applied successfully in combination with such reactions to measure level lifetimes.

1. Introduction

In order to investigate exotic nuclei far from the line of stability deep inelastic reactions or reactions with radioactive beams have been used successfully. Absolute transition probabilities which are crucial for the understanding of nuclear structure can be deduced directly from lifetimes of excited states measured with the recoil distance Doppler shift (RDDS) method or plunger technique. Therefore it is of great interest to apply the plunger technique in combination with such reactions and with fast radioactive beams. In this paper examples are presented which prove that the RDDS technique in combination with the above mentioned reactions can be very useful. For radioactive ion beam experiments only beam energies around 100MeV/u are considered here.

A general description of the plunger technique and recent developments can be found in [1].

With fast radioactive beams both Coulomb excitation and knockout reactions have been used.

Since the beam intensities which can be provided nowadays for experiments are still limited, it is often necessary to get some compensation by using thick targets (up to 1000 μm). At beam energies of about 100 MeV/u only one step Coulomb excitation occurs. That means that in such experiments only lower 2^+ states of even-even nuclei can be excited. This is a considerable limitation of the method but, on the other hand, it also means that the problem with unknown level feeding is practically eliminated.

2. RDDS after fast projectile Coulomb excitation

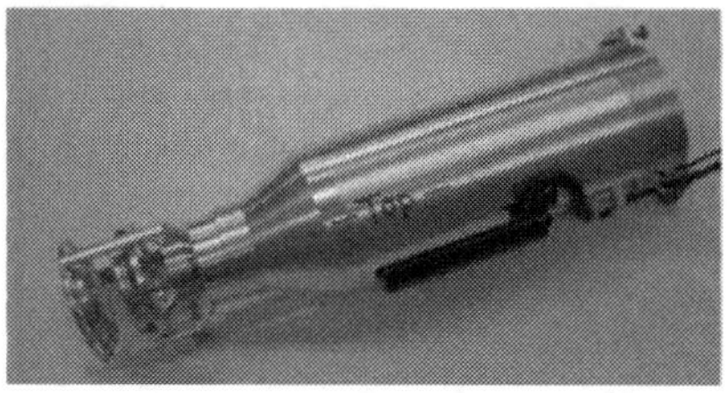

Fig. 1: plunger device for experiments with fast radioactive beams

As an example of a RDDS lifetime measurement after fast projectile Coulomb excitation I want to present the measurement aiming for the lifetime of the 2_1^+ states in the neutron rich ^{62,64,66}Fe nuclei [2]. This experiment was performed at the National Superconducting Cyclotron Laboratory at Michigan State University (NSCL/MSU) where the secondary Fe beams were produced by fragmentation of a primary ^{76}Ge beam at 130 AMeV. The generated fragments were purified by the A1900 [3]) fragment separator. Rather pure Fe beams were delivered and directed to the secondary target (300 μm thick Au) which was mounted in a dedicated plunger device [4], built at the Institute for Nuclear Physics of the University of Cologne (see Fig.1) and mounted at the target position of the S800 spectrometer [5]. Instead of the standard stopper foil a degrader foil was used (400 μm thick Nb) which enables a detection of the recoiling nuclei produced in the secondary reaction. The plunger device allows to set different target-degrader-separations. The S800 spectrometer was used to identify Coulomb excited Fe nuclei. The gamma spectra were measured with the SeGA array [6]. Spectra which have been measured at different target-degrader distances and with SeGA detectors positioned at different angles with respect to the beam axis are shown in Figure 2 after Doppler correction. The lifetimes were determined from a simulation which aims to reproduce the two components of the gamma transition of interest simultaneously in all spectra using the level lifetime as a free parameter. The resulting simulated spectra are shown as solid curve in Figure 2. The obtained results can be found in Table 1 of [2].

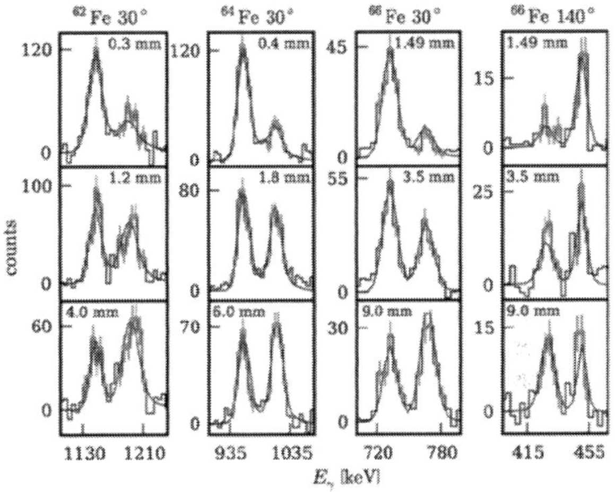

Fig1: Results from gamma-ray line-shape fits (blue line) to Doppler corrected spectra for the $2_1^+ \rightarrow 0_1^+$ transitions in ^{62,64,66}Fe

3. RDDS after knockout reactions using fast radioactive beams

Via knockout reaction higher lying levels with medium spins can be populated. This fact increases considerably the potential of the RDDS technique with fast radioactive beams. The first pioneering plunger experiment with fast radioactive beams using knockout reactions as secondary reactions was on ^{62}Zn and ^{64}Ge [7]. Unfortunately the level feeding cannot be neglected as in the case of one step Coulomb excitation and has to be considered in the data analysis. However, in contrast to standard compound fusion reactions it has been found that only transitions which have been observed in the experiment have to be considered [7]. Thus the level feeding is less complex than that of a standard compound fusion reaction where many long cascades feed the lower levels which are observed in the experiment. In a knockout reaction the feeding times are very close to the level lifetimes from which the feeding occurs since only one or two step feeding can be assumed. This makes the consideration of the feeding after knockout reactions much easier as compared to the compound fusion reaction. A recent experiment was performed at the NSCL/MSU aiming to measure lifetimes in neutron rich Cr isotopes. A primary ^{82}Se beam of 140 AMeV was fragmented at a beryllium production target. The secondary beams of ^{59,61,63}Mn were purified with the A1900 fragment separator and directed on beryllium plunger targets (288μm/^{58}Cr, 520μm/^{60}Cr, 1295μm/^{62}Cr). These targets with different thickness were used to compensate partially the different ^{59,61,63}Mn beam intensities.

In Fig. 3 examples of Doppler corrected γ-ray spectra of ^{58}Cr are shown. The intensities of the feeding transitions which are considered in the final data analysis were deduced from so called "target only" spectra which are generally of higher quality that the ones where a degrader foil has been used. The actual lifetime analysis was performed in two ways: via simulation of the spectra aiming to reproduce the fast and degraded components simultaneously in all spectra corresponding to different target-degrader separations and by using the extracted decay curves.

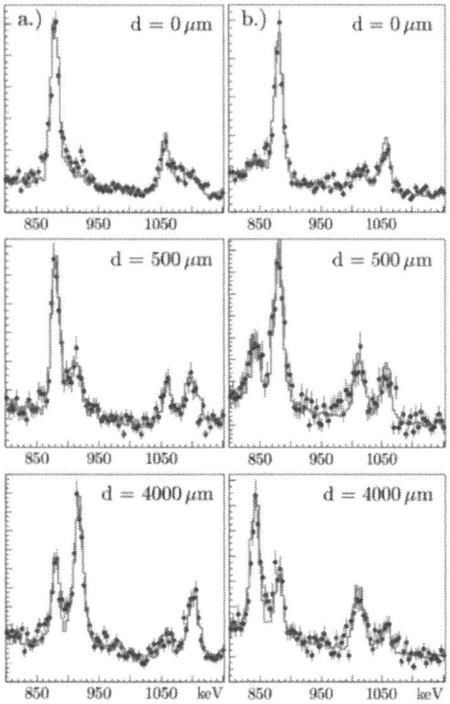

Fig.3: Doppler corrected γ spectra of ^{58}Cr for different target-degrader separations. The red line represents the simulation from which the level lifetimes were determined.

Table 1: Measured transition strengths of ^{58,60,62}Cr (preliminary) compared with LNPS calculations [8]. The B(E2;6$^+$→4$^+$) values were determined from effective lifetimes only. Therefore these values have to be considered as lower limits only.

A	$J^{\pi_2}_{\sigma_2}\rightarrow J^{\pi_1}_{\sigma_1}$	**this experiment** **(preliminary)**	**theory** **LNPS**
		B(E2↓) [e²fm]	
^{58}Cr	$2^+_1\rightarrow 0^+_1$	206^{32}_{24}	174
	$4^+_1\rightarrow 2^+_1$	206^{41}_{28}	206
	$6^+_1\rightarrow 4^+_1$	$\geq 183^{55}_{34}$	173
^{60}Cr	$2^+_1\rightarrow 0^+_1$	285^{37}_{30}	262
	$4^+_1\rightarrow 2^+_1$	345^{55}_{43}	372
	$6^+_1\rightarrow 4^+_1$	$\geq 173^{53}_{33}$	374
^{62}Cr	$2^+_1\rightarrow 0^+_1$	357^{65}_{47}	340
	$4^+_1\rightarrow 2^+_1$	531^{72}_{47}	507

Preliminary transition probabilities are consistent to previously measured B(E2;$2^+_1\rightarrow 0^+_1$ values) [9]. The experimental transition probabilities are presented in table 1 together with shell model calculations [8].

RDDS after deep inelastic reactions

An alternative way to produce exotic nuclei is offered by deep inelastic reactions. Since the cross sections are normally very low it is needed to clean the measured gamma spectra by demanding a coincidence with detected recoil nuclei in the focal plane of magnetic spectrometer like e.g. PRISMA [10] at the LN Legnaro/Italy or VAMOS [11] at GANIL/France. The reaction cross sections are strongly dependent on the emission angles of the nuclei produced in the reaction with respect to the beam axis. Therefore a special plunger device is needed which allows the recoils after leaving the target to pass the degrader foil at angles up to 70° and to enter into a spectrometer to detect and identify them. Such a plunger device was built at the Institute of Nuclear Physics of the University of Cologne and is

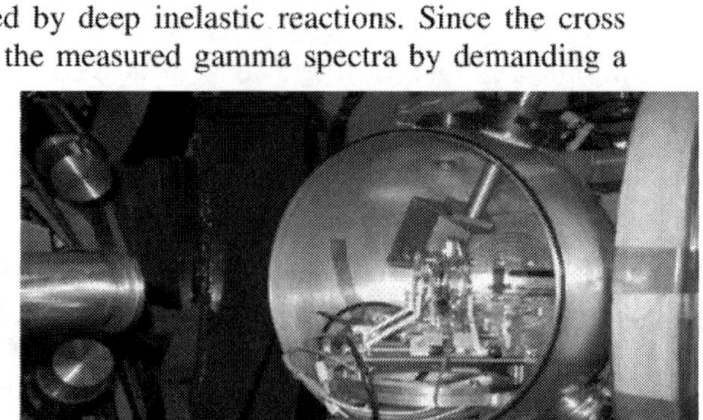

Figure 4: Plunger positioned in the target chamber in front of the PRISMA spectrometer

shown in Figure 4. Pioneering experiments were done at Legnaro [12,13] using the PRISMA spectrometer and the CLARA HP-Ge-detector array [14] for the gamma detection.

Recently the CLARA array was replaced by the gamma tracking spectrometer AGATA in its first implementations (AGATA-Demonstrator) [15,16]. With this setup a first plunger experiment was performed for ^{70,72,74}Zn [17]. Our group performed with this setup a lifetime measurement on ^{84,86}Se in collaboration with groups from IPNO/CNRS-University Paris Sud-11, IPHC/CNRS-University of Strasbourg and the INFN LN Legnaro and others participating in the experiment No. 10.41.

We used a ^{82}Se beam at 557 MeV which was impinging on the target at an angle of 15° with respect to the normal on the plane defined by the stretched target. The recoils were registered with the PRISMA spectrometer at an angle of 58° with respect to the beam axis. The target consisted of 2 mg/cm^2 ^{238}U evaporated on 1.2 mg/cm^2 thick Ta backing. As a degrader a ^{93}Nb foil with a thickness of 4 mg/cm^2 was used which was sufficient to reduce the recoil velocity such that two separated Doppler shifted components of the transitions of interest were obtained in the gamma spectra (see Figure 5). The analysis of the data is still ongoing.

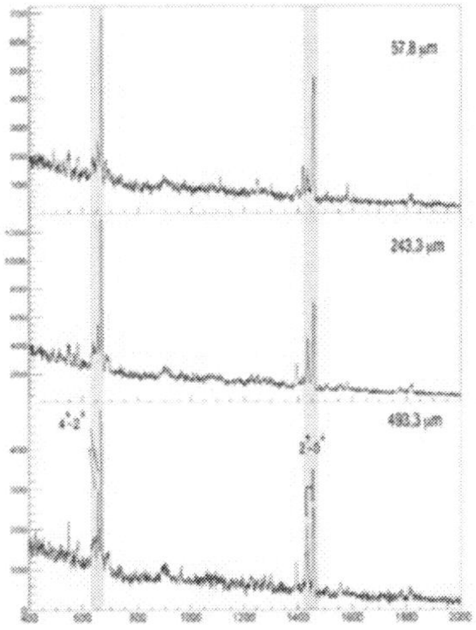

Figure 5: Spectra of ^{84}Se measured at different target-degrader separations (right).

Acknowledgements: I would like to thank all collaborators of the experiment E10021@ NSCL/MSU and H. Iwasaki for the excellent local support given especially by him. I would like to thank T. Braunroth for supplying me with his first results of the data analysis and for the preparation of figures and tables. I also have the thank all collaborators of the Se experiment (10.41) and especially C. Fransen the spokesperson for the part of the experiment which was proposed by the Cologne group as well as J. Litzinger who is working on the data analysis. On behalf of the Cologne group we want to express our gratitude for the excellent support given by the AGATA collaboration and the PRISMA group at Legnaro.

References

[1] A. Dewald et al., Prog. Part. Nucl. Phy. 67 (2012) 786
[2] W. Rother et al., Phys. Rev. Lett. 106 (2011) 022502
[3] D.J. Morrissey et. al., Nucl. Instr. Meth. Phys. Res. B 204 (2003) 90
[4] A. Dewald et al., Phys. Rev. C 78 (2008) 051302(R)
[5] D. Bazin et al., Nucl. Instr. Meth. Phys. Res. B 204 (2003) 629
[6] W.F. Mueller et al., Nucl. Instr. Meth. Phys. Res. A466 (2001) 492
[7] K. Starosta et al., Phys. Rev. Lett. 99 (2007) 042503
[8] S. M. Lenzi, et al., Phys. Rev. C 82 (2010) 054301
[9]) T. Baugher, et al., Phys. Rev. C 86 (2012) 011305(R)
[10] A. Stefanini, et al., Nucl. Phys. A 701 (2002) 217c
[11] S. Pullanhiotan et al., Nucl. Instr. Meth. Phys. Res. A593 (2008) 343
[12] A. Gadea, et al., Acta Phys. Polonica B38 (2007)193
[13] J.J. Valiente-Dobon, et al., Phys. Rev. Lett. 102 (2009) 242502
[14] A. Gadea, et al., Eur. Phys. J. A20 (2003) 193
[15] A. Gadea, at al., Nucl. Instr. Meth. Phys. Res. A 654 (2011) 88
[16] S. Akkoyun, et al., Nucl. Instr. Meth. Phys. Res. A 668 (2012) 26
[17] C. Louchart, et al., Phys. Rev. C78 (2013) 054302

Different collectivity in the two signatures of the $i_{13/2}$ stemming band in ^{167}Yb

P Petkov[1], K A Gladnishki[2], A Dewald[3], O Möller[4], I Deloncle[5], M Reese[4], C Fransen[3], M Hackstein[3], J Jolie[3], Th Pissulla[3], W Rother[3] and K O Zell[3]

[1]Bulgarian Academy of Sciences, Institute for Nuclear Research and Nuclear Energy, 1784 Sofia, Bulgaria
[2]Faculty of Physics, St. Kliment Ohridski University of Sofia, 1164 Sofia, Bulgaria
[3]Institut für Kernphysik, Universität zu Köln, D-50937 Köln, Germany
[4]Institut für Kernphysik, Technische Universität Darmstadt, D-64289 Darmstadt, Germany
[5]CSNSM, IN2P3/CNRS and Université Paris-Sud, F-91405 Orsay Campus, France

E-mail: petkov@inrne.bas.bg

Abstract.
Six lifetimes have been determined in the $5/2^+[642]$ band from $\nu i_{13/2}$ parentage in ^{167}Yb by means of Recoil distance Doppler-shift (RDDS) measurements carried out at the Cologne FN tandem. The deduced transition strengths and the level scheme are reasonably described by Particle plus triaxial rotor model (PTRM) calculations except for the behavior of the quadrupole collectivity in the two signatures of the $5/2^+[642]$ band. In that band, the quadrupole collectivity of the favored signature is appreciably larger than this of the unfavored signature. The effect increases with increasing the spin. Naturally, the rigid PTRM cannot explain these features, but the structure of its wave functions suggests a possible solution. It is associated with the enhanced contribution of low-Ω orbitals from $\nu i_{13/2}$ parentage in the favored signature compared to the unfavored one. This could selectively increase the deformation of the favored signature band members and give rise to a dynamic shape coexistence taking place between the two signatures which needs quantitative explanation by future theoretical work.

The nucleus ^{167}Yb is well studied after numerous experimental and theoretical investigations in the past. The experimental findings are summarized in [1]. Very recently [2], the collectivity of the band structures was investigated at low (I$\sim$11/2$\hbar$) and medium (I$\leq$29/2$\hbar$) spins using Recoil distance Doppler-shift lifetime measurements. In Figure1, we present a partial level scheme of ^{167}Yb presenting the highly perturbed $5/2^+[642]$ band from an $i_{13/2}$ parentage. It consists of two signatures, α=1/2 (favored) and α=-1/2 (unfavored). In the literature, there are theoretical indications [3, 4] that different deformations could characterize the signature partners. Prior to the study described in reference [2], this effect was not established experimentally. In the present paper, we present shortly these experiment and data analysis, and concentrate on the collectivity of the $5/2^+[642]$ band. To populate excited states in ^{167}Yb, we used the reaction ^{154}Sm(^{18}O,5n). The beam, with an energy of E = 80 MeV, was provided by the FN tandem of the Institut für Kernphysik of the Cologne University. Details on the plunger experiment can be found in [2], here we remind only some general features of the setup. Deexciting γ-rays were recorded in coincidence by five large volume Germanium detectors positioned symmetrically at the backward angle of 143° and a Euroball cluster [5] detector positioned at 0° with respect to

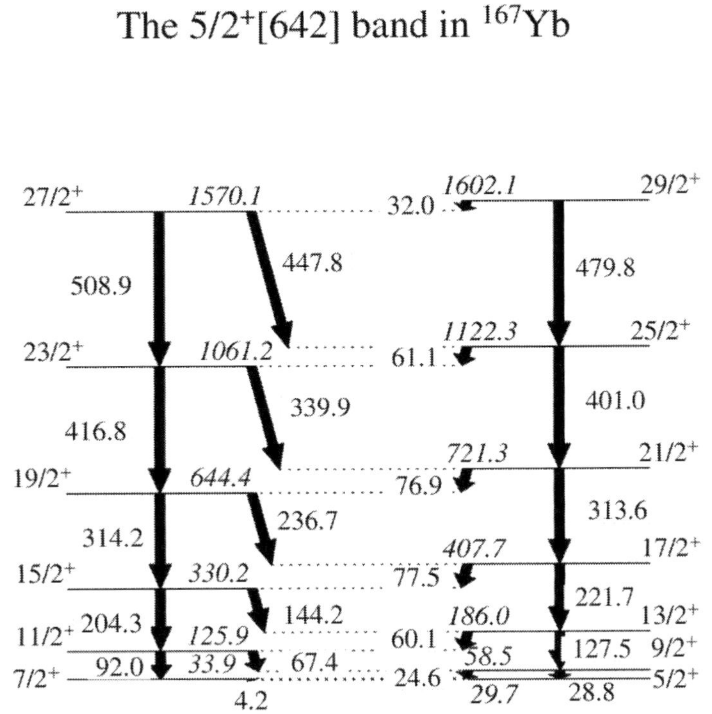

Figure 1. Partial level scheme of ^{167}Yb with the band investigated in the present work. The data are taken from reference [1].

the beam axis. The Germanium crystals were grouped in three rings, namely ring 0 (polar angle of 143° with respect to the beam axis), ring 1 (outer cluster segments at a polar angle of 29°) and ring 2 (central segment of the cluster at 0°). Data were taken for 21 target-to-stopper distances x in the range from 0.5 μm to 3001 μm. A total of about 8.0x10^9 unfolded double events were collected. After corrections for energy shifts and gain matching, the data were sorted into 4kx4k γ-γ coincidence matrices. Each matrix contains events where two γ-rays are registered by detectors belonging to a particular two-ring combination at a given distance. The normalization of the data taken at different target-to-stopper distances was performed using coincidence events corresponding to pairs of strong transitions in the yrast band of ^{168}Yb populated in the dominant 4n exit channel of the reaction ^{154}Sm+^{18}O (cf. reference [6]).

The RDDS method is a well established technique for the determination of picosecond lifetimes of excited nuclear states (see e.g. [7] and references therein). The data analyzing procedure used by us is described in references [8, 9]. It represents a concrete development within the framework of the Differential Decay Curve Method (DDCM) [10, 11]. In order to prepare spectra for further analysis, we used the procedure from reference [8], namely gates were set on the complete line shape (both shifted and unshifted components included) of the directly feeding transition and of the transition of interest, respectively. This approach solves the problem with the unknown (unobserved) feeding. In order to increase statistics, we summed up the spectra corresponding to gates set in the three independent rings.

In the procedure for the analysis, the background-subtracted line shapes corresponding to the transition of interest at all distances and the shifted decay curve

$$S_{ij}(t) = b_{ij} \int_0^t \lambda_i n_i(t')dt'. \qquad (1)$$

are fitted simultaneously. In eqs. 1, $n_i(t)$ is the population as function of time of the level of interest i, λ_i is its decay constant (the lifetime τ_i=1/λ_i) and b_{ij} is the branching ratio of the transition $i \to j$. The function $S_{ij}(t)$ is represented by second-order polynomials which are continuously interconnected at the borders of an arbitrarily chosen set of neighboring time-intervals. The fitting procedure consists in changing the borders of the time-intervals until the best reproduction of the spectra is achieved. More details can be found in references [8, 9] The final result for the lifetime for a particular ring combination is obtained by fitting a horizontal line through the points of the τ-curve (cf. references [10, 11]) within the region of sensitivity where the values are reliable. Then, these results are averaged while paying attention to possible systematic errors [9]. The derived lifetime values and the extracted reduced electromagnetic transition strengths are presented in [2].

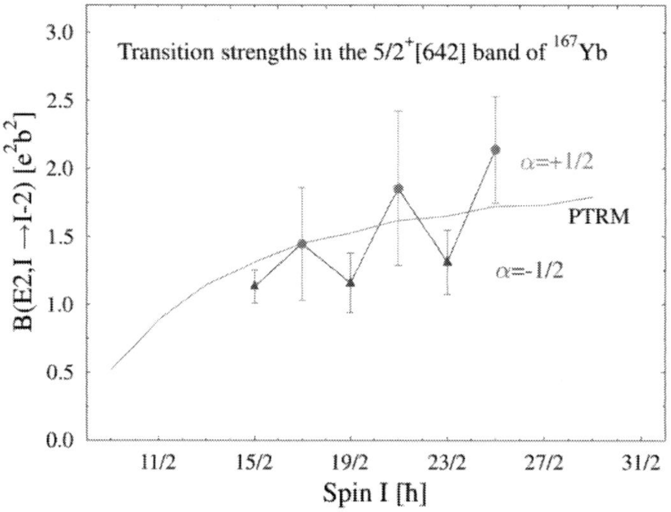

Figure 2. B(E2) transition strengths along the $5/2^+[642]$ band in ^{167}Yb compared to PTRM calculations. The solid line interconnecting the data points is to guide the eye.

Six lifetimes have been determined in both signature partners of the $5/2^+[642]$ band (three in each signature). Their knowledge allows to derive transition strengths which will be compared in the following to the results of PTRM calculations. The ytterbium nuclei in the A≈170 mass-region are believed to be well deformed and characterized by a stability against deformation changes induced by the rotation. Therefore till the back-banding region one may expect a reasonable description of the properties of ^{167}Yb by the PTRM. We used the version of the model described in reference [12] as implemented in the computer codes GAMPN, ASYRMO, PROBAMO and E1PROBAM presented in references [13, 14]. To describe the spectroscopic properties (excitation energies and transition strengths) of ^{167}Yb we varied the parameters ϵ, γ, ϵ_4 as well as $E(2_1^+)$ and the attenuation ζ of the Coriolis interaction. In this procedure, the best choice of the deformation parameters ϵ=0.26, γ=6°, ϵ_4=0.0 insured a reasonable reproduction of the positions of the band-heads observed at lower spin and of the intraband B(E2,I→I-2) values. The final results for the calculated level scheme were obtained by a fine tuning of the moments of inertia and the attenuation of the Coriolis interaction. In this way, the values $E(2_1^+)$=0.076 MeV and $\zeta(5/2^+[642])$=0.83 were obtained. An overall agreement between experiment and theory is observed, although the small spacings in the bottom of the $5/2^+[642]$ band are not completely reproduced (cf. [2]). In Figure2, the behavior of the B(E2) reduced transition probabilities within the $5/2^+[642]$ band is presented. The experimental quantities are compared to calculations within the PTRM. A striking feature of the data immediately emerges: the B(E2)'s in the two signatures of the band differ and this effect increases with increasing the spin. Although the relatively large error bars, the effect is systematic and very well seen at spins $23/2\hbar$ and $25/2\hbar$. At our knowledge, up to now there is not an experimental evidence for different deformation (collectivity) in the two signatures of a rotational band. In the literature, there are theoretical considerations predicting that the two signatures may have different quadrupole deformations (e.g. [3] for ^{79}Rb). For $\nu i_{13/2}$ based bands, this effect was studied systematically in [4] in relation with the signature splitting and predictions for the odd ytterbium isotopes can be found (Figure 4 in that work). The quadrupole deformation for ^{167}Yb is in agreement with the results from the present study, however, the unfavored signature is predicted to be a bit more deformed than the favored one, in disagreement with the experimental findings. It should be mentioned that the PTRM calculations of the present study predict some zig-zag behavior of the B(E2)'s, following the trend of the experiment, but the effect is very weak.

A possible way for explanation is suggested by the structure of the wave functions of the levels from the $5/2^+[642]$ band according to the PTRM calculations. The most interesting feature is the significant presence of the deformation-driving low-Ω orbitals ($1/2^+[660]$ and $3/2^+[651]$) in the wave functions of the favored signature, while their presence in the unfavored signature is much weaker. Thus, a calculation of the mean value of Ω for each level gives a nearly constant value $< \Omega > \sim 2.5\ \hbar$ in the unfavored signature. In the favored signature, $< \Omega >$ continuously decreases, and reaches about 2 $\hbar$ at spins 25/2 and 29/2. This involvement of low-Ω orbitals could lead dynamically to a larger deformation of the favored signature. In this sense, we could associate the observed behavior of the B(E2)'s in the $5/2^+[642]$ band of ^{167}Yb with a dynamic shape coexistence of the two signatures induced by the Coriolis interaction in the rotating nucleus. Of course, much more detailed theoretical considerations are needed to quantitatively explain the rotationally induced effect.

Acknowledgments

The authors are grateful to I.Ragnarsson and A.Gelberg for fruitful discussions. This work was supported by the Deutsche Forschung Gemeinschaft (DFG) within contracts No. Jo 391/8-1 and No. Jo 391/11-1. The authors are indebted for the support by the partnership agreement between the University of Cologne and University of Sofia financed by the DAAD. The authors are also grateful for the financial support of the Bulgarian Science Fund under contract DFNI-E 01/2.

References

[1] Baglin C M 2000 *Nuclear Data Sheets* **90** 431
[2] Gladnishki K A *et al* 2013 *Nucl. Phys.* A **903** 1
[3] Bengtsson R, Nazarewicz W, Skeppstedt Ö and Wyss R 1991 *Nucl. Phys.* A **528** 215
[4] Mueller W F, Jensen H J, Reviol W, Riedinger L L, Yu C H, Zhang J Y, Nazarewicz W and Wyss R 1994 *Phys. Rev.* C **50** 1901
[5] Eberth J, Thomas H G, von Brentano P, Lieder R M, Jäger H M, Kämmerling H K, Berst M, Gutknecht D and Henck R 1996 *Nucl. Instr. Meth. Phys. Res.* A **369** 135
[6] Dewald A *et al Nucl. Phys.* A **545** 822
[7] Dewald D, Möller O and Petkov P 2012 *Prog. Part. Nucl. Phys.* **67** 786
[8] Gladnishki K A, Petkov P, Dewald A, Fransen C, Hackstein M, Jolie J, Pissulla Th, Rother W and Zell K O 2012 *Nucl. Phys.* A **877** 19
[9] Petkov P *et al* 2000 *Nucl. Phys.* A **674** 357
[10] Dewald A, Harissopulos S and von Brentano P 1989 *Z. Phys.* A **334** 163
[11] Böhm G, Dewald A, Petkov P and von Brentano P 1993 *Nucl. Instr. Meth. Phys. Res.* A **329** (248
[12] Larsson S E, Leander G and Ragnarsson I 1978 *Nucl. Phys.* A **307** 189
[13] Ragnarsson I and Semmes P B 1988 *Hyp. Int.* **43** 425
[14] Semmes P B, Computer manual, presented at Nuclear structure theory workshop, Oak Ridge, TN (August 5-16, 1991)

XX International School on Nuclear Physics, Neutron Physics and Applications (Varna2013) IOP Publishing
Journal of Physics: Conference Series **533** (2014) 012035 doi:10.1088/1742-6596/533/1/012035

Observation of the Ground State Bands in ^{109}Pd and ^{111}Pd

E.A. Stefanova[1], S. Lalkovski[2], A. Korichi[3], T. Kutsarova[1], A. Lopez-Martens[3], F.R. Xu[4], H.L. Liu[4], S. Kisyov[2], A. Minkova[2], D. Bazzaco[5], M. Bergström[6], A. Görgen[7,8], F. Hannachi[3,9], B. Herskind[6], H. Hübel[7], A. Jansen[7], T. L. Khoo[10], Zs. Podolyák[11,12], G. Schönwasser[7]

Institute for Nuclear Research and Nuclear Energy, Bulgarian Academy of Science, 1784 Sofia, Bulgaria[1]
Faculty of Physics, University of Sofia "St. Kliment Ohridski", 1164 Sofia, Bulgaria[2]
CSNSM Orsay, IN2P3/CNRS, F-91405, France[3]
School of Physics, Peking University, Beijing 100871, China[4]
Dipartimento di Fizica e UNFN, Sezione di Padova, 1-35131 Padova, Italy[5]
The Niels Bohr Institut, Blegdamsvej 17, DK-2100 Copenhagen, Denmark[6]
ISKP, Universität Bonn, Nussallee 14-16, D-53115, Germany[7]
DAPNIA/SPhN, CEA-Saclay, Gif-sur-Yvette, France[8]
Centre d'Etudes Nuclaires de Bordeaux-Gradignan, Gradignan, France[9]
Physics Division, Argonne National Laboratory, Argonne, Illinois 60439, USA[10]
INFN, Laboratori Nationali di Legnaro, Italy[11]
Department of Physics, University of Surrey, Guildford, GU27XH, UK[12]

E-mail: `elenas@inrne.bas.bg`

Abstract.
The neutron-rich nuclei ^{109}Pd and ^{111}Pd were produced as fission fragments following the ^{30}Si + ^{168}Er reaction at a beam energy of 142 MeV. Using the identification based on the coincidences with the complementary fission fragments, the ground state bands in ^{109}Pd and ^{111}Pd were found. They are the only positive-parity bands observed so far in ^{109}Pd and ^{111}Pd. A band, built on top of the $5/2^+$ ground state exhibiting $\Delta I = 1$ energy-level staggering was observed in each of these nuclei. Both nuclei of interest, ^{109}Pd and ^{111}Pd, are suggested to lie in the transitional region of Pd isotopes of maximum γ-softness. The ground states of both nuclei are predicted by TRS calculations to be extremely γ-soft with shallow triaxial minima. The first crossing in the new bands is proposed to be due to an alignment of $h^2_{11/2}$ neutrons.

1. Introduction

The neutron-rich nuclei in the $A \approx 100$ ($38,40 \leq Z \leq 50$, $N > 56$) region are known to exhibit a variety of structural phenomena and shapes and were for example a subject of detailed theoretical study by the Nilsson-Strutinski calculations with cranked Woods-Saxon average potential in Ref. [1]. The calculations predict that the expected spherical-deformed shape transition goes through a region of γ-softness in the chain of Pd isotopes. The present study of ^{109}Pd and ^{111}Pd is intended to shed more light on the question of γ-softness in the chain of Pd isotopes.

Prior to the present work, a search for band structures in the neutron-rich nuclei ^{109}Pd and ^{111}Pd was performed using a fusion-fission reaction [2] and the negative-parity $\nu(h_{11/2})$ bands were identified. The first crossing was proposed to be a result of $\nu(h_{11/2})^2$ alignment, which stabilizes the prolate shape [2]. How to understand that in the light of the expected γ-softness? The situation was expressed in Ref. [3] as "a prolate, but still γ-soft shape is stabilized with increased rotational frequency" in ^{110,112,114,116}Pd isotopes. The basis of this assumption were cranked Woods-Saxon and total Routhian surface (TRS) calculations [3]. Most of the nuclei in the discussed region are neutron rich and are not accessible by means of conventional fusion-evaporation in-beam techniques. In the present work, positive-parity bands in ^{109,111}Pd are reported for the first time.

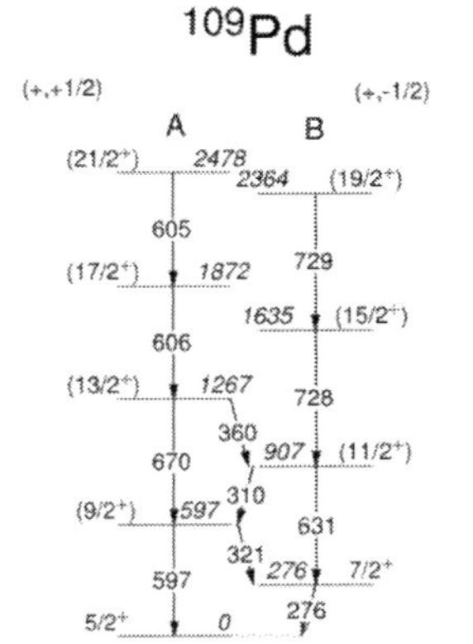

Figure 1. The positive-parity ground-state band of ^{109}Pd deduced from the present work. Only the $7/2^+$ state at an energy of 276 keV with the 276 keV transition depopulating it to the ground state and the 597 keV level (the transition depopulating the 597 keV level was not observed) were observed in previous studies [4].

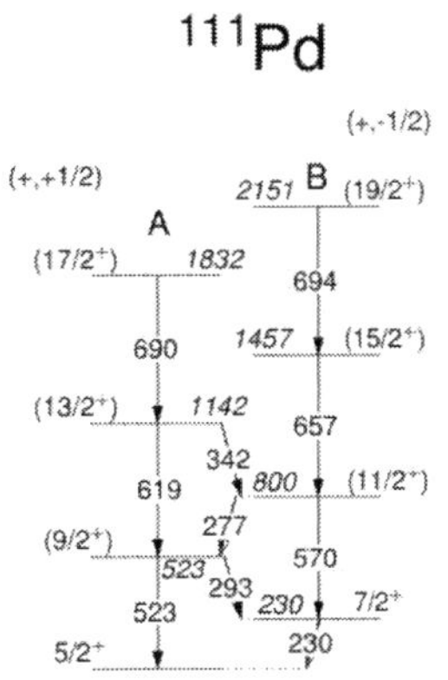

Figure 2. The positive-parity ground-state band of ^{111}Pd deduced from the present work. The levels at 230 keV and 523 keV as well as the 230, 293 and 523 keV transitions were previously known [5].

2. Experiment, Analysis and Experimental Results

Excited states in the neutron-rich nuclei ^{109}Pd and ^{111}Pd were populated via an induced fusion-fission channel of the ^{30}Si + ^{168}Er reaction. The beam of ^{30}Si at an energy of 142 MeV was provided by the XTU tandem accelerator at the Legnaro National Laboratory. Prompt γ rays emitted from the excited nuclei were detected with the EUROBALL III multidetector array. The ^{168}Er target foil of 1.15 mg/cm^2 thickness was evaporated on a 9 mg/cm^2 gold backing, in which the recoiling nuclei were slowed down and finally stopped.

In order to search for new transitions in ^{109}Pd and ^{111}Pd, we examined spectra double gated on low-lying transitions in ^{109,111}Pd and on low-lying transitions in the complementary fission fragments ^{84,82}Kr, respectively. Positive-parity rotational (quasi-rotational) bands were observed for the first time in ^{109}Pd and ^{111}Pd in the present work and are shown in Figures 1 and 2.

No band structures built on the two known low-lying positive-parity states were observed in ^{109}Pd before the present study. The first $7/2^+$ and $(9/2^+)$ states at energies of 276 keV

Journal of Physics: Conference Series **533** (2014) 012035 doi:10.1088/1742-6596/533/1/012035

and 597 keV, respectively, were observed previously as single not connected levels [4]. Based on these previous experiments, a positive parity can be assigned to the newly observed band structure. The 597 keV transition was not observed before the present study. No positive-parity rotational band has been observed in ^{111}Pd before the present experimental study. Only two excited positive-parity states (at 230 keV and 523 keV) were known in ^{111}Pd prior to the present study [5]. Based on the observed rotational (quasi-rotational) band structure (with $\Delta I = 2$ and $\Delta I = 1$ transitions as most likely), on the previously existed experimental data and on the systematics of the ground-state bands in the region, the spins and parities of the observed states can be proposed.

3. Discussion

The nuclei ^{109}Pd and ^{111}Pd lie in the transitional region between vibrational and rotational nuclei. It is known that this transition from sphericity to deformation in the Pd isotopes is more gradual than the ones in the Sr and Zr isotopes [6]. Nilsson-Strutinski calculations with cranked Woods-Saxon average potential [1] predict that this transition in palladium goes rather through γ-softness than just through softness to axial quadrupole deformation. The ratios of $E(4_1^+)/E(2_1^+)$ for ^{110}Pd and ^{112}Pd are also very close to the ones predicted by the IBM-2 model value of 2.5 for a γ-soft $O(6)$ rotor. The energy-level staggering in γ-bands in the chain of the Pd isotopes was analyzed in Ref. [7]. The levels of the γ-bands in ^{108}Pd [7], ^{110}Pd [7], ^{112}Pd [8] are grouped as 2^+, (3^+,4^+), (5^+,6^+) and etc., which is consistent with γ-soft behavior accordingly to the model of Wilets and Jean [9]. While the γ-bands are not so well developed in ^{102,104,106}Pd as in the heavier Pd isotopes, the amplitude of the staggering effect decreases in ^{114}Pd and ^{116}Pd with respect to that in ^{108,110,112}Pd, where it reaches its maximum [7]. It was concluded that ^{108}Pd, ^{110}Pd and ^{112}Pd are the most γ-soft Pd isotopes [7].

Thus, one may expect a significant degree of γ-softness at low energies and angular momenta for the nuclei of interest, ^{109}Pd and ^{111}Pd. Moreover, according to the behavior of the γ-bands in the even-even Pd isotopes, one may suggest that both nuclei lie in the region of maximum γ-softness. This suggestion is also supported by the TRS calculations [10] reported below.

In the present experiment, the ground-state positive-parity bands were observed in ^{109}Pd and in ^{111}Pd. While ^{107}Pd is predicted to be moderately γ-soft at its ground state, but still around prolate deformation by the TRS calculations, the TRS plots presented in Figure 3 for ^{109}Pd and ^{111}Pd draw a long valley from $\gamma = 60°$ till $\gamma = -120°$ revealing extreme γ-softness (especially for ^{111}Pd). Indeed, the nucleus ^{109}Pd looks considerably γ-soft at its ground state with a minimum at $\gamma = -20°$ (see Figure 3a) and extreme γ-softness with a minimum at $\gamma = -30°$ (see Figure 3e) is seen for the ground state of ^{111}Pd. Calculations were performed separately for both signature partners of these bands in ^{109}Pd and ^{111}Pd as shown in Figure 3. The signature partners have almost the same deformation up to the crossing.

A band crossing in ^{109}Pd occurs at around 0.35 MeV (the figure is not shown), which is at about the same frequency as in the even-even Pd neighbors. This behavior is consistent with an alignment of a pair of $h_{11/2}$ neutrons, as no blocking is expected in these bands. The first crossing cannot be seen in ^{111}Pd in the present data. In agreement, an alignment of $h_{11/2}^2$ neutrons is predicted by the TRS calculations for these bands in ^{109}Pd and ^{111}Pd. The calculations predict that the alignment of $h_{11/2}^2$ neutrons ($\hbar w \approx 0.300$ MeV) forces the α = -1/2 signature partners in ^{109}Pd (see Figure 3d) and ^{111}Pd (see Figure 3h) towards more γ-stable near prolate shape. However, the $\alpha = +1/2$ signature partners in both nuclei are predicted to preserve and actually to stabilize the triaxial deformation (see Figures 3b and 3f). Thus, a more stable triaxiality is predicted at higher spins. Experimental observation of the bands at higher spins in future experiments may allow to test this predictions.

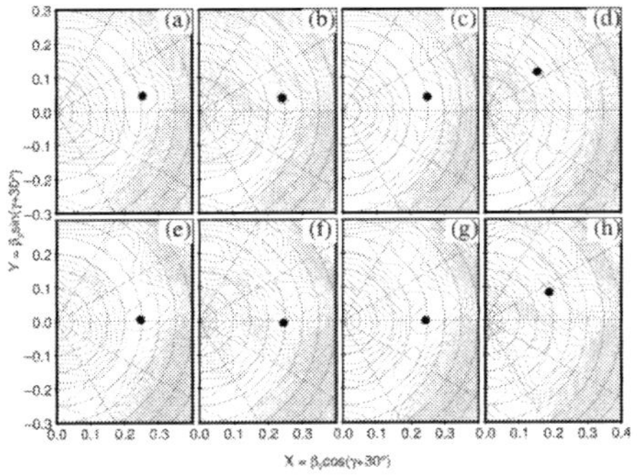

Figure 3. Total Routhian surface calculations for the positive-parity bands in ^{109}Pd and ^{111}Pd. Contour lines are in 200 keV increment. a) ^{109}Pd, (+,+1/2), $\hbar w$ = 0.100 MeV, β_2 = 0.256, γ = -19°; b) ^{109}Pd, (+,+1/2), $\hbar w$ = 0.350 MeV, β_2 = 0.245, γ = -21°; c) ^{109}Pd, (+,-1/2), $\hbar w$ = 0.100 MeV, β_2 = 0.253, γ = -20°; d) ^{109}Pd, (+,-1/2), $\hbar w$ = 0.350 MeV, β_2 = 0.191, γ = +7°; e) ^{111}Pd, (+,+1/2), $\hbar w$ = 0.100 MeV, β_2 = 0.246, γ = -29°; f) ^{111}Pd, (+,+1/2), $\hbar w$ = 0.350 MeV, β_2 = 0.244, γ = -31°; g) ^{111}Pd, (+,-1/2), $\hbar w$ = 0.100 MeV, β_2 = 0.244, γ = -30°; h) ^{111}Pd, (+,-1/2), $\hbar w$ = 0.350 MeV, β_2 = 0.204, γ = -9°.

4. Conclusion

Although few low-lying positive-parity states were known before the present study, no band structures built on them were observed. Positive-parity bands in ^{109}Pd and ^{111}Pd were observed for the first time in this work. It was concluded that ^{109}Pd and ^{111}Pd, together with ^{113}Pd, lie in the transitional Pd region where maximum γ-softness is expected to occur. The first band crossing in the observed positive-parity bands of this work is proposed to be due to the alignment of the $h^2_{11/2}$ neutrons. The calculations predict that this alignment drives one of the band signatures in both nuclei to a less γ-soft, near-prolate shape, while a stabilization of the triaxial shape is predicted for the other signature in both nuclei. Further experiments are definitely needed in order to test for these predictions.

Acknowledgements

The authors would like to acknowledge the useful discussions with N. Minkov. This work was partly supported from the Bulgarian National Science Fund, contract No: DMU02/1 and the collaboration agreement between Bulgarian Academy of Sciences-CNRS under contract No. 4847, and by the German BMBF under Contract No. 06BN109.

References

[1] Skalski J et al. 1999 *Nucl. Phys.* A **617** 282
[2] Kutsarova T et al. 1998 *Phys. Rev.* C **58** 1966
[3] Krücken R et al. 1999 *Phys. Rev.* C **60** 031302
[4] Blachot J 2006 *Nucl. Data Sheets* **107** 355
[5] Blachot J 2009 *Nucl. Data Sheets* **110** 1239
[6] Hotchkis M A C 1991 *Nucl. Phys.* A **530** 111
[7] Lalkovski S 2003 *Eur. Phys. J* A **18** 589
[8] Krücken R 2001 *Eur. Phys. J* A **10** 151
[9] Wilets L and Jean M 1956 *Phys. Rev.* **102** 788
[10] Wyss R 1988 *Phys. Lett.* B **215** 211

Journal of Physics: Conference Series **533** (2014) 012036 doi:10.1088/1742-6596/533/1/012036

Extensive Test of an SU(3)–based Partial Dynamical Symmetry

R.F. Casten

A.W. Wright Nuclear Structure Laboratory, Yale University, New Haven, CT USA

E-mail: richard.casten@yale.edu

Abstract. The concept of symmetries pervades much of our understanding of nature. In nuclear structure, the IBA embodies a framework with three dynamical symmetries U(5), O(6) and SU(3). Of course, most nuclei break these symmetries. Leviatan has discussed a concept of Partial Dynamical Symmetry (PDS) in which the states of the ground and gamma bands, only, are *exactly* described by SU(3) while all others are not. With an $E2$ operator which is not a generator of SU(3), this PDS gives a *parameter–free* description of γ to ground band relative $B(E2)$ values in ^{168}Er that is virtually identical to the best collective model (IBA) calculations with 2-3 parameters. We have carried out the first extensive study of this PDS, in 47 rare earth nuclei. Overall, the PDS works very well, and the deviations from the data are usually understandable in terms of specific kinds of mixing.

The study of nuclear structure is typically approached in two complementary ways—in terms of the interactions and orbits of the individual nucleons and in terms of the symmetries, quantum numbers and structure of the many-body system as a whole. Very broadly, the second approach is a powerful tool for telling us *what* nuclei do, and given a structure for a nucleus, for generating a wide variety of detailed predictions, while the first can tell us why the nuclei do what they do. This work proceeds from the second perspective, focusing on the symmetries and partial symmetries of well-deformed collective even-even nuclei.

The concept of dynamical symmetries has long been a powerful technique to interpret nuclear structure. The IBA model [1] codifies this in terms of three symmetries, U(5)–a vibrator-like structure, SU(3)—an axial rotor-like structure, and O(6)—a non-axial type of rotor. We are concerned here about SU(3). A dynamical symmetry description is characterized by specific degeneracies, quantum numbers, and selection rules. While few nuclei actually manifest an unbroken or nearly unbroken symmetry, these symmetries act as benchmarks, much like the magic numbers do for spherical nuclei. Most well-deformed nuclei somewhat resemble SU(3) but exhibit strong deviations from it [2, 3, 4]. Such nuclei have often been described in terms of numerical calculations involving a specific IBA Hamiltonian that includes symmetry breaking terms [3, 4]. One of the best studied is ^{168}Er [3, 5]. The IBA is able to account for the low lying intrinsic excitations as well as many absolute and relative $E2$ transition rates. Seemingly, this description identified ^{168}Er as a well deformed nucleus with strong SU(3)-symmetry breaking.

However, a number of years ago Leviatan [6] proposed the idea of a Partial Dynamical Symmetry (PDS) in which certain aspects of the parent symmetry are *exactly* preserved while others are not. In the application to ^{168}Er, the ground and intrinsic excitations are exactly described as pure SU(3) excitations while all other states involve significant SU(3) breaking and the mixing of SU(3) representations.

Journal of Physics: Conference Series **533** (2014) 012036 doi:10.1088/1742-6596/533/1/012036

His description of $E2$ transition rates was remarkably successful. Using a generalized $E2$ operator the predictions for relative to ground $B(E2)$ values are *parameter-free* but yet reproduce the data as well as the best parameterized broken-SU(3) calculations. Indeed, the PDS results are almost identical to those of the broken-SU(3) numerical IBA calculations. Of course, they are similar as well to a standard SU(3) calculation with the same operator for the γ-ground band transitions.

This surprising result raises many questions. However, before the theoretical issues are investigated it is important to know if this success is unique to ^{168}Er or if it characterizes other well-deformed nuclei. It is the purpose of this work to address this question. To this end, we have collected all relevant data on relative γ to ground band $B(E2)$ values in the rare earth region. In all, we studied 47 nuclei. In some, the data is insufficient for a substantial test—often relevant $E2$-$M1$ mixing ratios, or weak transitions are unknown or poorly known. However, in about 22 nuclei there are sufficient data. We present examples of the results below.

First, a word about the PDS calculations. They were carried out with the Hamiltonian of Eq. 1 of Ref. [6]. Here the Hamiltonian has two parameters, h_0 and h_2. As long as the Hamiltonian is written in the form of Ref. [6] the solutions satisfy an SU(3) PDS with pure SU(3) structure solely for the γ and ground representations. The calculations are parameter-free for relative γ to ground transitions (intra-band $E2$ transitions do require one overall parameter for each nucleus so we ignore them for the present purposes) but they do depend weakly on boson (valence nucleon) number. The PDS calculations were done in each case for the appropriate number of valence nucleons.

Examples of the results are shown in Figs. 1-3. Figure 1 includes ^{168}Er as well as several other well-deformed nuclei and show truly excellent agreement between the data and the PDS. The only notable discrepancies are for a few *spin-increasing* transitions. Figure 2 shows other nuclei in the same region with good agreement but slightly larger discrepancies—again primarily for the same transitions. Figure 3 shows two nuclei, in transitional regions, where there are huge disagreements. These latter are, of course, not surprising since an SU(3) PDS would not be expected to match the properties of transitional nuclei. However, it is nevertheless interesting to note that the discrepancies with the data, while much larger, are for exactly the same spin-increasing transitions (increasing relative to the normalized transition) as are the smaller discrepancies for well-deformed nuclei. Moreover, the discrepancies for spherical-deformed transitional nuclei at $N = 90$ are similar to those for nuclei in the prolate to soft transition seen in the Os nuclei. However, we note that, while ^{176,180,188}Os all show large deviations, ^{184}Os, with slightly larger $E_{4/2} = 3.20$, shows reasonable agreement, as seen in Fig. 2.

An interesting question is why the discrepancies are specifically and nearly always for the spin-increasing transitions, especially those such as $2_g \rightarrow 4_g$, $3_g \rightarrow 4_g$, $4_g \rightarrow 6_g$, and $5_g \rightarrow 6_g$. The reason is actually simple. Spin-increasing transitions always have small values in the case of a pure rotor with an uncoupled intrinsic γ vibration. This is seen in the Alaga rules [7] and has a simple physical origin that is easily seen if one decomposes the wave function into rotational and intrinsic parts. Here we explain this for the decay of the 4^+_γ level. The explanation is similar for higher spin band members. The 4^+ level of the ground band has rotational angular momentum $R = 4$. The 4^+_γ level has intrinsic angular momentum 2. It has rotational angular momentum components $R = 0$,2 and 4 but $R = 0$ dominates. Therefore an $E2$ transition to the 4^+ level of the ground band with $R = 4$ is highly hindered.

Actual deformed nuclei always show deviations from the Alaga rules, which are usually interpreted in terms of γ-ground bandmixing. Such bandmixing, depending on the relative signs of the direct and admixed components of the wave function can, in general, either increase or decrease the unperturbed $B(E2)$ value. However, for unperturbed transitions that are forbidden or very small, such as the γ-ground spin increasing transitions, the mixing can only increase the

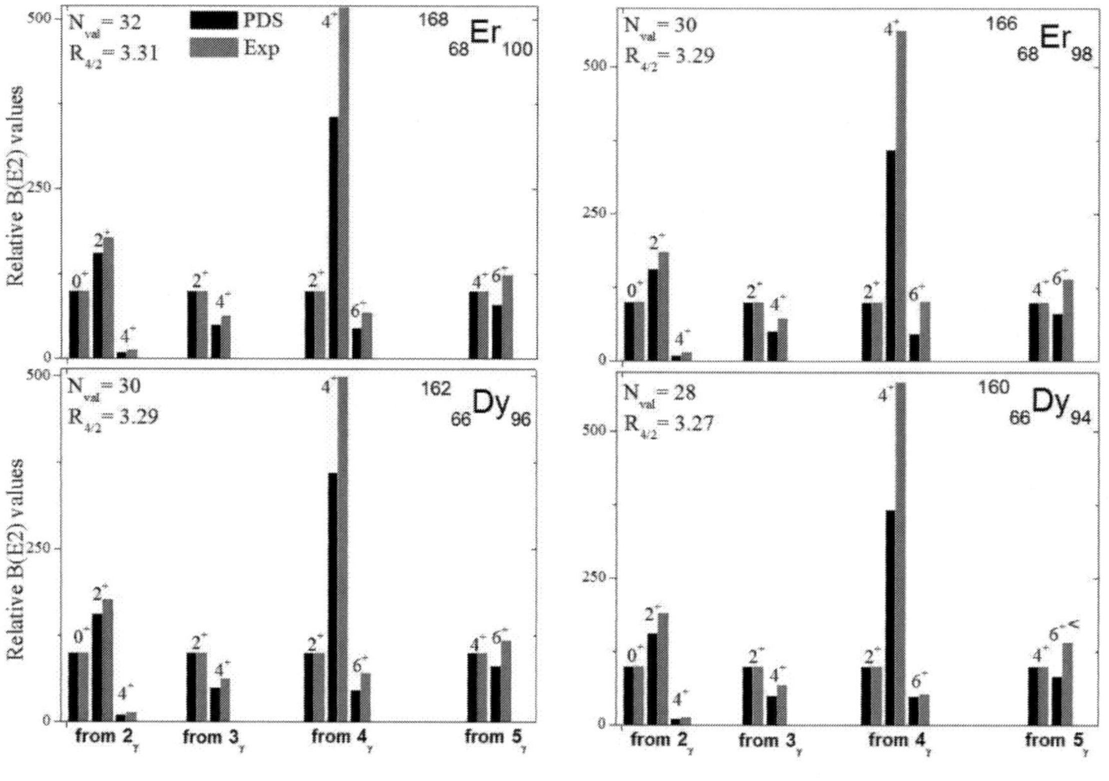

Figure 1. Comparison of the data on γ to ground band relative $B(E2)$ values in some well-deformed nuclei with the SU(3) PDS of Ref. [6]. For each initial level, the most spin-decreasing transition (the leftmost one) is normalized to 100. All data from NDS. The symbol $<$ next to a transition bar means that the height of the bar is an upper limit. Usually this is due to an unknown $M1$ component.

$B(E2)$ value since the sign of the mixed component gets squared. Therefore, one expects, and finds, that the spin increasing transitions are nearly always stronger than the Alaga rules. This is what one sees in the data and in comparison to the PDS as well. The extent of the deviations from the PDS seen in Figs. 1-3 measures approximately the amount of such bandmixing.

I am grateful to my collaborators R.B. Cakirli and A. Couture, and to A. Leviatan, P. Van Isacker, M. Macek, and N. Pietralla for discussions of the PDS and patient explanations of the physics behind it. Work supported by the US DOE under Grant No. DE-FG02-91ER-40609.

1. References

[1] Arima A and Iachello F 1974 *Phys. Lett.* **53** 309–12
[2] Casten RF and Warner DD 1988 *Rev. Mod. Phys.* **60** 389–469
[3] Warner DD and Casten RF 1982 *Phys. Rev. Lett.* **48** 1385–1389
[4] McCutchan EA and Casten RF 2006 *Phys. Rev.* C **74** 057302(1-4).
[5] Warner DD, Casten RF and Davidson WF 1980 *Phys. Rev. Lett.* **45** 1761–5
[6] Leviatan A 1996 *Phys. Rev. Lett.* **77** 818(1-4)
[7] Alaga G, Alder K, Bohr A and Mottelson BR 1955 *Mat. Fys. Medd. Dan. Vid. Selsk.* **29** No. 9

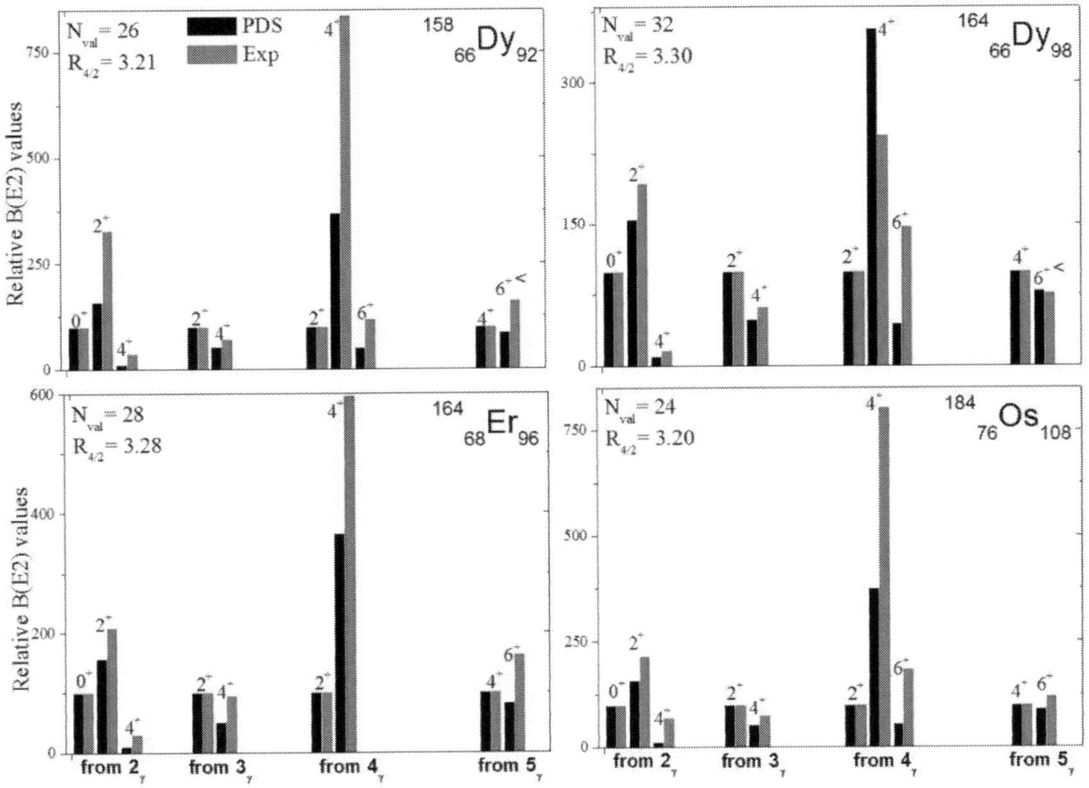

Figure 2. Similar to Fig. 1 for nuclei showing somewhat larger disagreements with the PDS.

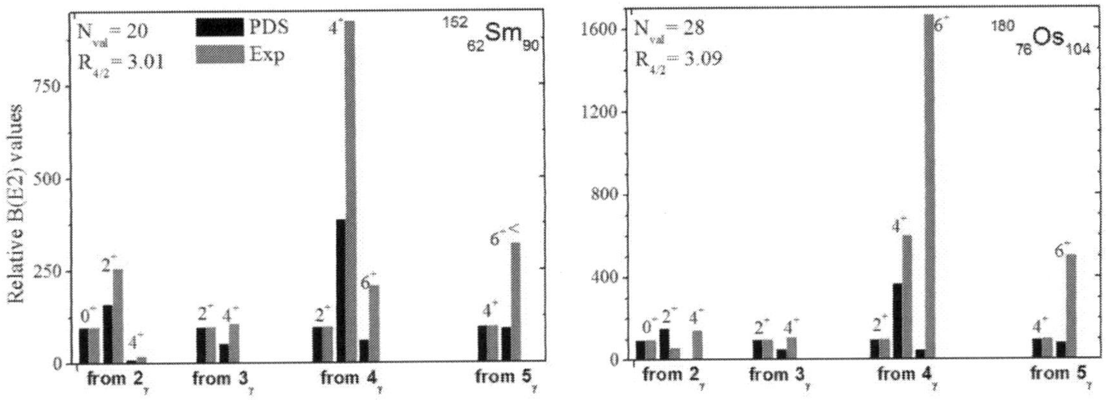

Figure 3. Similar to Fig. 1 for transitional nuclei which, not surprisingly, show large disagreements with the PDS.

Empirical p-n interactions, the synchronized filling of Nilsson orbitals, and emergent collectivity

R B Cakirli[1]
Department of Physics, Istanbul University, Istanbul, Turkey

E-mail: rburcu@istanbul.edu.tr

Abstract. The onset of collectivity and deformation, changes to the single particle energies and magic numbers and so on are strongly influenced by, for example, proton (p) and neutron (n) interactions inside atomic nuclei. Experimentally, using binding energies (or masses), one can extract an average p-n interaction between the last two protons and the last two neutrons, called δV_{pn}. We have studied δV_{pn} values using calculations of spatial overlaps between p and n Nilsson orbitals, considering different deformations, for the Z= 50-82, N= 82-126 shells, and comparison of these theoretical results with experimental δV_{pn} values. Our results show that enhanced *valence* p-n interactions are closely correlated with the development of collectivity, shape changes, and the saturation of deformation in nuclei. We note that the difference of the Nilsson quantum numbers of the last filled Nilsson p and n orbitals, has a special relation, 0[110], in which they differ by only a single quantum in the z–direction, for those nuclei where δV_{pn} is largest for each Z in medium mass and heavy nuclei. The synchronised filling of such orbital pairs correlates with the emergence of collectivity.

1. Introduction and empirical proton-neutron (p-n) interactions

Magic numbers, the onset of deformation, configuration mixing, single particle energies etc can be interpreted in terms of proton(p)-neutron(n) interactions. Heavy nuclei with many possible interactions between nucleons are more complicated than lighter nuclei. Thus, to simplify this, it is often easier to consider only (number of) valence nucleons outside of a closed shell, which is a good approximation to gain an idea about the origins of the structure of nuclei. In these proceedings, empirical p-n interactions for heavy nuclei will mainly be discussed by considering only the number of valence nucleons. Before this, how one can extract empirical p-n interactions using masses (binding energies) will briefly be explained.

Equation (1), a double difference of binding energies, filters most of the interactions between nucleons and gives an average p-n interaction between the last two protons and the last two neutrons, called δV_{pn}. Its interpretation close to closed shells is obtained by looking at the Shell Model orbits for the unpaired last proton and the last neutron, and the overlap between their wave functions by considering their shell model quantum numbers, nl_j. For more details see Refs. [1,2].

$$| \delta V_{pn}(Z,N) | = 1/4 [((B(Z,N) - B(Z,N-2)) - (B(Z-2,N) - B(Z-2,N-2))] \quad (1)$$

[1] To whom any correspondence should be addressed.

where B is the binding energy. To interpret δV_{pn} for deformed nuclei (near mid-shell), one has to take into account the deformed shell model, that is, Nilsson orbitals (see below).

2. The synchronized filling of Nilsson orbitals, and emergent collectivity

Figure 1 shows experimental δV_{pn} values against neutron number for light nuclei (left) and heavy nuclei (middle) and a color coded $R_{4/2}$ plot (right) for the Z=50-82 and N=82-126 shells. It is clearly seen that δV_{pn} has spikes at N=Z in the left panel where both proton and neutron orbits fill exactly the same orbits. These spikes with large p-n interactions have been interpreted [3] by Wigner's SU(4) spin-isospin symmetry [4]. Such a large p-n interaction is not expected in heavy nuclei due to the Coulomb force and spin-orbit force which brings unique parity orbits into each shell. However, although p-n interactions in heavy nuclei are reduced in magnitude, one can still see large p-n interactions at the same (or similar) number of valence protons (Z_{val}) and valence neutrons (N_{val}). One example is ^{168}Er which has 18 valence protons and 18 valence neutrons and it has a maximum in δV_{pn} in the N~100 region (see Refs. [5,6] for details and for more examples). Figure 1 (middle) has some sharp zig-zags. These comprise nuclei with both $N_{val} = Z_{val}$ and odd-Z, odd-N values due to the absence of pairing effects on the last nucleons. One example here is N=97 (N_{val}=15) for $_{65}$Tb (Z_{val}=15). The average p-n interaction has a peak when $N_{val} = Z_{val}$. Similar examples are also seen for Ho and other nuclei. The right panel of fig. 1 shows the locus of the maximum δV_{pn} values shown with white circles on an Z-N contour plot for $R_{4/2}$ which indicates the changes in structure from spherical (blue) through deformed (red). The black line is placed to indicate $N_{val} = Z_{val}$. As seen in this plot, the nuclei for which we have peaks in the experimental δV_{pn} values, lie where the saturation of collectivity starts so these maxima in δV_{pn} have a clear link to the evolution of collectivity [6].

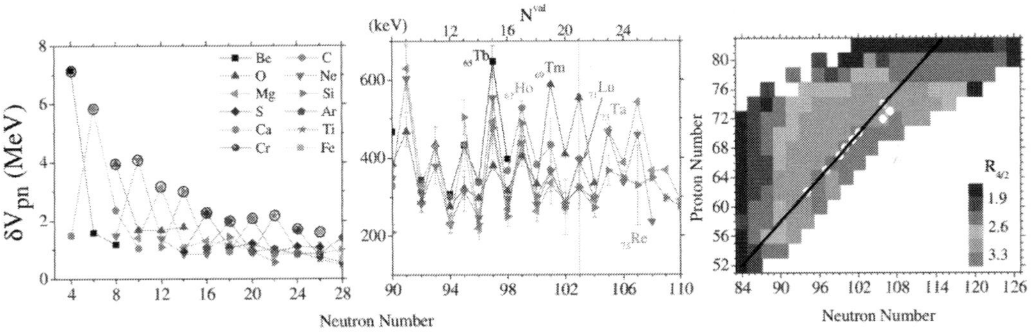

Figure 1. Experimental δV_{pn} values for light nuclei (Left) and heavy nuclei (Middle). (Right): Color coded $R_{4/2}$ values. White circles are shown for the nuclei for which there is a maximum in δV_{pn} in the middle panel. Based on Refs. [6,7].

As mentioned above, for these heavy nuclei, Nilsson orbitals, K[N,n_z,Λ], are considered for the interpretation. Quantum numbers for a given Nilsson orbital are K projection of the total angular momentum, N oscillator quantum number, n_z number of quanta in the z-direction and Λ projection of the orbital angular momentum. Reference [5] points out that large δV_{pn} values with known data in heavy nuclei appear when neutron and proton Nilsson orbital quantum number differences δK[δN, δn_z, δΛ]= 0[110]. In the same example as above, ^{168}Er is filling the 7/2[523] proton orbital and 7/2[633] neutron Nilsson orbitals. The difference between these orbitals is 0[110]. This occurs for almost all of these nuclei with the largest δV_{pn} values [5,6]. This 0[110] combination is special since proton and neutron Nilsson orbitals differ by only a single quantum in the z –direction.

The most important point is that this special 0[110] relation of proton and neutron Nilsson orbitals is (this has not been explicitly noted before) satisfied for each pair of orbits near the beginning of the shell as proton and neutron orbitals fill sequentially. This can be seen in fig. 2 where Nilsson orbitals are shown for protons (left) in the Z =50-82 shell and neutrons (right) in the N =82-126 shell. Since the overall shapes or patterns of the Nilsson diagrams are so similar this sequential filling of highly overlapping pairs of orbitals satisfying 0[110] gives large δV_{pn} values (see fig. 1 (middle)) along the line corresponding to saturation in $R_{4/2}$ (right) and following $N_{val} \sim Z_{val}$ line. even as the deformation increases. As a result, this synchronized filling of 0[110] relation is directly related to onset of deformation.

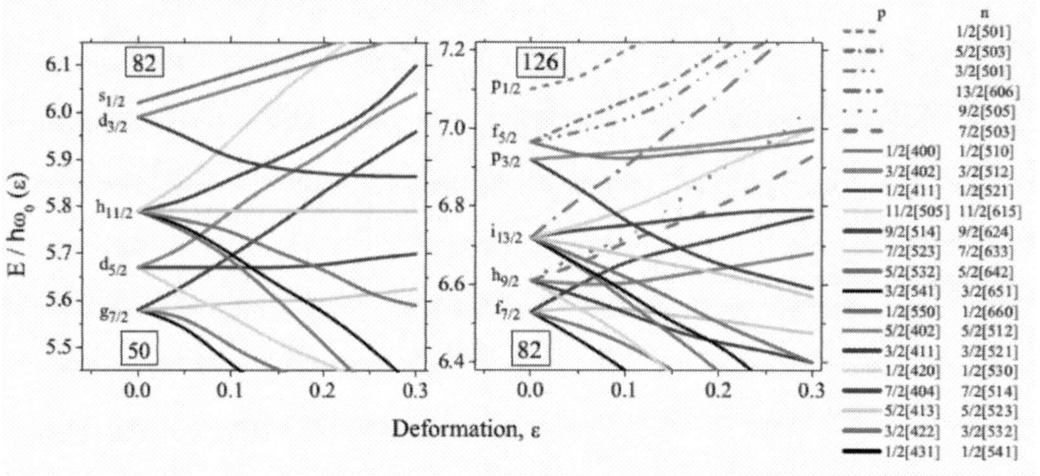

Figure 2. Nilsson diagrams for the proton Z= 50-82 and neutron N= 82-126 shells. Based on Refs. [6,7].

In order to understand these interactions in a simple way, we calculated [6] spatial overlaps of proton and neutron Nilsson wave functions with three deformations, namely ε= 0.05, 0.22, 0.3 [6,7]. The calculated overlaps are compared with the experimental δV_{pn} results in fig. 3. The left panel is experimental δV_{pn} values, theoretical spatial overlaps for the nuclei where the data exist are shown in the middle panel. All the calculated nuclei are illustrated in the right panel of fig. 3.

Since the proton shell has 32 nucleons and the neutron shell 44 nucleons, the figures are designed as square. Thus, along the diagonal line in fig. 3, nuclei will fill similar orbitals so large overlaps (large δV_{pn}) are expected along this line. The upper black line corresponds to $N_{val} = Z_{val}$ nuclei.

Overall agreement between the data and theory is very good. There are some disagreements such as in the upper right in fig. 3. However, this is not surprising because this disagreement may happen due to a γ-soft structure that our calculations do not take into account [7]. In addition, we obtain large δV_{pn} values with a 0[101] relation (not 0[110]) around Z~ 52-64 with N~ 92-108 (5/2[413] p - 5/2[512] n and 1/2[420] p - 1/2[521] n orbitals). Mass measurements in this region will help us to understand the p-n interactions more, in particular in the future at facilities such as FAIR or FRIB.

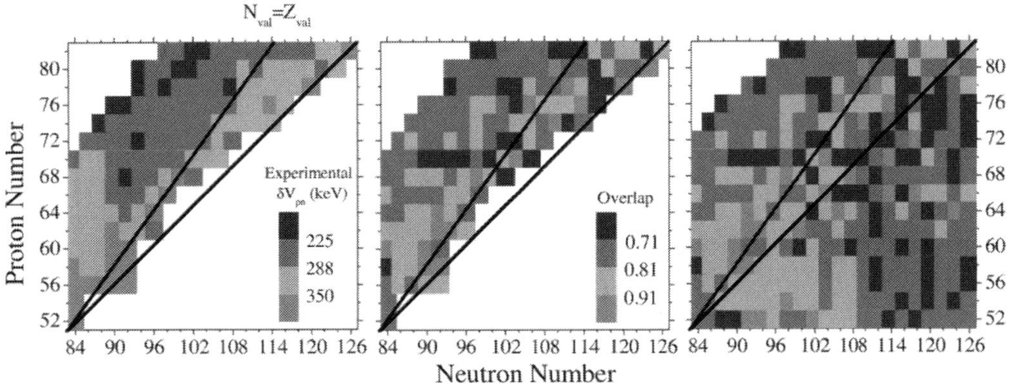

Figure 3. (Left): Color coded empirical δV_{pn} values for the Z= 50-82 and N= 82-126 shells. Large values have redder colors. (Middle): Similar as (Left) but for calculated overlaps for nuclei where empirical values of δV_{pn} are known. (Right): Calculated overlaps for the full major shells (excluding the nuclei beyond the proton dripline at far upper left). (see text for details). Based on Ref. [6,7].

3. Conclusion

Double differences of binding energies, δV_{pn}, extracted from high-precision mass measurements, are discussed from both experimental and theoretical perspectives by focusing on the Z= 50-82, N= 82-126 shells. Reflecting the clear spikes in δV_{pn} for light nuclei, studies of δV_{pn} for heavy nuclei are discussed in terms of the number of valence nucleons. We obtained a systematic behavior (large δV_{pn} values) when a nucleus has equal (or almost equal) numbers of valence protons and neutrons and noticed a consistent difference between proton and neutron Nilsson orbital quantum numbers, namely 0[110]. Calculated spatial overlaps between proton and neutron Nilsson orbitals are presented and compared with the experimental results. The sequential filling of orbitals with the 0[110] relation for equal numbers of valence protons and neutrons leads to a maximizing in the growth of collectivity along the $N_{val} \sim Z_{val}$ line.

4. References

[1] Garrett J D and Zhang J.-Y 1988 Cocoyoc, Book of Abtsracts ; Zhang J.-Y, Casten R F, and Brenner D S 1989 *Phys. Lett. B* **227**, 1
[2] Cakirli R B, Brenner D S, Casten R F, and Millman E A 2005 *Phys. Rev. Lett.* **94** 092501 (2005); 95 119903(E)
[3] Isacker P Van, Warner D D and Brenner D S 1995 *Phys. Rev. Lett.* **74** 4607
[4] Wigner E P 1937 *Phys. Rev.* **51**, 106 (1937).
[5] Cakirli R B, Blaum K, Casten R F 2010 *Phys. Rev. C* **82** 061304 (R) (2010)
[6] Bonatsos D, Karampagia S, Cakirli RB, Casten RF, Blaum K, Amon Susam L, submitted to *Phys. Rev. C* (Rapid Communication)
[7] D. Bonatsos (private communication)

Acknowledgments

We are grateful to D. Bonatsos, R. F. Casten, K. Blaum, S. Karampagia, L. Amon Susam for very useful discussions. This work was supported by the US DOE under Grant No. DE-FG02-91ER-40609, by the Max-Planck Society and by the Istanbul University Scientific Research Projects, Numbers. 26433. R.B.C. acknowledges support by the Max-Planck-Partner group.

Application of the interacting boson model and the interacting boson-fermion model to β decays

Nobuaki Yoshida

Faculty of Informatics, Kansai University, Takatsuki-shi, 569-1095, Japan

E-mail: yoshida@res.kutc.kansai-u.ac.jp

Abstract. The β decay is studied in the interacting boson model. The application to single-β decay is extended to two-neutrino double-β decay.

1. Introduction

The interacting boson model (IBM) [1, 2], including the interacting boson-fermion model (IBFM) [3] and the interaction boson-fermion-fermion model (IBFFM) [4], has been successful in describing the energy levels and the electromagnetic properties of various kinds of nuclei. One of the important applications is β decay [5, 6, 7, 8, 9, 10, 11, 12, 13, 14] because β decay is related to neutrino physics. In this school, I talk about the application of IBM to the β-decay. The decays include single-β decay from odd-A nuclei as well as double β-decay from even-even nuclei [15].

2. Description of β-decay in IBM

Single-β decay between odd-even nuclei has been studied in wide regions in IBM [5, 6, 8, 9, 10]. Double-β in IBM decay was studied by Scholten and Yu [7]. Recently, Barea, Kotila and Iachello studied extensively both neutrino-less double-β decay ($0\nu\beta\beta$) and two-neutrino double-β decay ($2\nu\beta\beta$) [12]. These works use the closure approximation in treating the intermediate states in odd-odd nuclei. I talk about the work on $2\nu\beta\beta$ with Iachello [15].

For $2\nu\beta\beta$, the Gamow-Teller (GT) $M_{2\nu}^{\mathrm{GT}}$ and the Fermi (F) matrix elements $M_{2\nu}^{\mathrm{F}}$ are calculated by [16]

$$M_{2\nu}^{\mathrm{GT}} = \sum_N \frac{\langle 0_F^+||t^+\sigma||1_N^+\rangle\langle 1_N^+||t^+\sigma||0_1^+\rangle}{\frac{1}{2}(Q_{\beta\beta}+2m_ec^2)+E_N-E_I}, \qquad M_{2\nu}^{\mathrm{F}} = \sum_N \frac{\langle 0_F^+||t^+||0_N^+\rangle\langle 0_N^+||t^+||0_1^+\rangle}{\frac{1}{2}(Q_{\beta\beta}+2m_ec^2)+E_N-E_I}, \tag{1}$$

where $t^{\pm}$ is the isospin increasing/decreasing operator, $\boldsymbol{\sigma} = 2\boldsymbol{s}$ is the Pauli spin matrix, while $Q_{\beta\beta}$ is the Q value of the double-β decay, and E_I and E_N are the energies of the initial and the intermediate states, respectively. The proton-neutron IBM is used in which the even-even core of the nucleus is treated as a system of proton bosons and neutron bosons of angular momentum zero (s-bosons) or angular momentum two (d-bosons), which represent proton pairs and neutron pairs outside the closed shell. The microscopic theory of IBM gives the images of the Fermi and

Gamow-Teller transition operators as [5]

$$t^{\pm} \longrightarrow O^{\mathrm{F}} = \sum_j -\sqrt{2j+1}\,[P_\pi^{(j)} P_\nu^{(j)}]^{(0)}; \qquad t^{\pm}\sigma \longrightarrow O^{\mathrm{GT}} = \sum_{j'j} \eta_{j'j}\,[P_\pi^{(j')} P_\nu^{(j)}]^{(1)} \quad (2)$$

where $P_\rho^{(j)}$ represents the boson-fermion image of the particle-transfer operator expanded in terms of fermion $a^\dagger_{jm}$ and boson $s^\dagger_{\pi,\nu}, d^\dagger_{\pi,\nu}$ operators: $P^{\dagger(j)}_{\rho,m} = \zeta_j a^\dagger_{jm} + \sum_{j'} \zeta_{jj'} s^\dagger_\rho [\tilde{d}_\rho \times a^\dagger_{j'}]^{(j)}_m$ with proper coefficients [17]. From these the $\log ft$ values can be calculated for β^- and β^+/EC transitions. In many calculations, the closure approximation is adopted in which the summation in (1) over the intermediate states N is replaced by the average. In the work that I am presenting, no closure approximation is made.

In the $2\nu\beta\beta$ decay from ^{128}Te to ^{128}Xe, the states in the intermediate nucleus ^{128}I are accounted for in the proton-neutron IBFFM. Some of the low-lying states are shown in Fig. 1. In the summation for the matrix elements, the states 1^+ up to 3 MeV in excitation energy

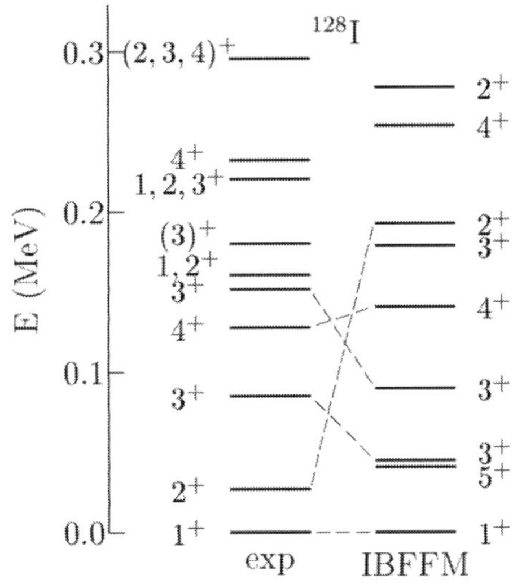

Figure 1. Energy levels in ^{128}I. The experimental data are from [18].

are included.

From ^{128}I, some of single-β decay (β^-, β^+/EC) are experimentally observed. Table 1 shows the $\log ft$ values of electron capture (EC) from ^{128}I, while those to ^{128}Xe are shown in Table

Table 1. The $\log_{10} ft$ values of EC from ^{128}I to ^{128}Te. The data are from [18].

transition	exp	cal	quenched
$1^+_1 \to 0^+_1$	5.049 (7)	3.836	5.15 (9)

2. Introducing a common hindrance factor: $h \approx 4.5$, which is equivalent to a quenched axial vector coupling constant: $g_{A,\mathrm{eff},\beta} = 1.269/h = 0.28$, we obtain a reasonable agreement, as shown as "quenched" in the Tables. The B(GT) values from ^{128}I have been also extracted from (^{3}He, t) reaction [19]. The related values are: ^{128}I, $B(\mathrm{GT})^{(^3\mathrm{He},t)}_{\mathrm{g.s.}} = 0.079$ (8); $\sum = 0.829$ (50). The corresponding IBM values are: ^{128}I$= 1.676$, $\sum = 15.09$. if we use the same quenched value of

Table 2. The $\log_{10} ft$ values of β^- decay from ^{128}I to ^{128}Xe. The data are from [18].

transition	exp	cal	quenched
$1_1^+ \to 0_1^+$	6.061 (5)	4.665	5.98 (9)
$1_1^+ \to 0_2^+$	7.748 (24)	5.262	6.57 (9)
$1_1^+ \to 0_3^+$	7.84 (6)	5.712	7.02 (9)
$1_1^+ \to 2_1^+$	6.495 (7)	5.212	6.52 (9)
$1_1^+ \to 2_2^+$	6.754 (9)	6.446	7.76 (9)

$g_{A,\text{eff},\beta} = 0.28$, then we have $B(\text{GT})[\text{IBM-quenched}] = 0.082$, $\sum[\text{IBM-quenched}] = 0.735$. These values are in good agreement with the (^{3}He, t) values.

Figure 2 shows the contributions from the intermediate states in ^{128}I to the GT matrix element in Eq. (1). The single-state dominance (SSD) discussed in Refs. [13, 14], namely, the

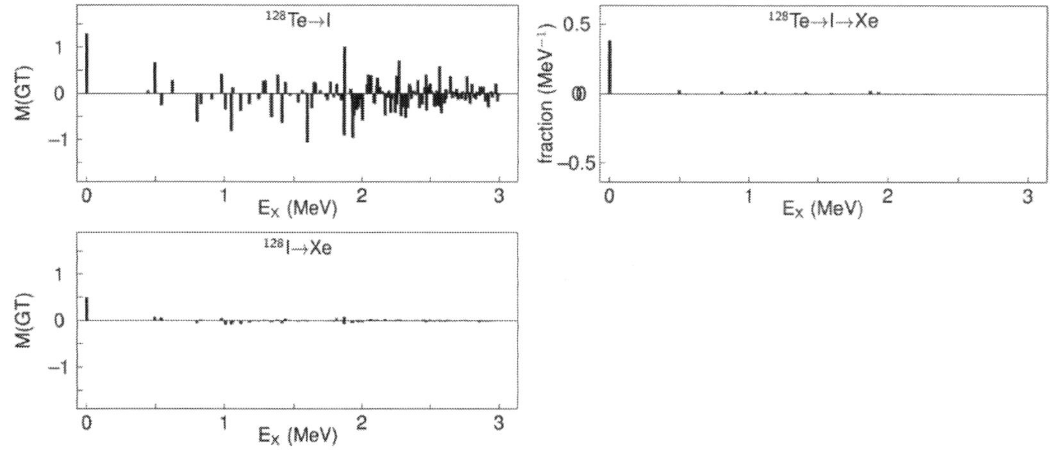

Figure 2. The values of $\langle 1_N^+||t^+\sigma||0_1^+\rangle$ (top-left), $\langle 0_1^+||t^+\sigma||1_N^+\rangle$ (bottom-left) and $\langle 0_1^+||t^+\sigma||1_N^+\rangle\langle 1_N^+||t^+\sigma||0_1^+\rangle/(\frac{1}{2}(Q_{\beta\beta}+2m_ec^2)+E_N-E_I)$ (top-right), for the double-β decay from the lowest 0^+ in ^{128}Te to the lowest 0^+ in ^{128}Xe through the intermediate 1^+ in ^{128}I, plotted as a function of the excitation energy.

dominance of 1_1^+ in the summation, is seen in the figures. Similar analysis has been made for the Fermi decay, as well as those from ^{130}Te. Table 3 shows the nuclear matrix elements calculated by (1) and a similar formula for the decay to a state 2^+. The inverse half-life of $0_1^+ \to 0_F^+$ can be calculated from

$$|M_{2\nu}^{\text{calc}}| = g_A^2 \left| M_{2\nu}^{\text{GT}} - \left(\frac{g_V}{g_A}\right)^2 M_{2\nu}^{\text{F}} \right| \tag{3}$$

by multiplying the lepton phase-space integral. Table 4 shows the thus obtained nuclear matrix element as "calu". By introducing the same quenched $g_{A,\text{eff},\beta\beta} = g_{A,\text{eff},\beta} = 0.28$ in

$$|M_{2\nu}^{\text{quenched}}| = g_{A,\text{eff},\beta\beta}^2 |M_{2\nu}^{\text{calc}}|, \tag{4}$$

while the ratio $g_V/g_A = 1/1.269$ in (3) is fixed, we obtain the values shown as "quenched" in Table 4, which are consistent with the experimental values.

Table 3. Nuclear matrix elements $M_{2\nu}^{\rm GT}$, $M_{2\nu}^{\rm F}$ of transitions from the ground state of ^{128,130}Te to some states in ^{128,130}Xe. The sign of $M_{2\nu}^{\rm GT}$ is chosen to be positive.

transition	^{128}Te→^{128}Xe	^{130}Te→^{130}Xe
GT		
$0_1^+ \to 0_1^+$	0.297	0.273
$0_1^+ \to 2_1^+$	0.00718	0.00639
$0_1^+ \to 0_2^+$		0.668
F		
$0_1^+ \to 0_1^+$	−0.0353	−0.0309
$0_1^+ \to 0_2^+$		−0.112

Table 4. Two-neutrino double-β decay matrix elements, $|M_{2\nu}|$ in IBFM.

	exp	calc	quenched
^{128}Te	0.044 (6)	0.514	0.040 (8)
^{130}Te	0.031 (4)	0.470	0.037 (8)

3. Conclusion

Use of a single value of $g_{A,{\rm eff},\beta} = g_{A_{\rm eff},\beta\beta}$ appears to describe well both single-β and double-β decay in a consistent way. The question of the small value of $g_{A,{\rm eff}}$ is a subject of further study.

References

[1] Arima A and Iachello F 1975 *Phys. Rev. Lett.* **35** p 1069
[2] Iachello F and Arima A *The Interacting Boson Model* (Cambridge: Cambridge University Press)
[3] Iachello F and Van Isacker P 1991 *The Interacting Boson-Fermion Model* (Cambridge: Cambridge University Press)
[4] Brant S and Paar V 1988 *Z. Phys.* A **329** p 151
[5] Dellagiacoma F 1988 Ph. D. thesis (New Haven: Yale University)
[6] Dellagiacoma F and Iachello F 1989 *Phys. Lett.* B **218** p 399
[7] Scholten O and Yu Z R 1985 *Phys. Lett.* B **161** p 131
[8] Yoshida N, Zuffi L and Brant S 2002 *Phys. Rev.* C **66** 014306
[9] Zuffi L, Brant S and Yoshida N 2003 *Phys. Rev.* C **68** 034308
[10] Brant S, Yoshida N and Zuffi L 2004 *Phys. Rev.* C **70** 054301
[11] Brant S Yoshida N and Zuffi L 2006 *Phys. Rev.* C **74** 024303
[12] Barea J and F. Iachello F 2009 *Phys. Rev.* C **79** 044301
[13] Kotila J and Iachello F 2012 *Phys. Rev.* C **85** 034316
[14] Barea J, Kotila J and Iachello F 2013 *Phys. Rev.* C **87** 014315
[15] Yoshida N and Iachello F 2013 *Prog. Theor. Exp. Phys.* **2013** 043D01
[16] Tomoda T 1991 *Rep. Prog. Phys.* **54** p 53
[17] Scholten O 1980 Ph. D. thesis (The Netherlands: University of Groningen)
[18] Kanbe K and Kitao K 2001 *Nucl. Data Sheets* **94** p 227
[19] Puppe P, Adachi T, Akimune H, Ejiri H, Frekers D, Fujita H, Fujita Y, Fujiwara M, Ganioğlu E, Grewe E -W, Hatanaka K, Hodak R, Iwamoto C, Khai N T, Lennarz A, Okamoto A, Okamura H, Povinec P P, Susoy G, Suzuki T, Tamii A, Thies J H and Yosoi M 2012 *Phys. Rev.* C **86** 044603

Journal of Physics: Conference Series **533** (2014) 012039 doi:10.1088/1742-6596/533/1/012039

Signature and parity splitting in rotational bands and chiral bands. Double minimum potential model.

R.V.Jolos

Joint Institute for Nuclear Research, 141980 Dubna, Russia

E-mail: jolos@theor.jinr.ru

Abstract. The effects of the signature and parity splitting in nuclear spectra and the properties of the chiral bands are analyzed basing on a one-dimensional Schrödinger equation with a double-minimum potential. Rotational bands in odd axial nuclei, alternating parity bands in even-even nuclei and the chiral bands in odd-odd nuclei are considered. The results obtained are discussed.

1. Introduction

Many phenomena in nuclear structure physics are analyzed basing on the intrinsic, i.e., body fixed frame. The nuclear intrinsic wave function lacks some symmetries of the total wave function, however, some discrete symmetries survive and we obtain the two-component wave function with positive or negative interference depending on the angular momentum. Because of the symmetry of the problem a potential also can have two symmetric minima. Two solutions of the Schrödinger equation $\psi^{(\pm)}$ with different symmetries with respect to $x \to -x$ transformation are splitted in energy. The splitting is determined by a behavior of the wave functions at the top of the barrier between two minima. The splitting in energy between the even and odd solutions introduces some irregularities in the spectra. Since double minimum potential depends on the angular momentum through the rotational energy term we obtain a possibility to extract an angular momentum dependence of the effective potential from the splitting data.

2. Odd axial nuclei. Single j case

In the case of an odd axial nucleus and an isolated single-j level the rotational state wave function takes the form

$$\Psi_{IM} = \sum_{K>0} \psi_I(K) \left(D^I_{MK} |jK\rangle + (-1)^{I-j} D^I_{M-K} |j-K\rangle \right) \tag{1}$$

For the most interesting cases (intermediate coupling and alignment) $\psi_I(K)$ can be treated as a continuous function of K. Then the finite difference equations for $\psi_I(K)$ can be transformed approximately to the ordinary differential equations [1], where K is a continuous variable

$$\left(-\frac{\hbar^2}{2\Im} \sqrt{(I^2-K^2)(j^2-K^2)} \frac{d^2}{dK^2} + U_I(K) \right) = E_I \psi_I(K). \tag{2}$$

Above

$$U_I(K) = \frac{\hbar^2}{2\Im}(I-j)^2(1+K^2/Ij) + \sqrt{\Delta^2 + (\kappa K^2 - \lambda)^2} \tag{3}$$

The potential is symmetric with respect to $K \to -K$ reflection. The first term in (3) comes from the core part of the total Hamiltonian. This term has a minimum at K=0 because in this case a single particle momentum is directed along the axis of the collective rotation, minimizing the rotational energy. The second term comes from the single quasiparticle part of the total Hamiltonian. This term has a minimum at the value of K, corresponding to the Nilsson state closest to the Fermi surface. If for I=0 the total potential has minima at $K = \pm K_0 \neq 0$ then the value of K_0 will decrease with I increase and for sufficiently large I $U_I(K)$ will have a minimum at $K = 0$ (complete alignment).

For a coupling scheme intermediate between the strong coupling and the complete alignment we obtain

$$E_{(-)} - E_{(+)} = \frac{\hbar^2}{\Im} Ij \left((\psi_I^{(+)}(1/2))^2 + (\psi_I^{(-)}(1/2))^2 \right) \tag{4}$$

Comparing (4) with a diagonal matrix element of the Coriolis interaction

$$-(-1)^{I-j}\frac{\hbar^2}{2\Im}(I+1/2)(j+1/2) \tag{5}$$

We see that (4) can be obtained multiplying the Coriolis matrix element by the square of the amplitude of the aligned component. Thus, in the axial symmetry case the signature splitting in the energy spectra can be used to extract the weight of the aligned component.

3. Parity splitting

To consider the parity splitting we assume that the main role in description of the alternating parity bands is played by the octupole mode with K=0 (β_{30}) which preserves the axial symmetry up to the high angular momentum. The corresponding Hamiltonian can be presented as

$$H_I = -\frac{\hbar^2}{2B}\frac{d^2}{d\beta_{30}^2} + V_I(\beta_{30}). \tag{6}$$

The wave function of the state belonging to the ground state alternating parity band with even I can be approximated to a good accuracy by a sum of two gaussians by making the following ansatz

$$\Psi_I^{(+)}(\beta_{30}) = \left(\frac{B\omega}{\pi\hbar}\right)^{\frac{1}{4}} \left[2\left(1+\exp\left(-\frac{B\omega}{\hbar}\beta_m^2(I)\right)\right)\right]^{-1/2}$$
$$\times \left(\exp\left(-\frac{B\omega\beta_m^2(I)}{2\hbar}(\beta_{30}/\beta_m(I)-1)^2\right) + \exp\left(-\frac{B\omega\beta_m^2(I)}{2\hbar}(\beta_{30}/\beta_m(I)+1)^2\right)\right), \tag{7}$$

where ω and $\beta_m(I)$ are the parameters defining the shape of the wave function. We see that $\Psi_I^{(+)}(\beta_{30})$ is symmetric with respect to the $\beta_{30} \to -\beta_{30}$ transformation.

This wave function contains two maxima which can overlap strongly and in this case we have a limit of octupole vibrations, or they are separated and we obtain a limit of the octupole deformation. The evolution from one limit to the other can be achieved by taking $\omega = const$ and β_m increasing with spin I. The experimental data on the E1 transitions in ^{240}Pu [2,3] indicate on a linear increase of $\beta_m(I)$ with I.

Having a parameterized wave function Eq. (7) we can substitute it into the Schrödinger equation with the Hamiltonian Eq. (6) and obtain the potential $V_I(\beta_{30})$. This potential has the form

$$V_I(\beta_{30}) = \frac{\hbar^2}{2B} \frac{\left(\frac{d^2\Psi_I^{(1,+)}(\beta_{30})}{d\beta_{30}^2}\right)}{\Psi_I^{(1,+)}(\beta_{30})}. \tag{8}$$

The results of the calculations of the excitation energies of the lowest negative parity states demonstrate the overall agreement of the calculated results with the experimental data.

4. Collective chiral mode

Since the chirality in nuclear physics was originally suggested [4] it has been one of the hot topics in the nuclear structure physics. For a rotating nucleus with triaxiality, the three angular momenta, including collective rotor angular momentum along the intermediate axis, together with angular momenta of the valence particles (holes) along the nuclear short (long) axis, can construct the chiral geometry. It has been demonstrated that the chiral doublet bands generally start from a left-right chiral vibrational mode and evolve into a static chiral mode with increase of the spin [5].

To describe the energy splitting between the chiral doublet bands, one has to go beyond the mean field approximation. It is necassary to search a unified method for studying both chiral rotation and vibration. In analogy to Bohr Hamiltonian, constructing a collective Hamiltonian provides one of the way out. Instead of the β and γ degrees of freedom in Bohr Hamiltonian a chiral degree of freedom should be introduced. Subsequently, a quantum tunneling in the region of chiral rotation can be described by considering the chiral fluctuations around the mean field minimum besides the region of chiral vibration naturally described by the collective Hamiltonian.

We have constructed the collective Hamiltonian for a system of one $h_{11/2}$ proton particle and one $h_{11/2}$ neutron hole coupled to a triaxial rigid rotor.

The cranking Hamiltonian for a considered system reads

$$\hat{h}' = \hat{h}_{def} - \omega \cdot \hat{\mathbf{j}}, \quad \hat{\mathbf{j}} = \hat{\mathbf{j}}_\pi + \hat{\mathbf{j}}_\nu, \tag{9}$$

$$\omega = (\omega \sin\theta \cos\varphi, \omega \sin\theta \sin\varphi, \omega \cos\theta), \tag{10}$$

where the Hamiltonian of the deformed field is $\hat{h}_{def} = \hat{h}^\pi_{def} + \hat{h}^\nu_{def}$ with the single j-shell Hamiltonian

$$\hat{h}^{\pi(\nu)}_{def} = \pm\frac{1}{2}C\left\{\left(\hat{j}_3^2 - \frac{j(j+1)}{3}\right)\cos\gamma + \frac{1}{2\sqrt{3}}\left(\hat{j}_+^2 + \hat{j}_-^2\right)\sin\gamma\right\} \tag{11}$$

The tilted axis cranking solutions are obtained selfconsistently by minimizing the total Routhian surface.

The quantized form of the collective Hamiltonian [6] is

$$\hat{H}_{coll} = -\frac{\hbar^2}{2\sqrt{B(\varphi)}}\frac{\partial}{\partial\varphi}\frac{1}{\sqrt{B(\varphi)}}\frac{\partial}{\partial\varphi} + V(\varphi). \tag{12}$$

The mass parameter is found following the procedure described in [7]. One obtains

$$B(\varphi) = 2\hbar^2 \sum_{l\neq 0} \frac{(E_l - E_0)\,|\frac{\partial\vec{\omega}}{\partial\varphi}\langle l|\hat{\mathbf{j}}|0\rangle|^2}{\left((E_l - E_0)^2 - \hbar^2\Omega^2\right)^2} \tag{13}$$

Above E_l is the eigenvalue of the Hamiltonian $\hat{h}'$ (9), where E_0 means the lowest eigenvalue. The vibrational frequence Ω is determined as follows: a) for the chiral rotation, it is approximately taken as zero because the barrier penetration between the left-handed and the right-handed states is low; b) for the chiral vibrations the potential energy $V(\varphi)$ can be approximated by the harmonic oscillator potential $\frac{1}{2}K\varphi^2$. The mass parameter is

$$B = \frac{K}{\Omega^2}. \tag{14}$$

Combining (13) and (14), the value of Ω can be obtained for chiral vibrations.

The calculated potential energy is symmetric about $\varphi = 0$. For the frequence $\hbar\omega \leq 0.15$ MeV, the potential $V(\varphi)$ is a harmonic oscillator type which has only one minimum at $\varphi = 0$. For the frequence $\hbar\omega \geq 0.20$ MeV, the potential $V(\varphi)$ has two symmetric minima. Due to the appearance of the potential barrier, the stable chiral solutions are found in the body-fixed frame. The heights of the barrier increases with the cranking frequency, e.g., from 1 keV at $\hbar\omega = 0.20$ MeV to about 2 MeV at $\hbar\omega = 0.50$ MeV.

The mass parameter $B(\varphi)$ is symmetric with respect to $\varphi = 0$ and increases dramatically when φ approaches $\pm\pi/2$. In the interior part, $B(\varphi)$ is increased remarkably with $|\varphi|$ for $\hbar\omega \geq 0.35$ MeV, while its dependence on φ is weak for $\hbar\omega = 0.25$ to 0.30 MeV.

The spectrum of the energy levels obtained by diagonalization of the collective Hamiltonian form the anharmonic vibrational spectrum for low $\hbar\omega$. However, with the $\hbar\omega$ increase the energy levels form the quasidegenerate doublets.

5. Conclusions

We have considered the signature splitting in the rotational bands of odd nuclei, the parity splitting in the ground state alternating parity bands of the even-even nuclei with strong octupole correlations, and the phase transition in the chiral bands. In all these cases the analysis is based on the one-dimensional Schrödinger equation with the potential which evolves from the harmonic oscillator to the double minimum potential. The signature, parity and the chiral splitting are determined by the closeness of the collective potential to the limit of the well separated two symmetric minima. Thus, the experimental data can be used to extract an information on the collective potential.

5.1. Acknowledgments

I am grateful to Profs. P. von Brentano, R. F. Casten, J. Jolie A. Gelberg, J. Meng and Drs. Q. B. Chen, S. Q. Zhao, P. W. Zhao for collaboration. This work was supported in parts by the DFG (Germany) under Contract Br 799/25-1, by the RFBR (Russia) under grant No. 10-02-00301 and by the Heisenberg-Landau program.

6. References

[1] Jolos R V and Gelberg A 1993 *Phys. Lett.* B **317** 495
[2] Wang X, Janssens R V F, Carpenter M P et al. 2009 *Phys. Rev. Lett.* **102** 122501
[3] Wiedenhover I, Janssens R V F, Hackman G et al. 1999 *Phys. Rev. Lett.* **83** 2143
[4] Frauendorf S, and Meng J 1997 *Nucl. Phys.* A **617** 131
[5] Qi B, Zhang S Q, Meng J et al. 2009 *Phys. Lett.* B **675** 175
[6] Chen Q B, Zhao S Q, Zhao P W, Jolos R V, Meng J 2013 *Phys. Rev.* C **87** 024314
[7] Rowe D J 1970 *Nuclear Collective motion. Models and Theory* (London: Methuen)

The mechanism of nuclear spontaneous time reversal symmetry breaking transfered to digital logic system

Ernest Grodner

Faculty of Physics, University of Warsaw, Institute of Experimental Physics, ul. Hoża 69, PL-00681 Warsaw, Poland.

E-mail: ernest.grodner@fuw.edu.pl

Abstract. Spontaneous time-reversal symmetry breaking in the system of an atomic nucleus is discussed on the basis of nuclear chirality phenomenon. The mechanism of nuclear chirality i.e. existence of three aplanar angular momenta vectors seems to be applicable in other fields of physics, for example in classical physics as coupling of three gyroscopes. Here, possible application of spontaneous time-reversal symmetry breaking in the design of a digital logic is discussed as an interesting curiosity.

1. Introduction

Possibility of time-reversal symmetry breaking is a fundamental question since the beginning of the quantum mechanics. Mechanisms of T-violation are not fully understood until today. There is room for T-violation in the field of high-energy physics since some elements of the KCM matrix may be complex values [1]. This however relates to heavy quark sector and no T-violation is expected to occur in low energy atomic physics. Search for time-reversal symmetry breaking in low energy collective modes of an atomic nucleus is a subject of extensive studies from theoretical as well as experimental standpoint since years. Variety of techniques, like gamma-gamma correlations [2, 3], Electric Dipole Moment measurement [4] or even mixing of chirality doublets [5, 6] were proposed in order to find T-violation in low energy nuclear physics. Finally, the possibility of time-reversal symmetry breaking in atomic and molecular physics could have large impact on industry due to its possible technical applications, for instance, an increase of security of digital network technologies. In contrast to fundamental T-violation it is relatively easy to find occurrence of spontaneous time reversal symmetry breaking. In the field of nuclear spectroscopy this phenomenon is known as nuclear chirality. Thus a question of a digital logic based on spontaneous T-symmetry breaking becomes an interesting subject that is briefly discussed here.

2. Nuclear chirality

In the most simple case, the phenomenon of nuclear chirality appears in odd-odd nuclei that are triaxially deformed. Such nuclei may be described as coupling of three elements: odd proton, odd neutron and an even-even triaxial core. The angular momentum vectors of the core, of the odd proton and of the odd neutron may be perpendicular to each other forming a system with specified handedness. The sign of the handedness is chosen spontaneously by the system and may be reversed by action of time-reversal operator, thus we deal with spontaneous time reversal symmetry breaking. The picture of nuclear chirality is based on coupling of three gyroscopes and seems to be applicable in other fields of physics. In ref.[7] it has been shown that such a coupling has similar behavior in nuclear as well as in classical physics. Treating the sign of the handedness as a bit of information one can obtain a logic that changes the bit values with the time-reversal. Such digital logic could be useful to prevent reverse computing or make isomorphic transformations being irreversible due to spontaneous T-violation.

3. Spontaneous time reversal symmetry breaking in the unit of information - rotating bit

The coupling of three gyroscopes in case of nuclear chirality ensures that the system is rotationally invariant. Therefore the sign of the handedness does not change with rotation of the system. Rotational invariance in the case of digital logic means that the behavior of the unit would be invariant with the change of its connectors. Modification of connections of the memory unit that is etched in a memory chip goes beyound the scope of this report as well as beyound the capabilities of a regular pc users. Therefore, the rotational invariance is not vital for spontaneous T-violation in the digital logic what reduces the problem of three to the problem of only one gyroscope. The rotation direction of a gyroscope changes to opposite with time-reversal. One gyroscope can easily be modeled by the logic of 2 bit counter that for instance changes its state on the down-slope of the clock signal - see fig. 1. Such a counter consists of

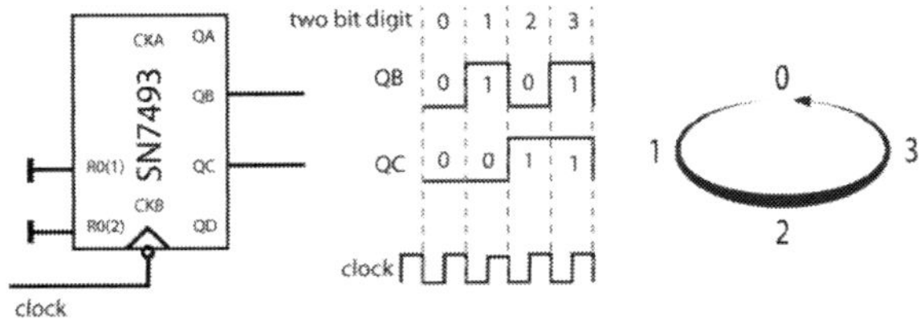

Figure 1. Digital logic model of a gyroscope as a two bit counter. Left: the two bit counter unit. Middle: input (clock) and output signals of the counter unit. Right: graphical representation of an analogy to a gyroscope.

two T-flip-flop units triggered by the down-slope of the clock signal. Going backward with time one can see reversed behavior of the flip-flops i.e. they change their states on the rising-slope of the clock signal - see fig.2. This modifies the behavior of the counter - after time-reversion it counts in opposite direction. Therefore, the counter is a good model of a physical gyroscope. The last thing that remains is to engineer an equivalent to the sign of the handedness of the chiral nucleus. The analog in the case of one gyroscope is the sign of rotation: +1 value for clockwise, -1 value for anticlockwise rotation. In the case of digital counter the analogy is the following: bit value 1 for counting up, 0 for counting down. Example of a logic that measures direction of counting consists of two D flip-flops, two EX-NOR gates and one triple triple AND gate connected to the same clock signal - see fig.3.

To check the direction of counting one has to compare the current with the previous state of the counter. Therefore the signals from counter are branched in two parts. First part is unchanged and keeps the present state of counter and the second part goes into the D flip-flops (triggered by the rising slope of the clock signal) that act as a delay. Signals leaving the D flip-flops in the second branch represent the previous state of the counter with respect to the signals in the first branch. The role of the EX-NOR and the triple AND gates is to compare the current and the previous state of the counter. The gates give clock signal in case of counting up and zero in case of counting down as an output signal. Using the analogy with the physical

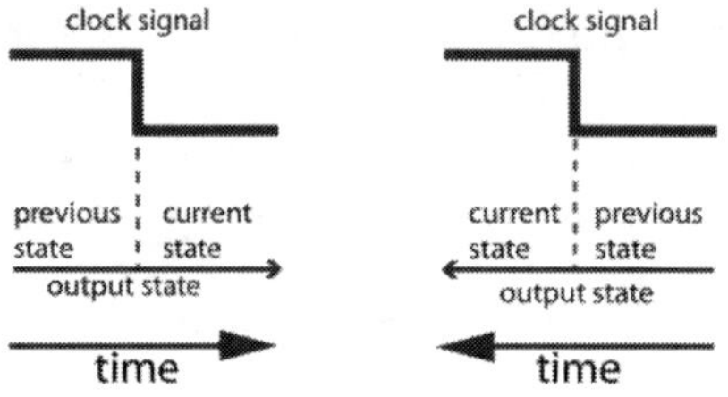

Figure 2. State change of the flip-flop unit with respect to clock signal. Left: flip-flop unit that changes state on the down-slope of the clock signal. Clock signal is plotted with thick solid line on top of the picture. Large arrow at the bottom shows the direction of time-flow. In between the change of the flip-flop state (from previous to the current state) is symbolically plotted with small arrow and dashed vertical line. Right: working of the flip-flop unit with reversed time. One can see that the triggering changes from the down-slope (left part) to the rising-slope (right part).

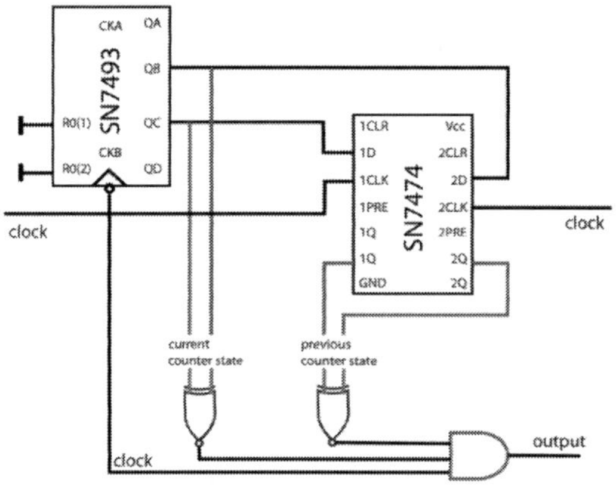

Figure 3. Rotating bit logic – a digital logic consisting of the two-bit counter and a part designed to check the direction of counting. The two bit counter (SN7493) is placed on the left. The output signals from the counter are branched in two parts. The first part remains unchanged (gray color) wile the second part goes into the SN7474 unit consisting of two flip-flops triggered by the rising-slope clock signal. The SN7474 unit works as a delay with respect to clock signal giving the previous counter state as an output. Two EX-NOR and the TRIPLE-AND gates compares the previous state of the counter with its current state determining the counting direction. It gives clock signal in case of counting up and zero in case of counting down as an output. See text fro more details.

gyroscope, the described logic represents a 'rotating bit' that gives the clock signal (1 value) for clockwise rotation and zero (0 value) for anticlockwise rotation.

In order to see the analogy of the 'rotating bit' and the physical systems one needs to verify the response of the digital logic to action of time reversal symmetry. In the chiral nucleus, time-reversal changes the sign of the handedness. To have a rotating bit that breaks spontaneously the time reversal symmetry we expect similar behavior: time reversal changes the logic output form +1 value (clock signal) to 0 value and vice-versa. The only thing the time-reversal changes in the presented logic is the behavior of the flip-flops used. As discussed above the flip-flops in the counter unit change their triggering form downslope to rising-slope of the clock signal.

Similarly the D flip-flops, used to compare the current and the previous state of the counter, change their triggering form rising-slope to down-slope of the clock signal. The remaining logic (including clock) is T-symmetric. The response of the logic output on time reversal is shown in fig.4 and in fact it corresponds to physical systems with spontaneously broken symmetry.

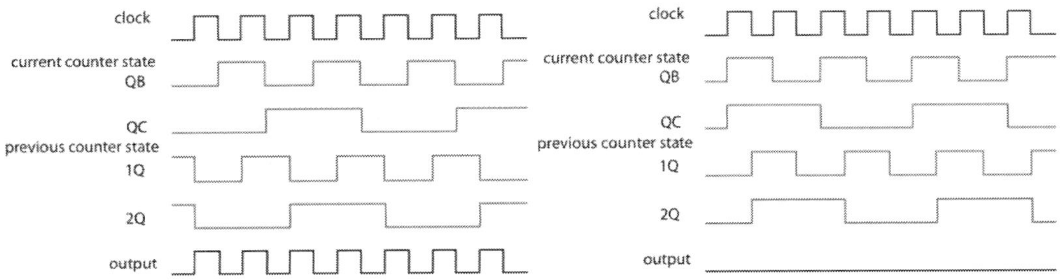

Figure 4. The signals in the rotating bit logic as a function of time. Left: the signals plotted for the counter set to counting up. The counter (SN7493) is triggered on the down-slope of the clock signal giving QB, and QC signals as an output. The SN7474 unit is triggered on the rising-slope of the signal giving 1Q and 2Q outputs. The output signal repeats the clock signal in that case. Right: The same logic in the time-reversed frame. Now the SN7493 is triggered on the rising-slope of the clock signal while the SN7474 on the down-slope giving zero as an output signal. This presents possible spontaneous time-reversal symmetry breaking of the logic.

4. Summary

Quest for time reversal symmetry breaking continues to be a subject of intensive study in various filed of the physics science. In the context of high energy physics the T-violation is a fundamental problem connected to the structure of the matter while in the low energy region the phenomenon relates to substantial possibilities of technical applications. Here the occurrence of spontaneous time reversal symmetry breaking was discussed, which is known since years in nuclear spectroscopy science. A transfer of this phenomenon to a model of the digital logic was shown as an interesting curiosity.

References

[1] Holstein B R 1989 *Weak Interactions in Nuclei* (Princeton Series in Physics) p 106
[2] Gimlett J L, Henrikson H E and Boehm F 1982 *Phys. Rev.* C**25** 1567
[3] Christenson C R 1987 *Phys. Rev. Lett.* **13** 138
[4] Auerbach N 2007 *Nucl. Phys.* A**787** 532c
[5] Grodner E *et al* 2006 *Phys. Rev. Lett.* **97** 172501
[6] Grodner E *et al* 2011 *Phys. Lett.* B**703** 46
[7] Olbratowski P, Dobaczewski J, Dudek J and Płóciennik W 2004 *Phys. Rev. Lett.* **93** 052501

In a search for a chiral symmetry in ^{102}Rh

M.S. Yavahchova[1], D. Tonev[1], N. Goutev[1], P. Petkov[1], G. de Angelis[2], R. K. Bhowmik[3], R.P. Singh[3], S. Muralithar[3], N. Madhavan[3], R. Kumar[3], M. Kumar Raju[4], J. Kaur[5], G. Mohanto[3], A. Singh[5], N. Kaur[5], R. Garg[6], A. Sukla[7], Ts. K. Marinov[1], S. Brant[8]

[1]Bulgarian Academy of Sciences, Institute for Nuclear Research and Nuclear Energy, 1784 Sofia, Bulgaria
[2]INFN, Laboratori Nazionali di Legnaro, I–35020, Legnaro, Italy
[3]Inter–University Accelerator Center, New Delhi, India
[4]Nuclear Physics Department, Andhra University, Visakhapatnam, India
[5]Department of Physics, Panjab University, Chandigarh, India
[6]Department of Physics and Astrophysics, Delhi University, New Delhi, India
[7]Department of Physics, Banaras Hindu University, Varanasi, India
[8]Department of Physics, Faculty of Science, University of Zagreb, Zagreb, Croatia

E-mail: m.yavahchova@gmail.com

Abstract. Excited states in ^{102}Rh were populated in the fusion-evaporation reaction ^{94}Zr(^{11}B, 3n)^{102}Rh at a beam energy of 36 MeV, using the INGA spectrometer at IUAC, New Delhi. The angular correlations and the electromagnetic character of some of the γ-ray transitions observed in ^{102}Rh were investigated in detail. A new candidate for a chiral twin band was identified in ^{102}Rh for the first time.

1. Introduction

Chirality is a phenomenon which is often found in nature. Examples of systems demonstrating chirality are present in chemistry, biology, high energy physics etc. A spontaneous breaking of the chiral symmetry can take place for configurations where the angular momenta of the valence protons, valence neutrons, and the core are mutually perpendicular [1]. The projections of the angular momentum vector on the axes of the intrinsic system can form a left- or a right-handed system. Since the chiral symmetry is dichotomic, its spontaneous breaking by the angular momentum vector leads to a pair of degenerate $\Delta I = 1$ rotational bands, called chiral twin bands. Pairs of bands, presumably due to the breaking of the chiral symmetry in triaxial nuclei, have been recently found in the mass regions $A \sim 130$ [2, 3], $A \sim 105$ [4, 5, 6, 7], $A \sim 195$ [8] and $A \sim 80$ [9]. However, only in few cases (^{126}Cs [10], ^{128}Cs [11]), the systematic properties [12] of the chiral bands, which originate from the underlying symmetry, were confirmed including the transition from chiral vibrations to static chirality in ^{135}Nd [3]. Thus, the yrast and side bands should be nearly degenerate. In many cases, the energy degeneracy of the chiral candidate bands was almost observed but the transition probabilities are different, like in the case of ^{134}Pr [13, 14]. The main goal of the present work was to check for the existence of chirality in the mass region A $\sim$ 100. According to the works [15] and [16], the nucleus of ^{102}Rh is one of the candidates to express multiple chirality. The goal of this work is to check for the existence of chirality in the mass region A $\sim$ 100.

2. Experiment

Excited states in ^{102}Rh were populated through the reaction ^{94}Zr(^{11}B,3n)^{102}Rh. The beam of ^{11}B with an energy of 36 MeV was provided by the 15-UD Pelletron accelerator at the Inter University Accelerator Centre (IUAC) in New Delhi, India. The target consisted of 0.9 mg/cm^2 ^{94}Zr, enriched to 96.5%, evaporated onto a 8 mg/cm^2 gold backing. The recoils were leaving the target with a mean velocity, v of about 0.9% of the velocity of light, c. The de-exciting gamma rays were registered with the Indian National Gamma Array (INGA), whose 15 Clover detectors were accommodated in 4π geometry [17]. Gain matching and efficiency calibration of the Ge detectors were performed using ^{152}Eu and ^{133}Ba radioactive sources before sorting the data in matrices and cubes.

3. Data analysis and results

To investigate the level scheme and electromagnetic properties of the transitions of interest in ^{102}Rh, we have performed four different types of data analyses as follows. The ordering of transitions in the level scheme was determined according to the γ-ray relative intensities, γ-γ coincidence relationships, and γ-ray energy sums. The character and multipolarity of the transitions was deduced from the investigation of linear polarization and angular correlations measurements, respectively. For this purpose, the clover detectors positioned close to 90° were used as a composite Compton polarimeter [18]. The efficiency with which the Compton scattering events were registered was determined with the ^{152}Eu and ^{133}Ba sources that emit γ rays of natural polarization isotropically. We have determined the spins, the levels and the multipolarity of the corresponding transitions using angular correlation analysis of data taken with the INGA spectrometer with the computer code CORLEONE [19]. In figure 1 the negative lines correspond to transitions of predominantly magnetic character while the positive lines correspond to transitions of predominantly electric character. The negative character of the lines 304, 306 and 363 in the spectrum, proves that the corresponding transitions are of predominantly magnetic character.

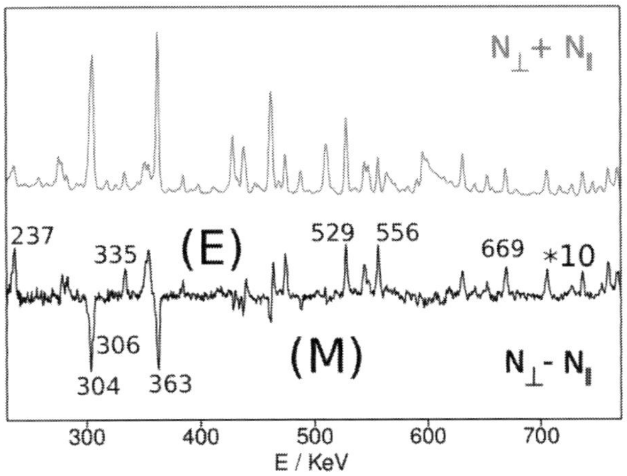

Figure 1. Coincidence spectra registered by the perpendicular and parallel arms of the compton polarimeter. Transitions with energies of 304, 306 and 363 keV have predominantly magnetic character.

The lifetimes were derived using the Doppler-shift attenuation methods. The analysis was carried out according to the procedure outlined in [20], where details about the Monte-Carlo simulation of the slowing down process, determination of stopping powers and fitting of line shapes can be found.

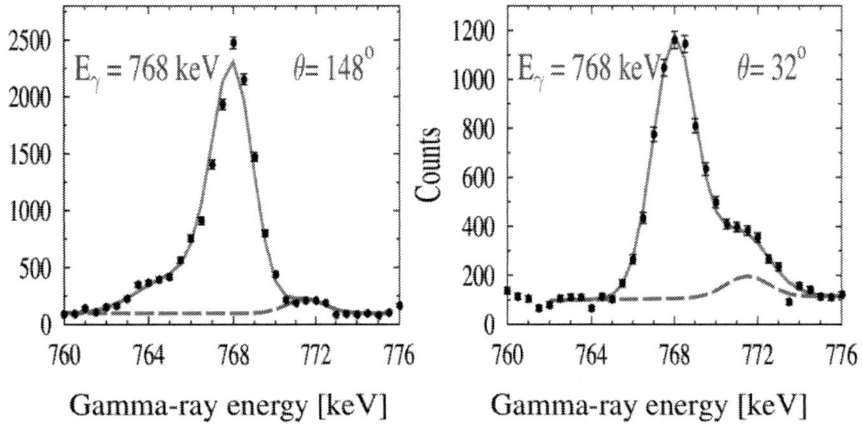

Figure 2. Fitted line-shape of transitions observed at 32° and 148°.

The spectra for every gated ring were summed up to increase the statistics. For each level, lifetimes were derived independently at the four rings where appreciable Doppler-shifts are observed. In figure 2 is shown an examples for such analysis and the Doppler-shifts are nicely seen. The gate is set from below of the transition of interest.

To describe the reduced transition probabilities $B(\sigma\lambda)$ and the level scheme in ^{102}Rh we performed two quasiparticles plus triaxial rotor (TQPTR) calculations within the approach given in [21]. Figure 3 illustrates the experimental level scheme of ^{102}Rh compared to TQPTR calculations.

The reduced transition probabilities $B(\sigma\lambda)$ deduced from the lifetime data are presented in Ref.[22]. The comparison between the experimental and calculated B(M1) transition strengths leads to the conclusion that the TQPTR calculations reproduce roughly the data in Band 1 and are consistent with the transition strength in Band 2 at spin 11 $\hbar$. The absence of an appreciable staggering of the data in Band 1 indicates that the expectations for the observation of a static chirality in ^{102}Rh are not realized [22]. If the chirality in ^{102}Rh exists it has mainly a dynamical character.

4. Summary and conclusions

For the investigation to the level-scheme of ^{102}Rh we have performed an experiment at the IUAC in New Delhi using the INGA spectrometer. To construct the level scheme of ^{102}Rh we used γ-γ coincidence data, relative γ-rays intensities and energy sums. For the first time we applied angular correlation analysis and polarization measurements for INGA spectrometer data. The results obtained from angular correlation and linear polarization measurements were essential for the spin and parity assignments. As a result from our analysis 4 new exited states in ^{102}Rh were determined for the first time. In addition 8 new lifetimes have been determined.

5. Acknowledgments

This research has been supported by Bulgarian Science Fund under contract DFNI-E 01/2 and by a NUPNET - NEDENSAA project funded by the Bulgarian Ministry of Education and Science. Thanks to the Pelletron Staff of IUAC, New Delhi for their perfect work for our experiment.

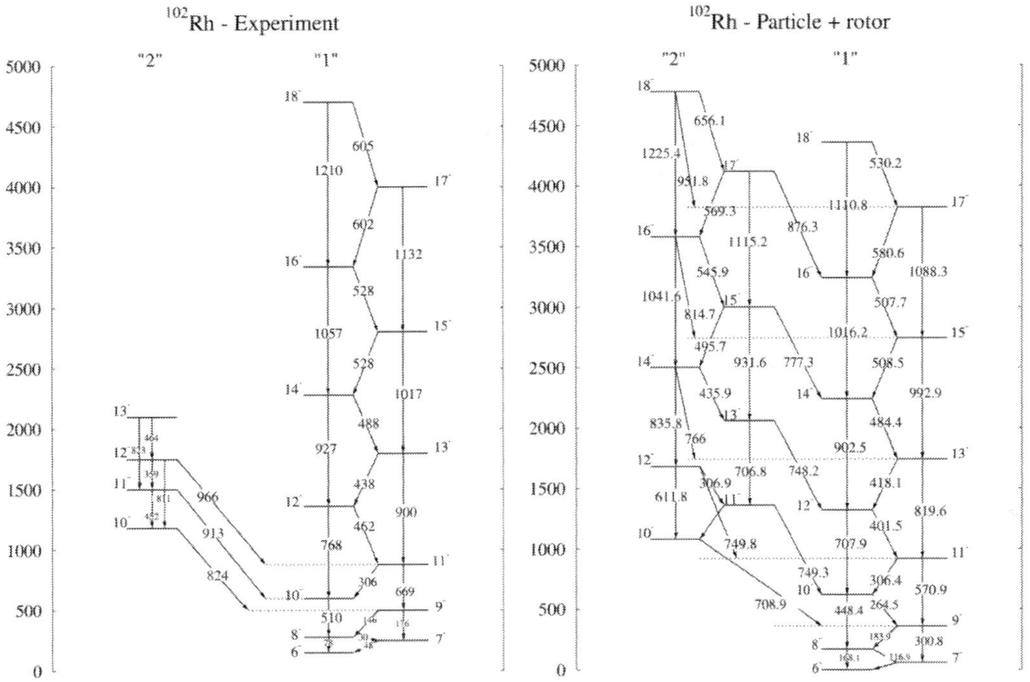

Figure 3. Comparison of the experimental and calculated within the TQPTR level schemes of ^{102}Rh.

References

[1] S Frauendorf, Jie Meng 1997 *Nucl. Phys.* A **617** 131
[2] K Starosta *et al.* 2001 *Phys. Rev. Lett.* **86** 971
[3] S Mukhopadhyay *et al.*, 2007*Phys. Rev. Lett.* **99** 172501
[4] J Timár *et al.* 2006 *Phys. Rev.* C **73** 011301
[5] C Vaman, D B Fossan, T Koike, K Starosta, 2004 *Phys. Rev. Lett.* **92** 032501
[6] T Suzuki *et al.* 2008 *Phys. Rev.* C **78** 031302
[7] P Joshi *et al.* 2004 *Phys.Lett.* B **595** 135
[8] E A Lawrie *et al.* 2008 *Phys. Rev.* C **78** 021305
[9] S Y Wang *et al.* 2011 *Phys. Lett.* B **703** 40
[10] E Grodner *et al.* 2011 *Phys. Lett.* B **703** 46
[11] E Grodner *et al.* 2006 *Phys. Rev. Lett.* **97** 172501
[12] S Brant, D Tonev, G De Angelis, A Ventura 2008 *Phys. Rev.* C **78** 034301
[13] D Tonev *et al.* 2006 *Phys. Rev. Lett.* **96** 052501
[14] D Tonev *et al.* 2007 *Phys. Rev.* C **76** 044313
[15] J Meng *et al.* 2006 *Phys. Rev.* C **73** 037303
[16] J Peng *et al.* 2008 *Phys. Rev.* C **77** 024309
[17] S Muralithar *et al.* 2008 *Proc. DAE-BRNS Symp. on Nucl. Phys.***53** 221
[18] N Goutev *et al.* 2012 *J. Phys.:Conf. Ser.* **366** 012021
[19] I Wiedenöver *et al.* 2008 *Proc. DAE-BRNS Symp. on Nucl. Phys.* **53** 221
[20] P Petkov *et al.* 1998 *Nucl. Phys.* A **640** 293
[21] I Ragnarsson and P Semmes 1988 *Hyp. Int.* **43** 425
[22] D Tonev 2014 *Phys. Rev. Lett.* **112** 052501

XX International School on Nuclear Physics, Neutron Physics and Applications (Varna2013) IOP Publishing
Journal of Physics: Conference Series **533** (2014) 012042 doi:10.1088/1742-6596/533/1/012042

Some recent experimental results related to nuclear chirality

J Timár[1], I Kuti[1], D Sohler[1], K Starosta[2], T Koike[3] and E S Paul[4]

[1] Institute for Nuclear Research, Hungarian academy of Sciences, H-4026 Debrecen, Hungary
[2] Department of Chemistry, Simon Fraser University, Burnaby, British Columbia V5A 1S6, Canada
[3] Graduate School of Science, Tohoku University, Sendai, 980-8578, Japan
[4] Oliver Lodge Laboratory, Department of Physics, University of Liverpool, Liverpool L69 7ZE, UK

E-mail: `timar@atomki.mta.hu`

Abstract. Detailed band structures of three chiral-candidate nuclei, ^{134}Pr, ^{132}La and ^{103}Rh have been studied. The aim of the study was twofold. First, to try to explore the reasons behind the contradiction between the theoretically predicted chirality in these nuclei and the recently observed fingerprints that suggest non-chiral interpretation for the previous chiral candidate band doublets. Second, to search for multiple chiral bands of different types in these nuclei. In ^{134}Pr a new $\pi h_{11/2}\nu h_{11/2}$ band has been observed besides the previously known chiral-candidate $\pi h_{11/2}\nu h_{11/2}$ doublet. This new band and the yrare $\pi h_{11/2}\nu h_{11/2}$ band show the expected features of a chiral doublet structure. This fact combined with the observed similarity between the band structures of ^{134}Pr and ^{132}La suggests that chirality might exist in these nuclei. The detailed study of the ^{103}Rh band structure resulted in the observation of two new chiral-doublet looking structures besides the previously known one. This is indicative of possible existence of multiple chiral doublet structure in this nucleus.

1. Introduction

Chiral-candidate doublet bands in the triaxial A≈105 and A≈130 nuclei attracted great interest in the past decade. Many such doublet structures corresponding to particular configurations have been identified in these mass regions. These doublets were subjects of numerous theoretical and experimental studies, which either confirmed or contradicted with the chiral interpretation of the actual bands. The performed studies have focused mainly on the observed chiral-candidate doublet band structures while a major part of the rotational bands has remained unexplored. However, recent theoretical studies indicated that study of the whole band structure in a nucleus can be important from chirality point of view.

It has been shown in Ref. [1] that the expected fingerprints in non-ideal chiral cases can change considerably. Thus, it is not enough to compare the observed properties with the fingerprints derived for ideal cases. Realistic calculations are needed to identify chirality, especially in cases of controversial observed chiral fingerprints. However, realistic calculations require knowledge on the whole band structure, not only on the particular doublet.

Also, recent theoretical studies have shown that chirality in nuclei can also be manifested in features that are beyond the properties of the firstly observed nearly-degenerate doublet bands.

It has been proposed that more than one configuration in a nucleus can have chiral properties [2], or more than one pair of degenerate bands corresponding to the same configuration can exist in case of chirality [3, 4].

Motivated by these predictions, we have studied the detailed band structures of several chiral-candidate nuclei in the A≈105 and A≈130 mass regions, of which we discuss here the cases of ^{134}Pr, ^{132}La and ^{103}Rh.

2. Possible chirality in ^{134}Pr and ^{132}La

^{134}Pr and ^{132}La were the first nuclei in which chiral candidate doublet band structures have been observed [5, 6]. These observations were in agreement with theoretical results that confirmed the existence of chirality in these nuclei. However, results of subsequent lifetime experiments for these doublet bands contradicted with the chiral interpretation [7, 8].

We have studied the detailed band structures of these nuclei based on high-statistics experiments. Excited states of ^{134}Pr were populated following the ^{116}Cd(^{23}Na,5n) reaction at a beam energy of 115 MeV and the gamma transitions were detected using the GAMMASPHERE spectrometer. The bands of ^{132}La were excited in the ^{100}Mo(^{36}S,p3n) reaction at 160 MeV beam energy, and studied with the EUROBALL IV spectrometer. In both cases the off-line analysis of coincident γ rays was carried out using the RADWARE software package and an angular-correlation analysis, based on the DCO (directional correlation of oriented states) was performed for γ-ray multipolarity assignments. In the case of the EUROBALL experiment, the clover detectors, placed at 90 degrees relative to the beam direction, also allowed of linear polarization analysis.

As a result, we have observed many new rotational bands in ^{134}Pr, assigned single-particle configurations to them, and identified one of the previously known two long-halflife isomers as the ground state [9]. One of the new bands (band 3 in Fig. 1) has $\pi h_{11/2}\nu h_{11/2}$ configuration and is linked by many transitions to the previously known $\pi h_{11/2}\nu h_{11/2}$ doublet (bands 1 and

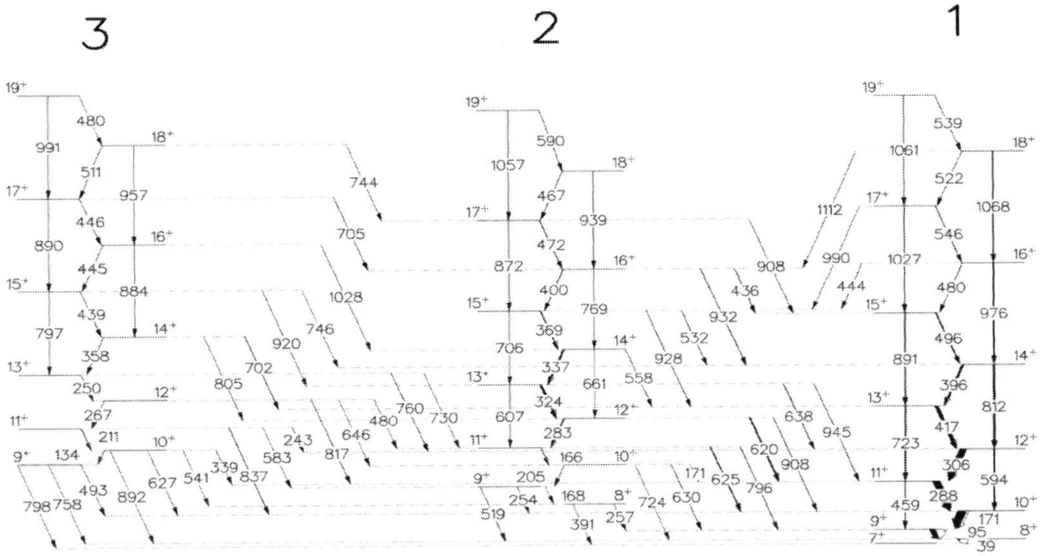

Figure 1. Partial level scheme of ^{134}Pr, showing the three $\pi h_{11/2}\nu h_{11/2}$ bands up to spin 19.

Journal of Physics: Conference Series **533** (2014) 012042 doi:10.1088/1742-6596/533/1/012042

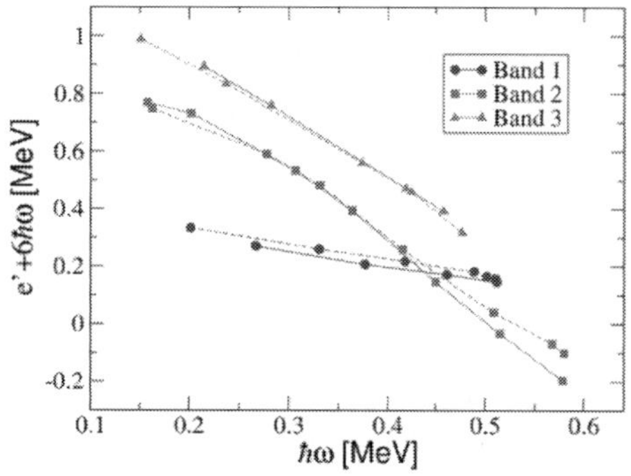

Figure 2. Routhian+$6\hbar\omega$ for the three $\pi h_{11/2}\nu h_{11/2}$ bands in ^{134}Pr as a function of the rotational frequency.

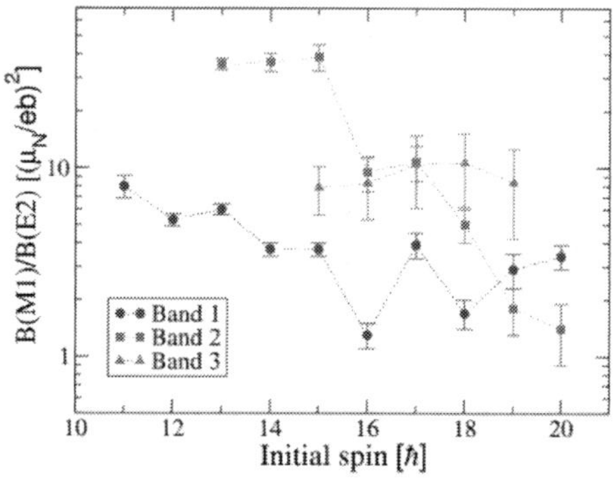

Figure 3. B(M1)/B(E2) values for the three $\pi h_{11/2}\nu h_{11/2}$ bands in ^{134}Pr as a function of the spin.

2 in Fig. 1). This new band has an approximately constant ~200 keV energy separation from band 2 and their Routhians have approximately the same slopes as a function of the rotational frequency, thus their single particle alignments are approximately equal (Fig. 2). We could not derive lifetime values for the levels of this new band; however it has been observed that the B(M1,$I \to I-1$)/B(E2,$I \to I-2$) ratios of bands 2 and 3 are close to each other in the 16 - 18 spin region (Fig. 3). These observations suggest that bands 2 and 3 may be chiral partner bands in a geometry corresponding to a chiral vibration. Therefore, the presence of chirality in this nucleus may not be excluded in spite of the lifetime results. We note that the occurrence of more than two bands close to each other corresponding to the chiral candidate configuration is predicted by calculations in this mass region [3]. In these predictions the first two and the second two bands are chiral partners of each other. However, the observed bands in ^{134}Pr are different from these predicted band structures because here the second and third bands show the chiral doublet features. In our case a possible scenario is that two shapes coexist in the $\pi h_{11/2}\nu h_{11/2}$ configuration, as it is discussed in Ref. [10], one of which is less triaxial while the other is close to the maximal triaxiality. In this case band 1 corresponds to the less triaxial shape, and is not chiral, while bands 2 and 3 correspond to the triaxial shape, and they are chiral.

As a byproduct of the lifetime experiment published in Refs. [11, 8], a new band has also been observed in ^{132}La, which is linked by many transitions to the yrast $\pi h_{11/2}\nu h_{11/2}$ band. Although the spins and the parity of the levels of this band could not be determined unambiguously, the authors tentatively assigned positive parity and spins that suggested a small energy separation between the new band and the previously known candidate chiral doublet. Moreover, the observed B(E2) values in the new band were found to be close to that of the yrast $\pi h_{11/2}\nu h_{11/2}$ band. Therefore, the scenario of the two bands being members of a chiral vibrational doublet, similarly to the case of ^{134}Pr, seemed to be possible. Our aim was to determine the spins and parities in order to confirm or reject this possibility. As a result of our studies, we built a corrected version of the new band and determined its parity and the spins of its levels [12]. The parity turned out to be negative, and the spins indicated a four-quasiparticle configuration. Thus, the possibility that this band is the chiral partner of the yrast band has been rejected. On the other hand, the results of our detailed study on the band structures of ^{134}Pr and ^{132}La revealed a strong similarity between both the low-lying states and the high-spin bands of the two nuclei. Based on this similarity, it is probable that ^{132}La also has more than two $\pi h_{11/2}\nu h_{11/2}$ bands close to each other in energy, and two of them may show the features of a chiral doublet. Although the observed hints for chirality are needed to be confirmed by higher-statistics experiments in case of ^{132}La, as well as by lifetime measurements and realistic calculations for both nuclei, the existence of chirality may not be excluded in ^{134}Pr and ^{132}La.

3. Possible multiple chiral doublet bands in ^{103}Rh

A chiral candidate doublet band structure, based on the three-quasiparticle $\pi g_{9/2}\nu(h_{11/2})^2$ configuration, has been observed previously in ^{103}Rh and ^{105}Rh [13]. In ^{105}Rh another chiral candidate doublet structure has been reported corresponding to the $\pi g_{9/2}\nu h_{11/2}(g_{7/2}, d_{5/2})$ configuration [14]. In agreement with these observations, diabatic and configuration-fixed constrained triaxial relativistic mean-field calculations predicted the existence of a new phenomenon, the multiple chiral doublet (MχD) in ^{105}Rh [15]. These calculations predicted two chiral doublet band structures with the $\pi g_{9/2}\nu h_{11/2}d_{5/2}$ and $\pi g_{9/2}\nu(h_{11/2})^2$ configurations.

We have studied the detailed band structure of the neighbouring ^{103}Rh to search for MχD in this nucleus, too. High-spin states in ^{103}Rh were populated using the ^{96}Zr(^{11}B,4n) reaction at 40 MeV beam energy. The emitted γ-rays were detected by the GAMMASPHERE spectrometer. A total of $\sim 9\times 10^8$ quadruple- and higher-fold events were accumulated using a thin target, and sorted off-line into 2-d, 3-d, and 4-d histograms.

As a result of the analysis, we have observed several new bands. According to the preliminary configuration assignments, four of the observed negative-parity bands have the $\pi g_{9/2}\nu h_{11/2}(g_{7/2}, d_{5/2})$ configuration and form two pairs which show the characteristics of chiral doublets. There is a very small energy difference between the members of the doublets (less then 100 keV), and their B(M1)/B(E2) ratios are also close to each other. This is the configuration for which chirality has been predicted in ^{105}Rh. If the calculations confirm the chirality for these configurations also in ^{103}Rh, then the existence of three chiral doublet bands will be probable including the previously known chiral candidate band in this nucleus.

Acknowledgments

The authors are indebted to the participants of the GAMMASPHERE and EUROBALL IV experiments. This work was supported in part by the Hungarian Scientic Research Fund, OTKA (Contract No. K100835), the Natural Sciences and Engineering Research Council of Canada under Contract No. SAPIN/371656-2010 and the UK Engineering and Physical Sciences Research Council.

References

[1] Lawrie E A et al. 2010 *Phys. Lett.* B **689** 66; Shirinda O et al. 2012 *Eur. Phys. J.* A **48** 118
[2] Meng J et al 2006 *Phys. Rev.* C **73** 037303; Peng J et al. 2008 *Phys. Rev.* C **77** 024309
[3] Chen Q B et al. 2010 *Phys. Rev.* C **82** 067302; Droste Ch et al. 2009 *Eur. Phys. J.* A **42** 79
[4] Hamamoto Ikuko 2013 *Phys. Rev.* C **88** 024327
[5] Petrache C M et al. 1996 *Nucl. Phys.* A **597** 106; Frauendorf S and Meng Jie 1997 *Nucl. Phys.* A **617** 131
[6] Starosta K et al. 2001 *Phys. Rev. Lett.* **86** 971
[7] Tonev D et al. 2006 *Phys. Rev. Lett.* **96** 052501
[8] Srebrny J et al. 2005 *Acta Physica Polonica* B **36** 1063
[9] Timár J et al. 2011 *Phys. Rev.* C **84** 044302
[10] Petrache C M et al. 2006 *Phys. Rev. Lett.* **96** 112502
[11] Grodner E et al. 2004 *International Journal of Modern Physics* E **13** 243
[12] Kuti I et al. 2013 *Phys. Rev.* C **87** 044323
[13] Timár J et al. 2006 *Phys. Rev.* C **73** 011301(R); Timár J et al. 2004 *Phys. Lett.* B **598** 178
[14] Alcántara-Núñez J A et al. 2004 *Phys. Rev.* C **69** 024317
[15] Li J, Zhang S Q and Meng J 2011 *Phys. Rev.* C **83** 037301

Journal of Physics: Conference Series **533** (2014) 012043 doi:10.1088/1742-6596/533/1/012043

Isomeric states in neutron-rich ^{129}In and the $\pi g_{9/2}^{-1} \otimes \nu h_{11/2}^{-1}$ multiplet

J. Taprogge[1,2], **A. Jungclaus**[1] **and G. Simpson**[3]

[1] Instituto de Estructura de la Materia, CSIC, E-28006 Madrid, Spain
[2] Departamento de Física Teórica, Universidad Autónoma de Madrid, E-28049 Madrid, Spain
[3] LPSC Grenoble, F-38026 Grenoble, France

E-mail: jan.taprogge@csic.es

Abstract. Within the RISING stopped beam campaign the neutron-rich indium isotopes with masses A=125-130 have been studied using the method of isomer spectroscopy. The decays of several isomeric states have been observed and here we compare our results for ^{129}In to previous measurements for this nucleus. The isomeric states were populated in the fragmentation of a ^{136}Xe beam at 750 MeV/u and in the relativistic fission of a ^{238}U beam at 650 MeV/u at the accelerator facility GSI (Darmstadt, Germany).

1. Introduction

In the last decade a large number of theoretical studies and experiments have been performed to investigate the region around the doubly magic ^{132}Sn. This region is of particular interest on one hand for testing the interactions used in shell model calculations and on the other, because its relevance for nuclear astrophysics, especially for nucleosynthesis calculations.

Results are presented for isomeric states in neutron-rich indium nuclei. Information on isomeric states in neutron-rich Ag, Cd, In and Sn isotopes, deduced from the same dataset, have already been presented in Refs. [1, 2, 3, 4, 5, 6, 7], including the first observation of a high-spin, core-excited isomer in ^{131}In. Here we focus on the results obtained for ^{129}In.

2. Experimental setup

The neutron-rich indium isotopes were studied with the RISING stopped beam setup employing the method of isomer spectroscopy. Two different reactions were used to produce these neutron-rich nuclei, namely the fragmentation of a ^{136}Xe primary beam at an energy of 750 MeV/u and relativistic fission of a ^{238}U beam at 650 MeV/u. The primary beams were provided by the accelerator facility at the Gesellschaft für Schwerionenforschung (Darmstadt, Germany) consisting of the linear accelerator LINAC and the heavy ion synchrotron SIS. Beryllium production targets with thicknesses of 4 and 1 g/cm^2, respectively, were used at the entrance of the fragment separator FRS.

The reaction products were separated and the ions of interest selected and identified on an event-by-event basis using the information from the different beam detectors, placed at the focal planes of the separator, and the $B\rho - \Delta E - B\rho$ method. The mass-to-charge ratio A/Z was determined from the measured time-of-flight through the second half of the fragment separator

Journal of Physics: Conference Series **533** (2014) 012043 doi:10.1088/1742-6596/533/1/012043

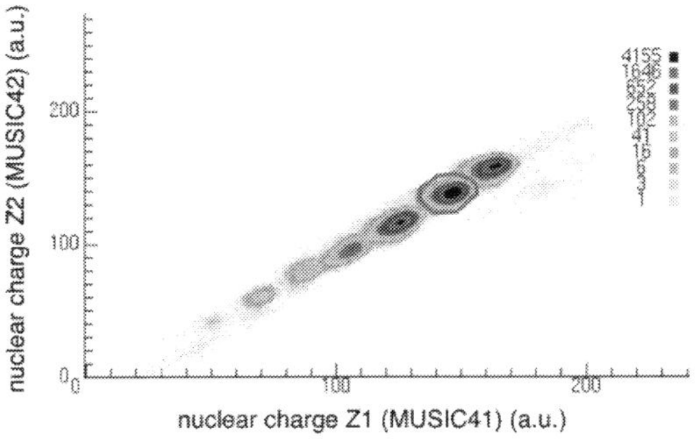

Figure 1. Calculated nuclear charge Z1 from the energy loss in MUSIC41 in comparison to Z2 from MUSIC42. The gate used for the offline analysis to select the Indium isotopes is shown in red.

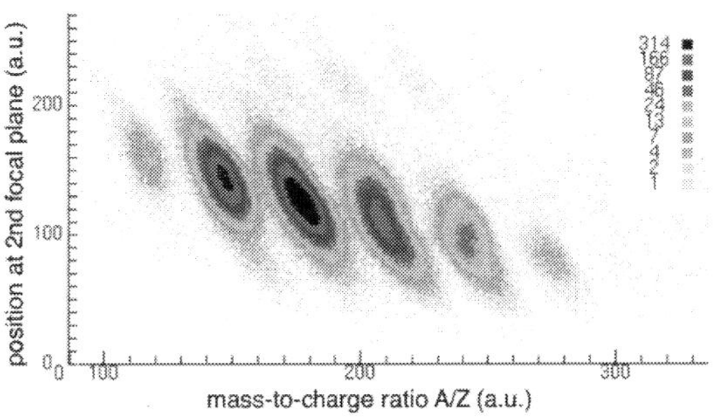

Figure 2. Measured position x2 at the second focal plane as a function of the mass-to-charge ratio A/Z gated on $Z = 49$ (Indium), as shown in Figure 1.

and several position measurements using plastic detectors and multi-wire proportional chambers (MWPC). The nuclear charge Z is determined from two independent energy-loss measurements performed with ionisation chambers called MUSIC. Figure 1 shows the nuclear charge measured in the first ionisation chamber, MUSIC41, plotted against the same quantity determined from the energy-loss in the second, MUSIC42. The different nuclear charges are clearly separated and a gate on Z=49 can be set. Figure 2 shows the different A/Z values, gated on Z=49, against the measured horizontal position x at the second focal plane to separate the different masses A and select the isotopes of interest in the offline analysis.

The ions which pass through the FRS are then stopped in a passive stopper at the final focal plane of the FRS which is surrounded by 15 former EUROBALL cluster detectors, each consisting of 7 encapsulated Ge-crystals to measure the γ-rays following isomeric decays. The Ge signals are split up into two separate electronic branches, one equipped with digital-gamma-finders (DGF) which processes the signal to obtain energy and time information, but have a limited time granularity of 25ns/bin. The other electronic branch consists of a conventional analog timing electronic chain with two different time-to-digital converters TDC with ranges of 1 and 800 μs and granularities of 0.31 ns/channel and 0.73 ns/channel, respectively.

3. Experimental results

In this section, the results obtained for the nucleus ^{129}In are presented. In ^{129}In, several isomeric states were known previous to this work. Fogelberg *et al.* [8] identified two ms isomeric states, which are connected by a 281-keV transition, and assigned spin/parity of $(29/2^+)$ and $(23/2^-)$ to the higher and lower isomeric state, respectively. Those isomeric states can not be observed in our experiment due to their long half-lifes of 110 and 700 ms, respectively. The presence of a μs isomer was first reported by Genevey *et al.* [9] who observed four γ-rays with energies of 334-, 359-, 995- and 1354-keV following the decay of this isomeric state to the ground state. Its half-life was measured in three independent experiments to 8.5(5) μs [9], 11(1) μs [10, 11] and 2.2(3) μs [12].

All four γ-rays reported by Genevey *et al.* [9] are observed in the delayed spectrum obtained in our experiment as shown in Figure 3 (time range 200 ns to 25 μs after implantation). The 334-, 359- and 995-keV transitions are observed in mutual prompt coincidence and in addition the 334-keV transition is found to be in prompt coincidence with the 1354 keV γ-ray. As an example the prompt coincidence spectrum for the 334-keV transition is presented in Figure 4.

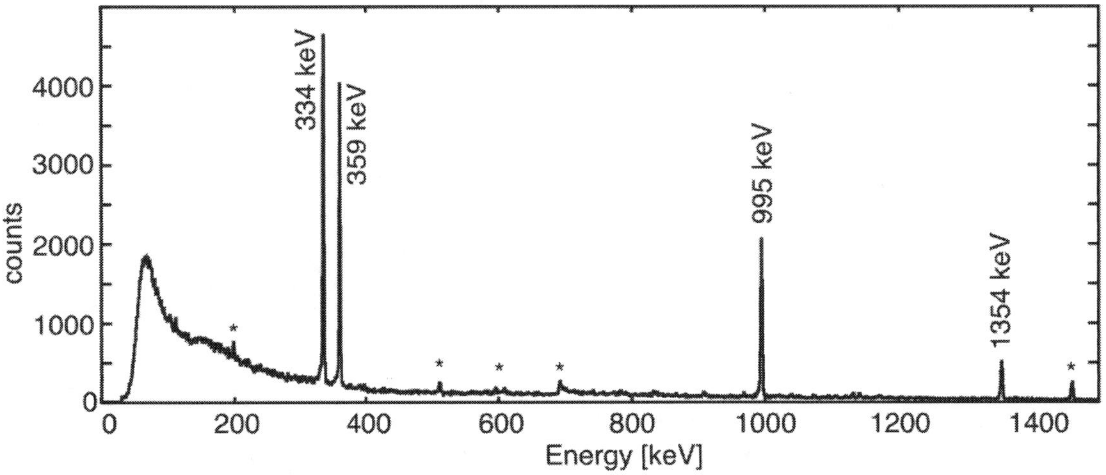

Figure 3. γ-ray spectrum in delayed coincidence with the implanted ^{129}In ions. The full time range of the DGF (25 μs) is used, excluding however the prompt flash after the implantation. The transitions marked with a star have been assigned to background radiation.

Note that the energy sum of the 359- and 995-keV transitions equals 1354-keV, the energy of the fourth transition, thus confirming the level scheme proposed in [9]. In that reference a spin and parity assignment of $17/2^-$ was made for the isomeric state on the basis of a conversion electron measurement for the 334-keV transition.

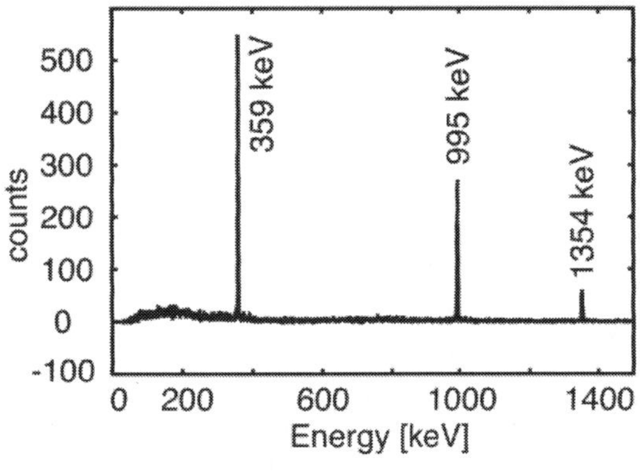

Figure 4. Prompt coincidence spectrum gated on the 334-keV transition in ^{129}In.

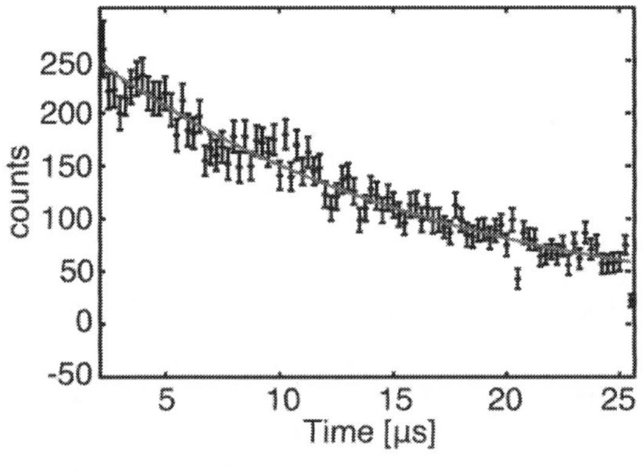

Figure 5. DGF time distribution gated on the 334-keV transition in ^{129}In and the least-square fit to the data (red line) to determine the half-life, $T_{1/2}$.

The half-life of the isomeric state was determined by least-square fits of the time distributions obtained for the four γrays. As an example the time distribution of the 334-keV transition is shown in Figure 5 . A mean half-life of $T_{1/2}$=11.2(2) μs was calculated from the individual values. This value agrees perfectly with the value published in [10, 11] and is only slightly higher as compared to the half-life reported in [9]. In contrast, no agreement is found with the value obtained in [12]. All measured half-lifes and the relative intensities of the four γrays are summarized in Table 1. The relative intensities are in good agreement with the ones quoted in [9].

Table 1. Summary of the γ-ray energies E_γ, half-lifes $T_{1/2}$, obtained from the individual fits, relative intensities I_γ and coincidence relations observed in this work for the γ-rays emitted in the decay of the $17/2^-$ isomer in ^{129}In. The relative intensities $I_{\gamma,lit}$ from [9] are listed for comparison.

E_γ [keV]	$T_{1/2}$ [μs]	I_γ [%]	$I_{\gamma,lit}$ [%]	obs. coinci.
333.6	11.5(3)	100(2)	100	358.9, 995.0, 1354.0
358.9	11.0(3)	82(3)	83	333.6, 995.0
995.0	11.3(4)	86(3)	82	333.6, 358.9
1354.0	10.7(9)	25(2)	28	333.6

4. Conclusion

The RISING stopped beam setup has been used to observe the decay of isomeric states in neutron-rich indium isotopes. Here we presented the re-measurement of the half-life of the $17/2^-$ isomeric state in ^{129}In. Our results are in good agreement with the previously published half-life values, coincidence relations and relative intensities.

Acknowledgments

We acknowledge financial support from the Spanish Ministerio de Ciencia e Innovación under Contracts No. FPA2009-13377-C02 and No. FPA2011-29854-C04.

References

[1] Górska, M. *et al.*, Physics Letters B **672**, (2009) 313-316.
[2] Lalkovski, S. *et al.*, Phys. Rev. C **87**, 034308 (2013).
[3] Naqvi, F., Phd thesis, Universität zu Köln, 2011.
[4] Naqvi, F. *et al.*, Phys. Rev. C **82**, 034323 (2010).
[5] Cáceres, L. *et al.*, Phys. Rev. C **79**, 011301(R) (2009).
[6] Jungclaus, A. *et al.*, Phys. Rev. Lett. **99**, 132501 (2007).
[7] Pietri, S. *et al.*, Phys. Rev. C **83**, 044328 (2011).
[8] Fogelberg, B. *et al.*, AIP Conf. Proc. No. 447, 191 (1998)
[9] Genevey, J. *et al.*, Phys. Rev. C **67**, 054312 (2003).
[10] Hellström, M. *et al.*, in Proceedings of the Third International Conference on Fission and Properties of the Neutron-Rich Nuclei, edited by Hamilton, J., Ramayya, A. and Carter, H. (World Scientific, Singapore, 2003), p. 22.
[11] Hellström, M. *et al.*, in Nuclear Structure and Dynamics at the Limits: Proceedings of the International Workshop XXXI on Gross Properties of Nuclei and Nuclear Excitations: Hirschegg, Austria, 2003, edited by Feldmeier, H. (Imprint Darmstadt: Gesellschaft fur Schwerionenforschung (GSI), 2003), p. 72.
[12] Tomlin, B., PhD thesis, Michigan State University, 2006.

Journal of Physics: Conference Series **533** (2014) 012044 doi:10.1088/1742-6596/533/1/012044

β-decay systematic trends of nuclei beyond ^{208}Pb

A.I. Morales[1,2] **and G. Benzoni**[1]
For the RISING collaboration
[1]Dipartimento di Fisica dell'Università degli Studi di Milano, Milano, Italy
[2]Istituto Nazionale di Fisica Nucleare, Sezione di Milano, Milano, Italy
E-mail: anaisabel.morales@mi.infn.it

Abstract. The β decay of several nuclei beyond ^{208}Pb has been investigated using the FRS+RISING set up at the GSI accelerator facilities, in Darmstadt. This contribution provides a compilation of the β-decay information obtained for neutron-rich Tl, Pb and Bi isotopes. On the basis of the new data, a general overview on the β-decay systematics in this mass region is provided. The study reveals the impact of the microscopic structure of these nuclei on the descriptive power of standard theoretical models. These are used to extrapolate gross properties towards the neutron-drip line, where the nuclei that concur in the formation of the third r-process abundance peak reside. Deficiencies in their predictive power are discussed.

1. Introduction

Since the beginnings of the 20th century, the β decay has been a powerful tool to obtain structural information on new exotic nuclei. This mechanism has also attracted great attention in astrophysics, since the abundance patterns and time scales of the nucleosynthesis processes leading to the formation of heavy nuclei, namely the s- and r-processes [1], are determined by gross β-decay properties, such as half-lives, Q_β values, and delayed emission probabilities. z In this work special attention is payed to the third r-process abundance peak. Since direct experimental access to r-process nuclei –lying near the neutron-drip line– is still not possible, r-process calculations must rely on nuclear theories. While their descriptive power near stability has already been tested and confirmed, their predictive power in nuclei at extreme conditions of isospin might be depleted by microscopic structural effects, such as deformation or shell quenching. Therefore information on more moderately neutron-rich nuclei is of extreme importance to probe their extrapolation capabilities.

2. Experimental technique and data analysis

The neutron-rich $^{211--213}$Tl, ^{215}Pb and $^{215--219}$Bi were produced in inverse kinematics, exploiting the in-flight fragmentation of a ^{238}U primary beam colliding with a thick Be target. The 1 GeV/nucleon beam was delivered by the UNILAC-SIS accelerator system at GSI. The heavy nuclei were sent to the FRagment Separator (FRS), a high-resolution magnetic spectrometer that separates the heavy nuclei and their atomic charge states following the Bρ-ΔE-Bρ method [2]. At the exit of the spectrometer a decay station, consisting of an active stopper detection system surrounded by the RISING γ array in its "Stopped Beam Configuration" [3], was set up. The energy, time, and position of all the nuclear species implanted in the active

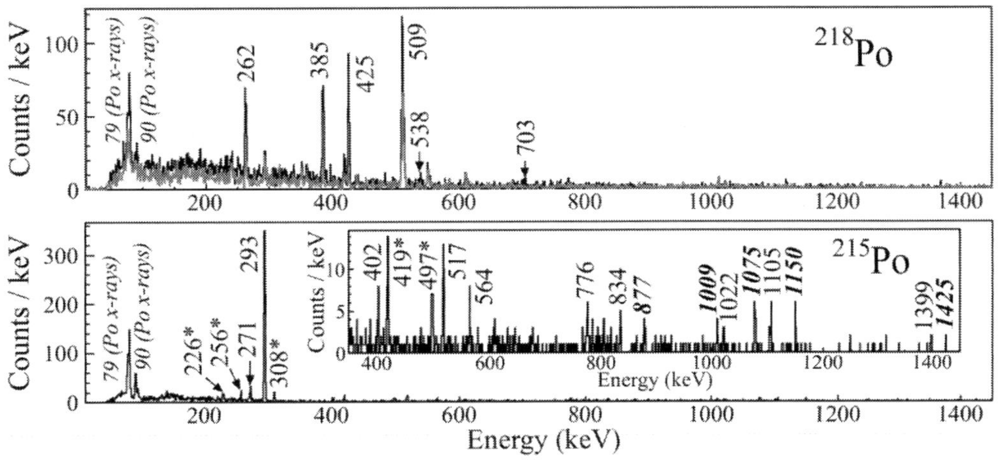

Figure 1. *(Color online) β-delayed γ spectra of the decays* $^{218}Bi \rightarrow ^{218}Po$ *(upper panel) and* $^{215}Bi \rightarrow ^{215}Po$ *(lower panel). The upper panel includes the ion-β correlations where the electron was detected in the implantation pixel. The background is shown in red. The ion-β correlations included in the lower panel, instead, are labeled with the α particle emitted by* ^{215}Po.

stopper were recorded on an event-by-event basis. This information was also collected for α and β particles, and for coincident γ rays.

Different types of time and position correlations were defined for the extraction of β-decay information. Depending on the background conditions, only ion-β correlations or additional tags to the subsequent α particles were applied. As an illustration, Fig. 1 shows the γ spectra of the decays ^{218}Bi$\rightarrow^{218}$Po and ^{215}Bi$\rightarrow^{215}$Po. In the first case, the γ spectrum includes the ion-β correlations where the electron was detected in the implantation pixel. Due to the long correlation times involved, the reverse-time background technique was applied [4, 5]. In the second case, the γ spectrum incorporates correlations with an additional tag on the α particle emitted by the daughter ^{215}Po. The short half-life of this α emitter, 1.78(5) ms [6], allowed for the extension of the β-α correlations to a greater area in the active stopper, including also the neighboring cells. In both cases, the extracted γ-rays agree with previous β-decay studies [7, 8], thus confirming the correlation procedures. Further half-life measurements were performed exploiting these α and β correlations. Ad-hoc techniques of data analysis, explicitly developed for experiments with complex background conditions [9, 10, 4], were employed. Half-lives for $^{211-213}$Tl, ^{215}Pb, $^{215-219}$Bi, $^{215-217}$Po and ^{217}At could be extracted [10, 6].

Partial β-decay level schemes have been built from β-γ and β-γ-γ spectra. The level orderings and proposed spin-parities are supported by systematics and, where possible, by shell-model calculations. As an example, Fig. 2 shows the level schemes of $^{211-213}$Pb. The newly observed states and γ rays are indicated in bold red fonts. β feedings expressed as % are also shown in the figure. These have been obtained assuming $E2$ multipolarities for all transitions, except for the 362-, 438-, 1190-, and 1395-keV γ rays, for which an $M1$ character has been considered [5]. In general, the feeding patterns show a strong population to final levels of equal parity, thus indicating an allowed Gamow-Teller nature for the decay of this isotopic chain. This behavior can be understood in terms of the shell-model structures of mother and daughter nuclei: Direct single-particle β decays are completely blocked since the ground states of the $N > 126$ Tl isotopes have a proton hole in the $s_{1/2}$ orbital, while the valence neutrons lie within the $g_{9/2}$ and $i_{11/2}$ sub-shells. Such large spin differences imply that the β transitions must be driven by the Gamow-Teller $(\nu i_{11/2}, \pi i_{13/2})$ and first-forbidden $(\nu g_{9/2}, \pi h_{9/2})$ single-particle components.

Journal of Physics: Conference Series **533** (2014) 012044 doi:10.1088/1742-6596/533/1/012044

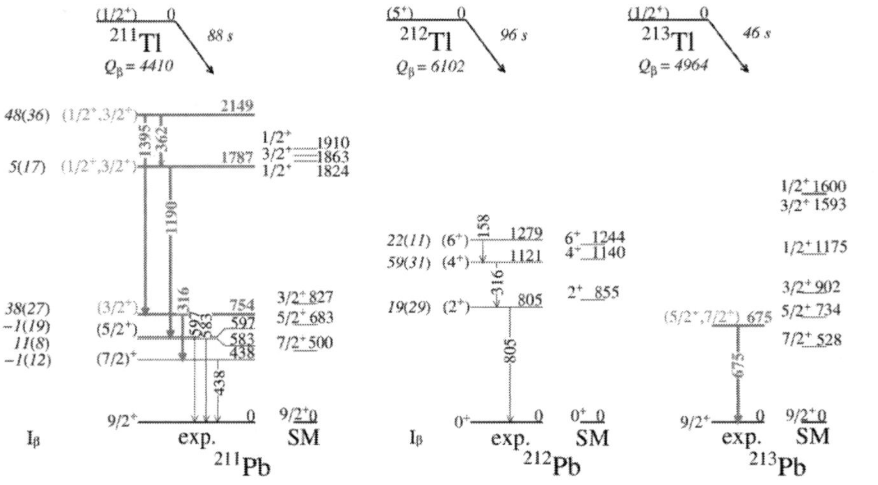

Figure 2. *(Color online) Experimental level schemes (exp.) of the $Z = 82$ isotopes $^{211-213}Pb$. Shell-model calculations (SM) are show on the right-hand side for each nucleus.*

Shell-model calculations show that the allowed part of the β-decay strength function is favored: On the one hand, the occupation number of the $i_{11/2}$ neutron sub-shell is nearly one for the Tl ground states, and, on the other, the calculations indicate that the $\pi i_{13/2}$ single-particle orbital is fragmented over several states in the Pb daughters due to correlations between nucleons, with an increasing presence for larger excitation energies. This leads to strong allowed β transitions to higher-lying states, in line with the β-decay feeding patterns observed. For instance, in the decay ^{211}Tl$\rightarrow^{211}$Pb, a 38% of the strength spreads to the first $(3/2^+)$ 754-keV state, while a 48% of β transitions populate the higher-lying $(1/2^+, 3/2^+)$ 2149-keV candidate. The situation reverses once the $Z = 82$ magic shell is filled, i.e., for the Pb and Bi nuclei with $N > 126$. Here, the main sub-shells mediating in the β decay are the $\pi h_{9/2}$ and $\nu g_{9/2}$ orbitals. As a consequence, strong first-forbidden β-decay transitions, mainly driven by the single-particle $(\nu g_{9/2}, \pi h_{9/2})$ configuration, appear.

3. β-decay systematics of trans-lead nuclei

The nuclei here discussed are of special interest to trace the systematics of β-decay half-lives in nuclei beyond ^{208}Pb. Though not extremely neutron-rich, these are important probes for the nuclear models, needed of experimental constraints that can help to strengthen their predictive power towards the exotic nuclei lying in the unexplored r-process region. Indeed, predictions of r-process time scales and production rates can vary over orders of magnitude depending on the nuclear framework employed. Whereas standard microscopic-macroscopic models emphasize the importance of the allowed Gamow-Teller β-decay strength [11], self-consistent microscopic treatments stress the role of first-forbidden transitions in the reduction of predicted lifetimes [12, 13]. The latter are expected to influence significantly the low-energy spectra of the r-process nuclei due to the presence of intruder single-particle states with different parity near the Fermi surface. The stringent constraints found in the $Z < 82$, $N < 126$ quadrant of the nuclear chart [14, 4] already provide evidence of the importance of first-forbidden transitions in the modeling of the underlying nuclear structure.

The aim of this work is to extend the former systematic studies of β-decay half-lives to the $N > 126$ region. Comparison of measured and predicted half-lives in neutron-rich Tl and Bi isotopic chains is shown in Fig. 3. For the sake of clarity, the experimental-to-theoretical ratios

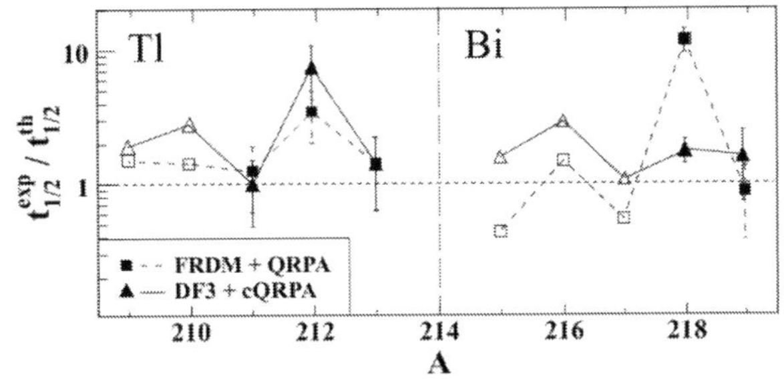

Figure 3. *Ratio between experimental and theoretical β-decay half-lives for the neutron-rich Tl and Bi nuclei studied in this work. The predictions shown by squares are obtained using the FRDM+QRPA approach while the ones shown by triangles by the DF3+cQRPA model.*

are displayed. Two models are discussed here. The first is the FRDM+QRPA approach [11], based on the microscopic Quasi-particle Random Phase Approximation upon the macroscopic Finite Range Droplet Mass model to account for the Gamow-Teller part of the β-strength function. The first-forbidden contribution, instead, is treated as a perturbation of the main Gamow-Teller strength, and thus is inserted macroscopically, employing the Gross Theory framework. The second is the DF3+cQRPA model, a full microscopic approach that implements the first-forbidden strength within the continuum Quasi-particle Random Phase Approximation, thus giving an equal treatment of Gamow-Teller and first-forbidden decays. Predictions of these models are consistent with the decay-feeding pattern derived from the γ-spectroscopy analysis: Whereas the hybrid theory gets closer to the measured values for the one-proton-hole Tl nuclei (whose β decay is mainly driven by Gamow-Teller transitions), predictions deviate up to a factor ten in the Bi isotopic chain (for which the first-forbidden transitions dominate the low-energy part of the β-decay strength distribution). Consistently, the fully microscopic model gets a better description of this isotopic chain. In general, both models underestimate the measured half-lives by an average factor of two, and predict an odd-even staggering that is not seen in the experimental data [12, 5].

The results extracted from this work have paved the way to model the underlying nuclear structure in the region spanned by the *r*-process during the freeze-out stage. They provide an important guide for the theoretical approaches used to extrapolate β-decay inputs in nucleosynthesis calculations. This study already represents a breakthrough in the understanding of the *r*-process matter flow to heavier fissioning nuclei, however, further measurements will be needed to confirm the persistence of the observed tendencies in the unexplored *r*-process region.

References

[1] Burbidge E M, Burbidge G R, Fowler W A and Hoyle F 1957 *Rev. Mod. Phys.* **29** 547
[2] Morales A I *et al.* 2011 *Phys. Rev. C* **84** 011601
[3] Regan P *et al.* 2008 *Int. J. Mod. Phys. E* **17** 8
[4] Morales A I *et al.* 2013 *Phys. Rev. C* **88** 014319
[5] Morales A I *et al. Submitted to PRC*
[6] Benzoni G *et al.* 2013 *In preparation*
[7] De Witte H *et al.* 2004 *Phys. Rev. C* **69** 044305
[8] De Witte H *et al.* 2013 *Phys. Rev. C* **87** 067303
[9] Kurtukián-Nieto T *et al.* 2008 *Nucl. Instrum. Methods A* **589** 472
[10] Benzoni G *et al.* 2012 *Phys. Lett. B* **715** 285
[11] Möller P, Pfeiffer B and Kratz K L 2003 *Phys. Rev. C* **67** 055802
[12] Borzov I N 2011 *Phys. Atom. Nucl.* **74** 1435
[13] Suzuki T, Yoshida T, Kajino T and Otsuka T 2012 *Phys. Rev. C* **85** 015802
[14] Kurtukián-Nieto T *et al.* 2009 *Nucl. Phys. A* **827** 587

XX International School on Nuclear Physics, Neutron Physics and Applications (Varna2013) IOP Publishing
Journal of Physics: Conference Series **533** (2014) 012045 doi:10.1088/1742-6596/533/1/012045

Beta decay and isomer spectroscopy in the ^{132}Sn region: New results from EURICA

A. Jungclaus[1], J. Taprogge[1,2,3], G.S. Simpson[4], G. Gey[4,5,3], S. Nishimura[3], P. Doornenbal[3], G. Lorusso[3], P.-A. Söderström[3], T. Sumikama[6], Z. Xu[7], H. Baba[3], F. Browne[8,3], N. Fukuda[3], N. Inabe[3], T. Isobe[3], H.S. Jung[9], D. Kameda[3], G.D. Kim[10], Y.-K. Kim[10,11], I. Kojouharov[12], T. Kubo[3], N. Kurz[12], Y.K. Kwon[10], Z. Li[13], H. Sakurai[3,7], H. Schaffner[12], H. Suzuki[3], H. Takeda[3], Z. Vajta[14,3], H. Watanabe[3], J. Wu[13,3], A. Yagi[15], K. Yoshinaga[16], S. Bönig[17], J.-M. Daugas[18], F. Drouet[4], R. Gernhäuser[19], S. Ilieva[17], T. Kröll[17], A. Montaner-Pizá[20], K. Moschner[21], D. Mücher[19], H. Nishibata[15], R. Orlandi[22], K. Steiger[19] and A. Wendt[21]

[1] Instituto de Estructura de la Materia, CSIC, E-28006 Madrid, Spain
[2] Departamento de Física Teórica, Universidad Autónoma de Madrid, E-28049 Madrid, Spain
[3] RIKEN Nishina Center, RIKEN, 2-1 Hirosawa, Wako-shi, Saitama 351-0198, Japan
[4] LPSC, Université Joseph Fourier Grenoble 1, CNRS/IN2P3, Institut National Polytechnique de Grenoble, F-38026 Grenoble Cedex, France
[5] Institut Laue-Langevin, B.P. 156, F-38042 Grenoble Cedex 9, France
[6] Department of Physics, Tohoku University, Aoba, Sendai, Miyagi 980-8578, Japan
[7] Department of Physics, University of Tokyo, Hongo 7-3-1, Bunkyo-ku, 113-0033 Tokyo, Japan
[8] School of Computing, Engineering and Mathematics, University of Brighton, Brighton BN2 4JG, United Kingdom
[9] Department of Physics, Chung-Ang University, Seoul 156-756, Republic of Korea
[10] Rare Isotope Science Project, Institute for Basic Science, Daejeon 305-811, Republic of Korea
[11] Department of Nuclear Engineering, Hanyang University, Seoul 133-791, Republic of Korea
[12] GSI Helmholtzzentrum für Schwerionenforschung GmbH, 64291 Darmstadt, Germany
[13] School of Physics and State key Laboratory of Nuclear Physics and Technology, Peking University, Beijing 100871, China
[14] Institute of Nuclear Reserach of the Hungarian Academy of Sciences (ATOMKI), Debrecen, H-4011 Hungary
[15] Department of Physics, Osaka University, Machikaneyama-machi 1-1, Osaka 560-0043 Toyonaka, Japan
[16] Department of Physics, Faculty of Science and Technology, Tokyo University of Science, 2641 Yamazaki, Noda, Chiba, Japan
[17] Institut für Kernphysik, Technische Universität Darmstadt, D-64289 Darmstadt, Germany
[18] CEA, DAM, DIF, 91297 Arpajon cedex, France
[19] Physik Department E12, Technische Universität München, D-85748 Garching, Germany
[20] Instituto de Física Corpuscular, CSIC-Univ. of Valencia, E-46980 Paterna, Spain
[21] IKP, University of Cologne, D-50937 Cologne, Germany
[22] Instituut voor Kern- en StralingsFysica, K.U. Leuven, B-3001 Heverlee, Belgium

E-mail: `andrea.jungclaus@csic.es`

Journal of Physics: Conference Series **533** (2014) 012045 doi:10.1088/1742-6596/533/1/012045

Abstract. The first EURICA campaign with high intensity Uranium beams took place at RIKEN in November/December 2012. Within this campaign experiment NP1112-RIBF85 was performed dedicated to the study of the isomeric and beta decays of neutron-rich Cd, In, Sn and Sb isotopes towards and beyond the N=82 neutron shell closure. In this contribution we will first provide information about the status of the analysis of the extensive data set obtained in this experiment and close with a short outlook.

1. Introduction

The region around doubly-magic ^{132}Sn is of great importance for nuclear structure physics because it is the only region around a heavy doubly-closed shell nucleus far-off stability (8 neutrons relative to the last stable isotope ^{124}Sn) for which detailed spectroscopic information can be obtained using modern state-of-the-art techniques. It therefore plays an essential role in testing the shell model and serves as input for any reliable future microscopic nuclear structure calculations towards the neutron drip line. In addition, this region is also relevant for nuclear astrophysics, in particular nucleosynthesis calculations, due to the close relation between the N=82 shell closure and the A≈130 peak of the solar r-process abundance distribution.

The main goal of experiment NP1112-RIBF85 was to extend the current knowledge on excited states in very neutron-rich Cd, In, Sn and Sb isotopes. In particular we were aiming i) for first experimental information on excited states in ^{136,138}Sn via the search for 6^+ seniority isomers in these isotopes in anology to the one known in ^{134}Sn [1], ii) for the first observation of transitions within the $\pi g_{9/2}^{-1} \otimes \nu f_{7/2}$ multiplet in ^{132}In populated in the β-decay of ^{132}Cd and iii) to follow the evolution of the $\pi g_{7/2} \otimes \nu f_{7/2}$ multiplet in ^{136,138}Sb.

The results of this experiment should serve to test the predictions of shell-model calculations in a very neutron-rich, medium-heavy region of the chart of nuclides. Indeed, these nuclei, with just a few neutrons beyond ^{132}Sn, are very sensitive to deficiencies in current state-of-the-art shell-model interactions.

2. Experimental setup

The exotic nuclei of interest were produced by the in-flight fission of a 345 MeV/nucleon ^{238}U beam from the RIBF facility, impinging on a 3-mm thick Be target. The ions of interest were separated from other reaction products and identified on an ion-by-ion basis by the BigRIPS in-flight separator [2]. The particle identification was performed using the ΔE-TOF-$B\rho$ method in which the energy loss, (ΔE), the time of flight (TOF) and the magnetic rigidity ($B\rho$) are measured and used to determine the atomic number, Z, and the mass-to-charge ratio, A/q, of the fragments. Details about the identification procedure can be found in Ref. [3]. The identified ions are transported through the ZeroDegree spectrometer (ZDS) and finally implanted into the WAS3ABi (Wide-range Active Silicon Strip Stopper Array for β and Ion detection) Si array positioned at the focal plane of the ZDS (F11). The WAS3ABi detector [4] consists of eight DSSSD with an area of 60×40 mm^2, a thickness of 1 mm and a segmentation of 40 horizontal and 60 vertical strips each. A sketch of the experimental facility together with an identification plot of the isotopes implanted into WAS3ABI during experiment NP1112-RIBF85 is shown in Fig. 1. To detect γ radiation emitted in the decay of the implanted radioactive nuclei 12 large-volume Ge Cluster detectors [6] from the former EUROBALL spectrometer [7] were arranged in a close geometry around the WAS3ABi detector.

The combination of the unprecedented high intensity of the primary Uranium beam (on average 8-10 pnA) and the high efficiency of the setup for both the detection of γ rays (7% at 1 MeV) and particles allowed to perform detailed decay spectroscopy in a region of the chart of nuclides which has not been accessible for this type of studies before.

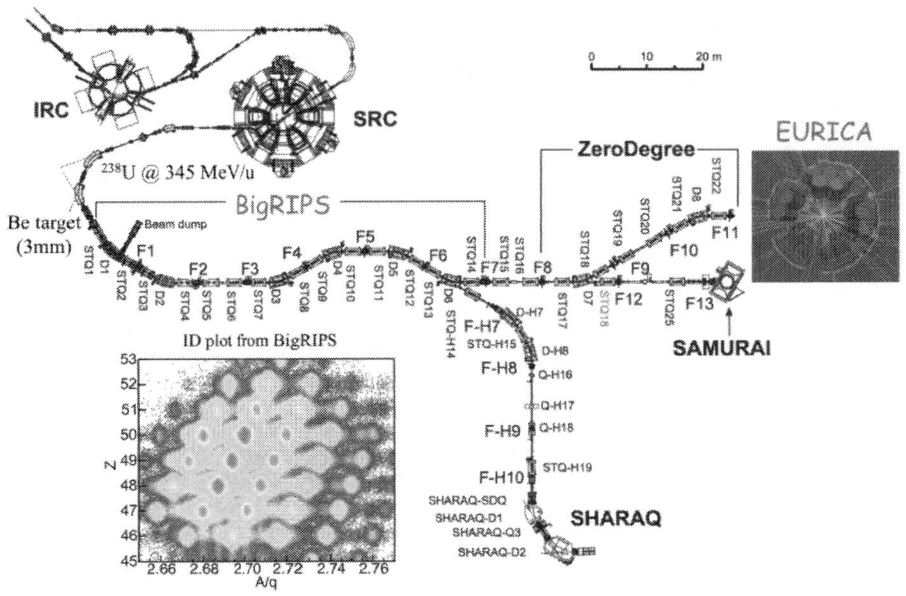

Figure 1. Sketch of the experimental facility (adopted from Ref. [5]) including the identification plot of the isotopes implanted into WAS3ABi during experiment NP1112-RIBF85.

3. First results

To study the γ radiation emitted in the β-decay of and after β-delayed neutron emission from the neutron-rich isotopes produced in this experiment, spectra were constructed including all the γ rays which were detected in prompt coincidence with decay events. Different conditions were applied on the correlations, both in space and time, between the implantation and the successive decay signals registered in the Si detectors of WAS3ABi. In the analysis process, we first concentrated on the decays of the Cd isotopes produced in our experiment (A=128-133). The high-statistics data obtained for the β-decay of ^{130}Cd, which had already previously been studied in detail [8], served as an excellent test case. In a next step, a β-decay scheme for ^{129}Cd was established for the first time. It comprises more than forty γ transitions connecting about twenty excited states. Only four of them, including a β-decaying $1/2^-$ isomer, were known previously. Finally, from the decay of the most exotic Cd isotopes, first information about excited states in nuclei in the quadrant south-east of ^{132}Sn was obtained.

In addition to the γ spectroscopy after β-decay, also information on isomeric decays is comprised in the data set. Several isomeric states with half-lifes in the μs and ms ranges are observed for the first time. One example are the neutron-rich Sn isotopes. Delayed γ rays are observed in coincidence with both ^{136}Sn and ^{138}Sn. They constitute the first observation of the decay of excited states in these very neutron-rich, semi-magic nuclei. Indeed, together with ^{128}Pd [9], they are the nuclei with the highest N/Z ratio in this region for which excited states are known and their semi-magic nature allows just the neutron-neutron part of the shell model interactions to be probed. Three delayed transitions have been observed for each nucleus and these have been assigned as E2 transitions from the 6^+, 4^+ and 2^+ states, by analogy with γ rays of similar energies observed from the decay of a 6^+ isomer in ^{134}Sn [1]. The small spacing between the 6^+ and 4^+ states, and their relatively pure $\nu(f_{7/2})^2$ configuration, are responsible for the isomerism.

We found that the energies of the 2^+, 4^+ and 6^+ levels remain fairly constant as the number of neutrons increases from N=84 to N=88. This agrees with the predictions of shell-model calculations performed using state-of-the-art interactions, e.g. the CD-Bonn bare nucleon-nucleon potential, renormalized using G-matrix [10] and V_{low-k} [11] prescriptions. In contrast calculations performed using empirical interactions (SMPN) deviate from the experimental data [12], despite the simple nature of these nuclei. These data serve as useful input for astrophysical r-process calculations as the path of this reaction includes these nuclei. A low excitation energy of the 2_1^+ state can change the effective half-lives of nuclei participating in this reaction at high temperatures.

4. Outlook

In the future, the energies of excited states will be established for the first time also in many other nuclei besides the cases discussed here. In particular, the data on the Sb and In isotopes will allow very sensitive tests of the shell-model predictions to be performed. Information on the excited states of these simple odd-Z nuclei is particularly important as the neutron-proton part of shell-model interactions is the most difficult part to reproduce. However, the data set obtained from the present experiment is not only of interest from the spectroscopic point of view. For many of the produced nuclei shown in Fig. 1 basic information such as β-decay half-lifes will become available for the first time, many of them being important ingredients for r-process calculations.

To conclude, a very rich data set has been obtained from experiment NP1112-RIBF85 which took place in December 2012 during the first EURICA campaign with high intensity Uranium beams at RIKEN. Exciting results with respect to the structure of neutron-rich Cd, In, Sn and Sb isotopes, some of them also relevant for nuclear astrophysics, will be presented in the near future.

Acknowledgments

We thank the staff of the RIKEN Nishina Center accelerator complex for providing stable beams with high intensities to the experiment. We acknowledge the EUROBALL Owners Committee for the loan of germanium detectors and the PreSpec Collaboration for the readout electronics of the cluster detectors. This work was supported by the Spanish Ministerio de Ciencia e Innovación under contracts FPA2009-13377-C02 and FPA2011-29854-C04.

References

[1] A. Korgul et al., Eur. Phys. J. A **7**, 167 (2000).
[2] T. Kubo, Nucl. Instr. Meth. B204 (2003) 97.
[3] T. Ohnishi et al., J. Phys. Soc. Jpn. **79**, 073201 (2010).
[4] P.-A. Söderström et al., Nucl. Instr. Meth. B, in press
[5] T. Kubo et al., Prog. Theor. Exp. Phys. 2012, 03C003.
[6] J. Eberth et al., Nucl. Instrum. Methods Phys. Res., Sect. A **369**, 135 (1996).
[7] J. Simpson, Z. Phys. A **358**, 139 (1997).
[8] I. Dillmann et al., Phys. Rev. Lett. **91**, 162503 (2003).
[9] H. Watanabe et al., Phys. Rev. Lett. **111**, 152501 (2013).
[10] M. P. Kartamyshev, T. Engeland, M. Hjorth-Jensen, and E. Osnes, Phys. Rev. C 76, 024313 (2007).
[11] A. Covello, L. Coraggio, A. Gargano, and N. Itaco, J. Phys.: Conf. Ser. 267, 012019 (2011).
[12] S. Sarkar and M. Saha Sarkar, Eur. Phys. J. A 21, 61 (2004); Phys. Rev. C 78, 024308 (2008); Phys. Rev. C 81, 064328 (2010).

Nuclear moments and nuclear structure near ^{132}Sn

Andrew E. Stuchbery

Department of Nuclear Physics, The Australian National University, Canberra, ACT 0200, Australia

E-mail: andrew.stuchbery@anu.edu.au

Abstract. Several g-factor measurements have been performed recently on nuclei near the neutron-rich, double-magic nucleus ^{132}Sn. The focus here is on ^{134}Te, the $N = 82$ isotone which has two protons added to ^{132}Sn. Comparisons are made with other nuclei that have two protons outside a double-magic core. The extent to which ^{132}Sn is an inert core is discussed based on these comparisons.

1. Introduction

The recoil in vacuum (RIV) method has proved to be a powerful method to measure the g factors (gyromagnetic ratios) of the first 2^+ states in neutron-rich isotopes near double magic ^{132}Sn. Following on from the pioneering work on ^{132}Te [1, 2], RIV g-factor measurements have also been published recently for the isotopes ^{124}Sn, ^{126}Sn and ^{128}Sn [3]. The measurements were performed at the Holifield Radioactive Ion Beam Facility (HRIBF). The quality of the radioactive beams, which were produced by the ISOL method and accelerated through the 25UR tandem, together with the segmentation and angular coverage of the CLARION [4] and bare HYBALL [5] detector arrays contributed to the success of these measurements.

In execution, an RIV g-factor measurement on a radioactive beam is identical to a measurement of $B(E2; 0_1^+ \to 2_1^+)$ by safe Coulomb excitation. In simplified terms, the $B(E2)$ is determined by the total intensity of the γ radiation emitted, whereas the g factor is determined by examining the spatial pattern of the γ radiation. As a consequence, a statistically precise $B(E2)$ is measured along with the g factor. The measurements are complementary as the g factor probes single-particle aspects of the wavefunction whereas the $B(E2)$ probes collectivity.

This paper will focus on ^{134}Te [6], which has two valence protons added to ^{132}Sn. Comparisons will be made with other nuclides that have two valence protons added to a double magic core.

2. Magnetic moments and the structure of ^{134}Te

Figure 1 shows the experimental [1, 2, 6–8] and theoretical [9–12] $g(2_1^+)$ systematics for the Te isotopes near $N = 82$. There is quite good agreement between theory and experiment for ^{130}Te and ^{132}Te, however the theories do not agree for ^{134}Te where the experimental g factor [6] falls between the predictions of the QRPA calculation [11] and the two shell model calculations [9, 10]. Elsewhere we have shown that these three calculations in fact predict similar wavefunctions, dominated by the $\pi(g_{7/2})^2$ configuration [6]. Once the differences in the $M1$ operator are considered, the QRPA can be brought into agreement with the shell model. The Monte Carlo Shell Model (MCSM) calculation [12], however, used the same $M1$ operator as the QRPA

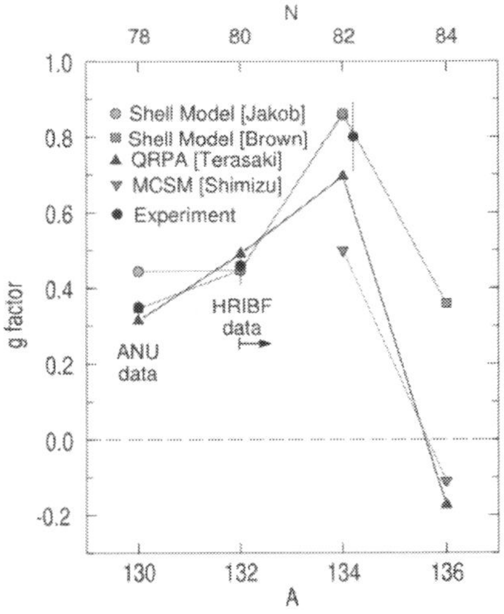

Figure 1. Theoretical g factors in the even Te isotopes near $N = 82$ compared with experimental data [1, 2, 6–8]. Jacob *et al.* [9] and Brown *et al.* [10] performed shell model (SM) calculations. Terasaki *et al.* [11] used the Quasiparticle Random Phase Approximation (QRPA), and Shimizu *et al.* [12] used a Monte Carlo Shell Model (MCSM) approach.

calculations. Thus the difference in the predicted g factors for these two models implies that there is a significant difference between their wavefunctions, which requires further investigation, particularly as the MCSM is two standard deviations from the experimental $g(2_1^+)$ value.

The g factors of the longer-lived 4_1^+ and 6_1^+ states in ^{134}Te have been measured [13, 14], as have the moments of the ground states of ^{133}Sb and ^{135}I [15]. To the extent that these states can be associated with pure $\pi(g_{7/2})^n$ configurations, their g factors should be the same. Shell model calculations, using OXBASH [16], have been performed for ^{134}Te and its neighbors. The interactions and model space were those of Brown *et al.* [10] (the two protons can occupy $g_{7/2}$, $d_{5/2}$, $s_{1/2}$, $h_{11/2}$, $d_{3/2}$), but the empirical $M1$ operator of Jakob *et al.* [9] was used. The calculated g factors are compared to the experimental data in Table 1. It is evident that the shell model predicts only small differences between the g factors of the states considered. Experimentally, the g factor of the 6_1^+ state in ^{134}Te is the same, within uncertainties, as that of the ground-state of ^{133}Sb, which represents the single-proton case. The $g(2_1^+)$ value is also consistent with the g factor of the single-proton $g_{7/2}$ state. Overall, the g-factor data point to rather pure $\pi(g_{7/2})^2$ configurations for the yrast 2_1^+, 4_1^+ and 6_1^+ states in ^{134}Te.

Table 1. Shell model and experimental g factors of few-proton states near ^{132}Sn.

Nuclide	J^π	Main configuration	g factor		
			Brown [10]	Present	Experiment
^{133}Sb	$7/2_1^+$	$\pi g_{7/2}$		0.807	0.857(3)
^{134}Te	2_1^+	$\pi(g_{7/2})^2$	0.833	0.824	0.76(9)
	4_1^+		0.830	0.819	$0.70^{+0.55}_{-0.38}$
	6_1^+		0.838	0.829	0.847(25)
^{135}I	$7/2_1^+$	$\pi(g_{7/2})^3$		0.808	0.840(1)

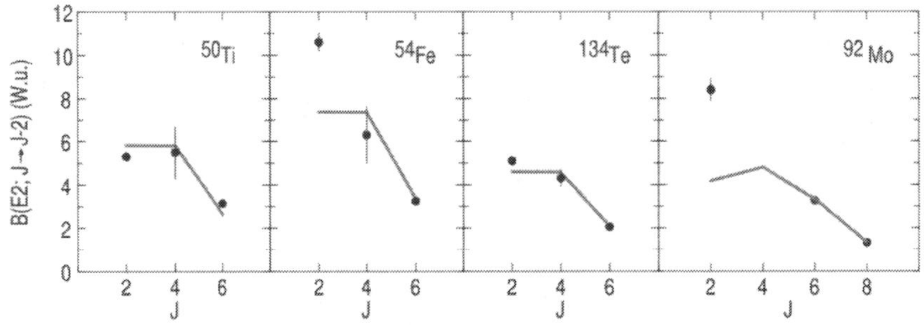

Figure 2. $B(E2; J \to J-2)$ in nuclei with two valence protons.

3. Comparison of nuclei with two protons added to a double-magic core

We have seen that the g factors of the 2_1^+, 4_1^+ and 6_1^+ states in ^{134}Te are consistent with these states being predominantly due to the $\pi(g_{7/2})^2$ configuration. Cases where two protons are added to a closed-shell, and both the 2_1^+-state and the maximum-spin state with $J_{\rm max} = 2j-1$ (usually an isomer) have known g factors are rare. Aside from ^{134}Te these are: $^{50}_{22}$Ti$_{28}$ (two $f_{7/2}$ protons added to ^{48}Ca), $^{54}_{26}$Fe$_{28}$ (two $f_{7/2}$ proton holes in ^{56}Ni), and $^{92}_{42}$Mo$_{50}$ (two $g_{9/2}$ protons added to ^{90}Zr). Along with the discussion of the g factors, it is useful also to consider the $E2$ transition rates. For transitions between the states of the pure j^2 configuration, the $B(E2)$ values are related to the single-particle matrix element $\langle j||T(E2)||j\rangle$, by

$$B(E2; J_i \to J_i - 2) = 4(2J_i - 3) \left\{ \begin{array}{ccc} j & J_i - 2 & j \\ J_i & j & 2 \end{array} \right\}^2 |\langle j||T(E2)||j\rangle|^2. \qquad (1)$$

The $B(E2)$ data for ^{50}Ti, ^{54}Fe, ^{92}Mo, and ^{134}Te are compared with Eq. (1) in Figure 2.

In order to judge the degree to which the $E2$ properties of these nuclei are consistent with the predictions for a pure πj^2 configuration, we define the double ratio

$$R_{E2} = \frac{B(E2; 2 \to 0)_{\rm exp}}{B(E2; J_{\rm max} \to J_{\rm max} - 2)_{\rm exp}} \times \frac{B(E2; J_{\rm max} \to J_{\rm max} - 2)_{j^2}}{B(E2; 2 \to 0)_{j^2}}. \qquad (2)$$

Thus R_{E2} is unity if the experimental ratio of $B(E2)$ values is consistent with the pure πj^2 model, and it exceeds unity if there is additional quadrupole collectivity in the 2_1^+ state.

Table 2 shows the R_{E2} and $g(2_1^+)/g(J^\pi_{\rm max})$ ratios for the nuclei of interest. It is clear from Figure 2 and Table 2 that ^{54}Fe and ^{92}Mo show additional $E2$ collectivity in the 2^+ state. These

Table 2. $B(E2)$ and g factor ratios for closed-shell nuclei plus two protons.

nuclide	core	j^2	$J^\pi_{\rm max}$	$R(E2)$	$g(2_1^+)/g(J^\pi_{\rm max})$
^{50}Ti	^{48}Ca	$f_{7/2}$	6^+	0.77(4)	0.93(11)
^{54}Fe	^{56}Ni	$f_{7/2}$	6^+	1.48(6)	0.77(5)
^{134}Te	^{132}Sn	$g_{7/2}$	6^+	1.13(5)	0.90(11)
^{92}Mo	^{90}Zr	$g_{9/2}$	8^+	2.04(13)	0.81(11)

two cases also have $g(2_1^+)/g(J^\pi_{\rm max})$ significantly less than unity (see Table 2). This behavior can be associated with added quadrupole collectivity in the 2_1^+ state, which has the effect of reducing $g(2_1^+)$ toward $Z/A \sim 0.4$. Both ^{56}Ni and ^{90}Zr are known to be soft cores. In the case of ^{54}Fe, for example, large-basis shell model calculations in the *pf* shell, which effectively include excitations of a ^{56}Ni core, can account for the observed $B(E2)$ and g-factor data [17].

The $B(E2)$ and g-factor ratios in Table 2 enable an assessment of the extent to which the core nuclei are inert. Elsewhere [6] we have concluded that there is additional quadrupole collectivity in the 2_1^+ state of ^{134}Te that is not accounted for by large-basis shell model calculations which assume an inert ^{132}Sn core. We have also shown that coupling the valence $\pi g_{7/2}^2$ configuration to a core vibration with the properties of the first-excited state in ^{132}Sn can readily account for the observed $2_1^+ \to 0_1^+$ transition strength in ^{134}Te [6]. It was found that the wavefunctions of the 2_1^+, 4_1^+ and 6_1^+ states of ^{134}Te nevertheless remain dominated by the $\pi g_{7/2}^2$ configuration. Combining these insights with the discussion above, it can be concluded that ^{132}Sn is a relatively inert shell-model core, roughly comparable to ^{48}Ca.

4. Concluding comments

The g factor and $B(E2)\uparrow$ for the first-excited state of the neutron-rich $N = 82$ isotone ^{134}Te have been measured at HRIBF [6]. The precision achieved for these radioactive beam measurements is remarkable: a comparison of the precision of the $B(E2)$ and g-factor ratios in Table 2 shows that the radioactive beam case of ^{134}Te matches the precision of the three stable-beam cases.

There is evidence of additional quadrupole collectivity in the 2_1^+ state of ^{134}Te due to coupling between the valence protons and excitations of the ^{132}Sn core. However, the electromagnetic properties of the low-excitation states of ^{134}Te are generally well described by the shell model, even in the approximation that the two protons are restricted to the $g_{7/2}$ orbit. The power of combined $B(E2)$ and RIV g-factor measurements on radioactive beams has been demonstrated.

Acknowledgments

The author is grateful to many colleagues who have contributed to the work discussed here, especially J.M. (Mitch) Allmond (Oak Ridge National Laboratory, USA). This research was supported in part by the ARC Discovery Scheme, Grant No. DP0773273.

References

[1] Stone N J *et al* 2005 *Phys. Rev. Lett.* **94** 192501
[2] Stuchbery A E and Stone N J 2007 *Phys. Rev. C* **76** 034307
[3] Allmond J M *et al* 2013 *Phys. Rev. C* **87** 054325
[4] Gross C *et al* 2000 *Nucl. Inst. Meth. in Phys. Res. A* **450** 12
[5] Galindo-Uribarri A 2010 *AIP Conf. Proc.* **1271** 180; www.phy.ornl.gov/hribf/research/equipment/hyball
[6] Stuchbery A E, Allmond J M, Galindo-Uribarri A, Padilla-Rodal E, Radford D C, Stone N J, Batchelder J C, Beene J R, Benczer-Koller N, Bingham C R, Howard M E, Kumbartzki G J, Liang J F, Manning B, Stracener D W and Yu C H 2013 *Phys. Rev. C* **88** 051304(R)
[7] Stuchbery A E *et al* 2007 *Phys. Rev. C* **76** 034306
[8] Danchev M *et al* 2011 *Phys. Rev. C* **84** 061306
[9] Jakob G *et al* 2002 *Phys. Rev. C* **65** 024316
[10] Brown B A, Stone N J, Stone J R, Towner I S and Hjorth-Jensen M 2005 *Phys. Rev. C* **71** 044317
[11] Terasaki J, Engel J, Nazarewicz W and Stoitsov M 2002 *Phys. Rev. C* **66** 054313
[12] Shimizu N, Otsuka T, Mizusaki T and Honma M 2004 *Phys. Rev. C* **70** 054313
[13] Goodin C *et al* 2008 *Phys. Rev. C* **78** 044331
[14] Wolf A and Cheifetz E 1976 *Phys. Rev. Lett.* **36** 1072
[15] Stone N J 2011 **INDC(NDS)-0594** http://www-nds.iaea.org/publications/indc/indc-nds-0594/
[16] Brown B A *et al* 2004 Oxbash for windows pc (*Preprint* Michigan State University, report no. MSU-NSCL 1289)
[17] Honma M, Otsuka T, Brown B A and Mizusaki T 2004 *Phys. Rev. C* **69** 034335

Journal of Physics: Conference Series **533** (2014) 012047 doi:10.1088/1742-6596/533/1/012047

Analysis and results of the ^{104}Sn Coulomb excitation experiment

G Guastalla[1], D D DiJulio[2], M Górska[3], J Cederkäll[2], P Boutachkov[1,3], P Golubev[2], S Pietri[3], H Grawe[3], F Nowacki[4], A Algora[5,6], F Ameil[3], T Arici[7,3], A Atac[8], M A Bentley[9], A Blazhev[10], D Bloor[9], S Brambilla[11], N Braun[10], F Camera[11], C Domingo Pardo[5], A Estrade,[3] F Farinon[3], J Gerl[3], N Goel[3,1], J Gręebosz[12], T Habermann[3,13], R Hoischen[2], K Jansson[2], J Jolie[10], A Jungclaus[14], I Kojouharov[3], R Knoebel[3], R Kumar[15], J Kurcewicz[16], N Kurz[3], N Lalović[3], E Merchan[1,3], K Moschner[10], F Naqvi[3,10], B S Nara Singh[9], J Nyberg[17], C Nociforo[3], A Obertelli[18], M Pfützner[3,16], N Pietralla[1], Z Podolyák[1], A Prochazka[19], D Ralet[3,1], P Reiter[10], D Rudolph[2], H Schaffner[3], F Schirru[19], L Scruton[19], T Swaleh[9], J Taprogge[1]0, R Wadsworth[9], N Warr[10], H Weick[3], A Wendt[10], O Wieland[11], J S Winfield[3] and H J Wollersheim[3]

[1] Institut für Kernphysik, Technische Universität Darmstadt, Darmstadt, Germany
[2] Department of Physics, Lund University, Lund, Sweden
[3] Helmholtzzentrum für Schwerionenforschung GmbH (GSI), Darmstadt, Germany
[4] IPHC, IN2P3-CNRS et Université de Strasbourg, Strasbourg, France
[5] IFIC (CSIS-Univ. Valencia) Valencia, Spain
[6] Institute of Nuclear Research of the Hungarian Academy of Sciences, Debrecen, Hungary
[7] Institute of Science, Istanbul University, Turkey
[8] Department of Physics, Ankara University, Ankara, Turkey
[9] Department of Physics, University of York, York, United Kingdom
[10] Institut für Kernphysik, Universität zu Köln, Köln, Germany
[11] Dipartimento di Fisica, Università di Milano, and INFN Sezione Milano, Milano, Italy
[12] The Institute of Nuclear Physics PAN, Kraków, Poland
[13] Department of Physics, Goethe University, Frankfurt am Main, Germany
[14] Instituto de Estructura de la Materia, CSIC, Madrid, Spain
[15] Inter-University Acceleration Center, New Dehli, India
[16] The Institute of Experimental Physics, Warsaw, Poland
[17] Department of Physics and Astronomy, Uppsala University, Uppsala, Sweden
[18] CEA, Centre de Saclay, IRFU/Service de Physique Nucléaire, Gif-sur-Yvette, France
[19] Department of Physics, University of Surrey, Guildford, United Kingdom
[20] Departamento de Física Teórica, Universidad Autónoma de Madrid, Madrid, Spain

E-mail: `g.guastalla@gsi.de`

Abstract. The analysis of the Coulomb excitation experiment conducted on ^{104}Sn required a strict selection of the data in order to reduce the large background present in the γ-ray energy spectra and identify the γ-ray peak corresponding to the Coulomb excitation events. As a result the B(E2; $0^+ \rightarrow 2^+$) value could be extracted, which established the downward trend towards ^{100}Sn and therefore the robustness of the N=Z=50 core against quadrupole excitations.

1. Introduction

The N=Z=50 nucleus ^{100}Sn, located at the proton-drip line, is predicted by the shell model to be the heaviest self-conjugated doubly magic nucleus. Experimental evidence has indeed confirmed the presence of a large shell gap [1] in this nucleus. However, at the present nuclear facilities, it is still impossible to measure excited states in ^{100}Sn which will give a direct proof of the stability of the shell closure. This aspect has been therefore studied by looking at the evolution of the shell structure in the tin isotopic chain. A sensitive tool to get information about the polarization of the core is the measurement of the reduced E2 transition strength from the ground state, the B(E2; $0^+ \rightarrow 2^+$) value (in the following called B(E2)). The results obtained for this quantity for Sn isotopes from N=64 down to N=56 indicated, despite the large uncertainties, a constant trend. This was at variance with the shell model predictions (see [2]) and from this behavior it was not possible to exclude a *softening* of the N=Z=50 shell closure. Therefore the measurement of the B(E2) value in the next even-even isotope toward ^{100}Sn, i. e. ^{104}Sn, was needed. An important aspect is also that this nucleus could be more precisely described by shell model calculations.

The experiment has been performed at GSI using the FRagment Separator (FRS) [3] to produce the exotic beam and the EUROBALL-PreSPEC setup [4] to measure the E2 transition strength, as will be described in the following paragraph. To excite the rare isotopes, the Relativistic Coulomb excitation technique in inverse kinematics has been used. A B(E2) value measurement has been performed in similar conditions also for the stable ^{112}Sn, for the purpose of normalization of the B(E2) value for ^{104}Sn. In this way one could neglect in the extraction of the B(E2) value possible systematic errors as well as feeding pattern from states above the 2^+ and the detection efficiency of the gamma-ray detector setup. This procedure assumes however a similar structure of the first excited states in both nuclei.

The analysis of the data collected in this experiment is particularly challenging because of the large contribution of background radiation in the γ-ray energy spectra and the low intensity of the exotic beam. In this paper, the analysis steps applied to extract the γ-ray energy peak associated to the $2_1^+ \rightarrow 0^+$ transition are briefly described.

2. The experiment

In the experiment, a stable ^{124}Xe beam of 793 MeV/A delivered by SIS-18 impinged on a ^{9}Be (4 g/cm^2 thick) target, which was positioned at the entrance of the FRS. Isotopes produced by fragmentation reactions in the target were separated by the FRS using the Bρ-ΔE-Bρ method and identified with several detectors placed at the intermediate and at the final focal plane. The proton number (Z) of the ions has been determined from the energy loss signals of the MUltiple Sampling Ionization Chamber (MUSIC), while the mass-over-charge ratio (A/q) was obtained from a Time-Of-Flight (TOF) measurement, using plastic scintillators in the second half of the separator. The position of the ions was also measured at the aforementioned focal planes in Time Projection Chambers (TPCs).

In order to induce Coulomb excitation of the selected fragments, a secondary target made of ^{197}Au (386 mg/cm^2 thick) was positioned at the end of FRS. The γ-rays emitted in the reactions in the target were detected by the RISING array [4], which comprised 15 EUROBALL Cluster detectors. To take advantage of the Lorentz boost, occurring due to the relativistic energy of the beam, these detectors were placed in three rings at small angles (up to 40°) relative to the beam direction. Finally, the recoiling particles were identified and tracked with the Lund York Cologne Calorimeter (LYCCA) placed downstream of the target position [5].

3. The analysis

The isotopes have been identified and selected before and after the secondary target in order to select mainly events arising from the Coulomb scattering channel. At the final focal plane, after

Journal of Physics: Conference Series **533** (2014) 012047 doi:10.1088/1742-6596/533/1/012047

gating on Z=50, the isotope of interest could be uniquely identified by plotting the position of the ion against the A/q value [6]. For the selection of the Sn isotopes after the target, the ΔE-E plot from the LYCCA calorimeter was used [6].

The next step of the analysis was to reduce contributions from nuclear reactions and the atomic background in the γ-ray energy spectra and to identify the peak associated with the $2_1^+ \rightarrow 0^+$ transition. Due to the relativistic energy of the beam and the high velocity of the particles after the target, a precise Doppler correction of the γ-ray energy signals was required. Event-by-event velocity information was acquired from the TOF measurements in the FRS combined with a constant offset taking into account the unknown layers of matter in the final focal plane. The best offset was chosen based on the peak-to-background ratios observed in the Doppler corrected energy spectra when using different velocity values. A selection of the scattering angles of the ions after the secondary target was carried out in order to reduce contributions from nuclear reactions and elastic scattering (see Fig.1). This parameter was calculated with the incoming and the outgoing angles of the ions, measured before and after the target using a TPC detector at the final focal plane, the target DSSSD and the LYCCA Wall detectors. The chosen range from 15 mrad to 40 mrad is compatible with the 15 fm calculated as a *safe* impact parameter for our experimental setup [4] and visibly improves the peak-to-background ratio. A narrow time window of about 15 ns has been also applied to select the prompt γ radiation.

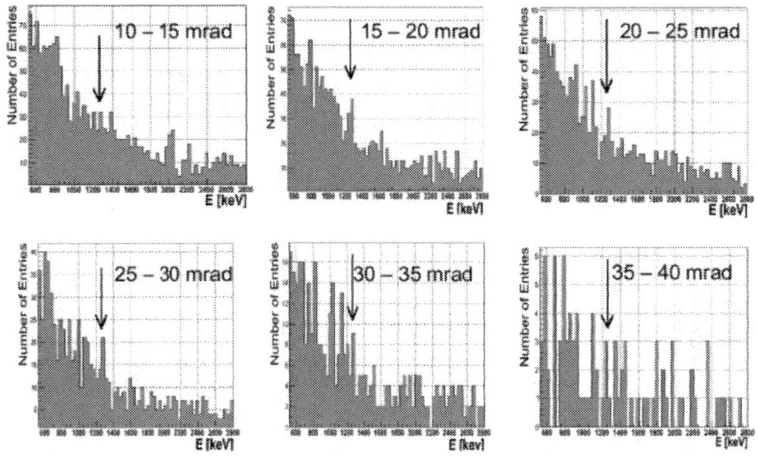

Figure 1. Examples of γ-ray energy spectra produced gating on different ranges of the scattering angle for the ^{112}Sn case. Below 15 mrad the contribution from elastic scattering is covering the peak corresponding to the $2_1^+ \rightarrow 0^+$ transition.

Due to the poor peak-to-background ratio of the ^{104}Sn data, the analysis gates described above have been optimized for the ^{112}Sn case and applied to ^{104}Sn. This procedure is adequate because of the assumption that the two isotopes have a similar structure and therefore an identical excitation process can be expected for both.

The analysis led to clear identification of the $2_1^+ \rightarrow 0^+$ transition in the final Doppler corrected energy spectra [6].

The reference value of B(E2) = 0.242(8) e^2b^2 for ^{112}Sn, measured in the most recent sub-barrier Coulomb excitation experiment [7], was used for normalization.

4. Conclusions

The analysis procedure described in this work allowed us to identify the Coulomb excitation γ-ray peak of the transition of interest, despite the scarce statistics and the large background in the γ-ray energy spectra. The method to determine the velocity of the ions after the secondary target by using event-by-event FRS velocity with a constant offset was appropriate for the Doppler correction of the γ-ray energy signals. The accurate choice of a narrow range for the time in the Germanium detectors and for the scattering angle of the ions after the secondary target also improved considerably the peak-to-background ratio.

As a result for ^{104}Sn the B(E2; $0^+ \rightarrow 2^+$)=0.10(4) e^2b^2 has been obtained and a decreasing trend of B(E2) values towards ^{100}Sn has been established for the first time. This value is in agreement with LSSM calculation performed without a significant truncation of the model space and it infers the stability of the N=Z=50 shell closure [6].

5. Aknowledgments

This work was supported by the Helmholtz International Center for FAIR (HIC for FAIR) within the LOEWE program by the State of Hesse, the BMBF under grant No. 05P12RDFN8, and the Swedish Research Council through contract No. 2009-3939. A. J. would like to thank the Spanish Ministerio de Ciencia e Innovación for financial support under Contract No. FPA2011-29854-C04. The support of the German BMBF under Contract No. 06KY9135I is acknowledged. Zs. D. and D. S. acknowledge the support of the Hungarian OTKA under Contract No. K100835.

References

[1] Blazhev A et al., 2004, *Phys. Rev.* C **69**, 064304.
Boutachkov P et al., 2011, *Phys. Rev.* C **84**, 044311.
Hinke C et al., 2012, *Nature* (London) **486**, 341.
[2] Banu A et al., 2005,*Phys. Rev.* C **72**, 061305.
J. Cederkall et al., 2007, *Phys. Rev. Lett.* **98**, 172501.
C. Vaman et al., 2007, *Phys. Rev. Lett.* **99**, 162501.
A. Ekstrom et al., 2008, *Phys. Rev. Lett.* **101**, 012502.
[3] Geissel H et al., 1992,*Nucl. Instrum. Methods Phys. Res.*, Sect. B **70**, 286.
[4] Wollersheim H J et al., 2005, *Nucl. Instrum. Methods Phys. Res.* A **537**, 637.
Simpson J, 1997 *Phys.* A **358**, 139.
Eberth J et al., 1996, *Nucl. Instrum. Methods Phys. Res.* A **369**, 135.
[5] Golubev P et al., 2013, *Nucl. Instr. Meth.*, Nucl. Instr. Meth. A **723**, 55 .
Hoischen R et al., 2011,*Nucl. Instrum. Methods Phys.Res.* A **654**, 354.
Taylor M et al., 2009, *Nucl. Instrum. Methods Phys. Res.* A **606**, 589.
[6] Guastalla G et al., 2013, *Phys. Rev. Lett.* **100**,172501.
[7] Kumar R et al., 2010, *Phys. Rev.* C **81**, 024306.

Journal of Physics: Conference Series **533** (2014) 012048 doi:10.1088/1742-6596/533/1/012048

Possibilities of research for on-line mass separator with heavy ion reactions

I. Siváček[1,2], J. Kliman[1,2], A.M. Rodin[2], Ľ. Krupa[2,3], A.V. Belozerov[2], A.V. Podshibyakin[2], V.S. Salamatin[2], S.V. Stepantsov[2], V.Yu. Vedeneev[2]

[1] Institute of Physics, Slovak Academy of Sciences, Bratislava, Slovakia
[2] Joint Institute for Nuclear Research, Joliot-Curie 6, 141 980 Dubna, Russia
[3] Institute of Experimental and Applied Physics, Czech Technical University, Prague, Czech Republic

E-mail: ivan.sivacek@savba.sk

Abstract. Mass Analyser of Superheavy Atoms is ISOL – type setup created for direct mass measurement heavy ions. Hot catcher and ECR ion source combination allows effective formation of secondary beams of volatile elements. Powerful magnetic analysing system offers possibility to achieve mass resolution M/ΔM > 1000 in the focal plane silicon strip detector. The efficiency, time characteristics and detection system properties are described. Two applications of setup in different fields of research are presented together with methodology of experiments and data analysis.

1. Introduction

To study the properties of nuclei at the limits of nucleon stability the most accurate separation of isotopes of exotic nuclei is needed. For this purpose, mass separators of heavy ions are widely used. One of them is the Mass Analyzer of Superheavy Atoms – MASHA, built in Flerov Laboratory of Nuclear Reactions (FLNR) at the Joint Institute for Nuclear Research (JINR) in Dubna, Russian Federation.

2. Separator characteristics

MASHA is the ISOL-type of setup consisting of target block with hot catcher, electron cyclotron resonance (ECR) ion source, ion – optical system of magnetic separator and focal plane with detectors (figure 1). Primary beam is delivered from U400M cyclotron (up to ≈ 500 pnA ^{48}Ca). Exotic nuclei can be produced in nuclear reactions on thick or thin targets. Reaction products are thermalized in plate made of special fiber-like graphite heated up to 2000 °C by tantalum heater and transported into 2.45 GHz ECR ion source developed in FLNR [1]. Reaction products are ionized to 1^+ state and formatted into secondary beam with efficiency 84% for ^{nat}Xe (ionization efficiency is proportional to Z). Mass resolution for ^{nat}Xe is M/ΔM = 1300. For details of magnetic analyser properties see [2]. Main part of focal plane detection system is the silicon multi-strip detector for alpha and fission fragments spectroscopy. Focal plane is divided into 192 strips with another 160 strips placed around in well-shape geometry, as can be seen on figure 2. Thickness of silicon strips is 300 μm with insensitive surface layer < 50 nm. Specifications can be found in [3]. Secondary beam of 40 keV is unable to pass through dead layer, so there is no information about implantation of incoming ions.

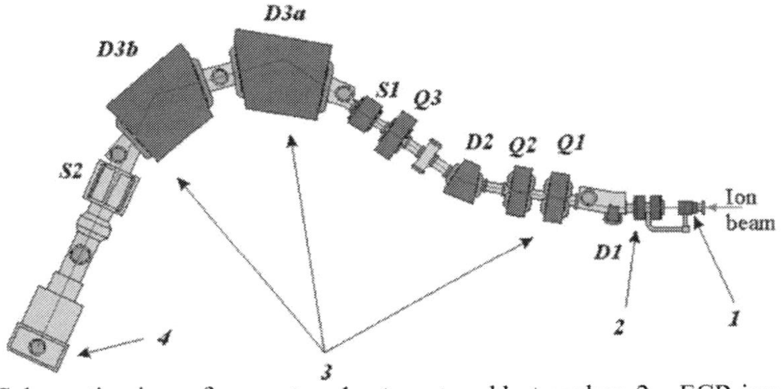

Figure 1: Schematic view of separator. 1 – target and hot cacher, 2 – ECR ion source, 3 – magnetic separator (D–dipole, Q–quadrupole and S–sextupole magnets), 4 – focal detectors.

Detector calibration is provided using ^{226}Ra spectrometric source. Energy resolution is 25 keV for 5.5 MeV alpha particles. Due to thin TiO_2 foil covering source and detector dead layer, peak energy correction is needed. Monte Carlo simulation in GEANT 4 showed that alpha particles lose 32 – 68 keV on their way to sensitive volume depending on geometrical conditions. Experimentally measured energies of alpha particles in model experiments agreed with tabular values with uncertainty ± 5 keV. Simulation of geometrical efficiency for decay in focal plane showed that 92 – 95 % of alpha particles are registered in first decay, depending on the implantation point. Alpha decay of daughter nucleus has chance to be registered about 70 %. Signal from each strip has its own independent spectrometric tract. Time necessary for signal processing is < 5 μs depending on pulse height. Secondary beams of stable currents are detected by true copy of focal plane made of copper structure on glass fibre laminate (mainly used for measurement of noble gasses currents).

Time characteristics of particular separation processes were estimated from measurement of calibrated leaks of noble gasses into the ECR ion source and hot catcher chambers, experiments on ^{40}Ar beam without target and by measuring decays of Hg isotopes created in reaction ^{144}Sm (^{40}Ar, xn)$^{184-x}$Hg. Total time of separation for isotopes $^{179-181}$Hg was assessed to 1.8 ± 0.3 s (from creation in nuclear reaction to implantation in focal plane detector). Transport efficiency estimated from overall yield of 2n, 3n and 4n channel of this reaction is 7 ± 1.4 %. Procedure of those measurements can be found in [4].

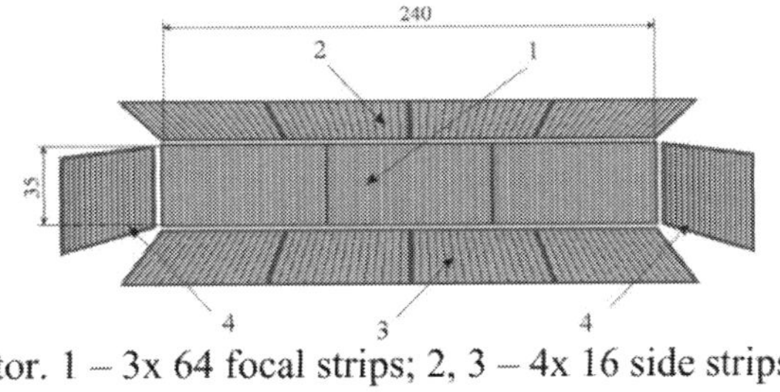

Figure 2: Multi-strip detector. 1 – 3x 64 focal strips; 2, 3 – 4x 16 side strips; 4 – 2x 16 side strips.

3. Fields of research

Main purpose of MASHA separator is direct mass measurement of isotopes of superheavy elements (SHE) on the principle of ion trajectory bending in magnetic field. Yet known information about masses of SHE were estimated from systematic of alpha decay. MASHA's aim is to confirm those data by method based on different principles. Isotopes of SHE will be synthetized in fusion-evaporation reactions of ^{48}Ca projectiles with actinide targets. Decay of all SHE from Z = 112 to Z = 118 created in this hot fusion method were observed on another setup at the FLNR – Gas Filled Recoil Separator (GFRS) [5]. First proposed is mass measurement of isotope ^{283}Cn created in reaction ^{238}U(^{48}Ca,3n)^{283}Cn.

Specific properties of hot catcher – ECR ion source combination allows to form secondary beams of volatile elements. Spectroscopic investigations of neutron shell closures N = 126, 152 can be performed on neutron-rich isotopes of radon and mercury. Their efficiency has been verified in model experiments [4]. Possibility of multi nucleon transfer reactions of ^{48}Ca with thick target ^{232}Th target in pyrolytic carbon matrix of catcher leads to production of neutron-rich isotopes of radon with $A \leq 232$. Estimated cross section for this reaction was evaluated using method described in [6] to predict the possibility to access the Rn isotopes reaching, even surpassing the presently known most neutron rich Rn isotopes. Isotopes ^{227}Rn – ^{232}Rn have from 3 to 5 β^- decays in chain before accessing the long lived alpha radioactive daughter. A GEANT 4 simulation was carried out to estimate possibility to measure decay of subjected isotopes by silicon detector. Since electrons are less ionizing particles than alphas, a 300 μm silicon thickness is not enough for complete stopping.

4. Events registration

When the magnetic fields inside separator magnets are set to requested values, secondary beam ions are focused to mass spots in the focal plane. Masses of radioactive nuclei are identified by their alpha decay in corresponding strips. In the experiments where a good statistics can be achieved, exact atomic mass up to 10^{-5} a.m.u. can be assessed from the mass peak position on the focal plane.

Neutron-rich Rn isotopes emit beta particles with maximum energy up to 5 MeV. Beta particles are less ionizing than alphas, so their track in silicon is longer resulting to incomplete energy deposition of particle in sensitive volume. Simulations in GEANT 4 showed that energy spectrum of beta particles in silicon block of 14 x 14 x 0.3 mm dimensions is strongly deformed, with most probable energy deposition about 120 keV and maximum about 1 MeV (figure 3a). Corresponding track length achieves most probable value about 340 μm (figure 3b). In lower part of spectrum noise-like background consisting of x-rays occurs. While x-ray part of spectrum takes place in low energies, β part of spectrum starts to rise above 60 keV. Since beta decays cannot be identified from energy spectrum, crucial parameter is probability of observing all β^- particles from one decay chain with suitable energies. Isotopes of interest are primary the heaviest with A = 229 – 232. All of them have at least $4\beta^-$ in decay chain before accessing long living alpha emitters. Probabilities are shown in table 1.

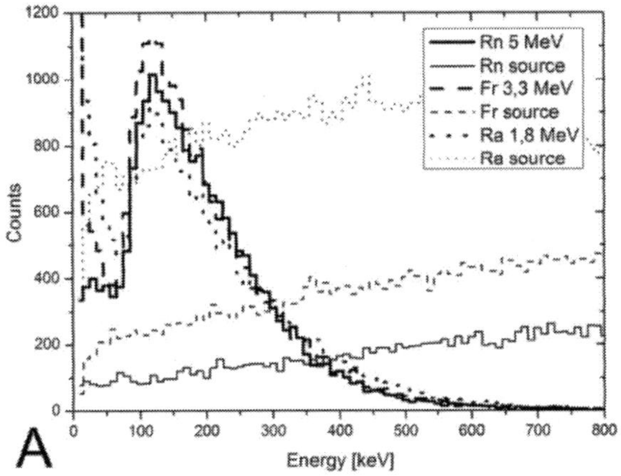

Figure 3a: Energy spectra of electrons in silicon – simulation.

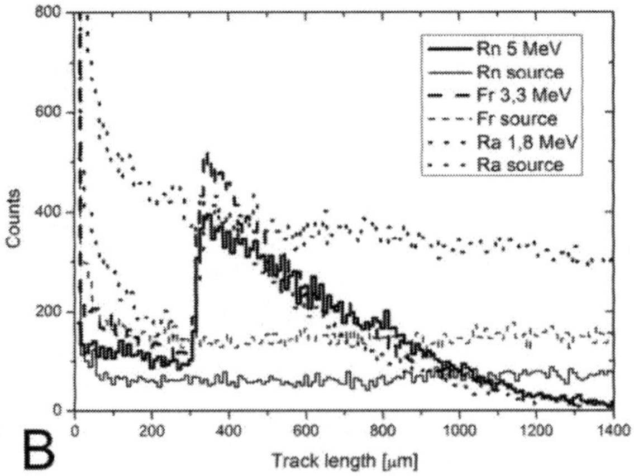

Figure 3b: Track lengths of electrons in silicon – simulation.

Table 1: Coincidence detection probabilities of neutron-rich Rn isotopes decay chains.

Isotope	^{227}Rn	^{229}Rn	^{230}Rn	^{231}Rn	^{232}Rn
Beta particles in coincidence	3β	4β	4β	4β	4β
Event detection probability [%]	0.654	0.187	0.149	0.223	0,2

5. Monte Carlo simulations of particle registering processes

Monte Carlo simulations in GEANT 4 were performed on the software suite Geant4 9.6 patch01 on RedHat Scientific Linux 5.8 platform for VMware [7]. Detector construction represents 352 silicon strips with exactly the same dimensions as real detector. Thickness of strips was 300 μm and on the surfaces of all detector planes were added a 50 nm thick plate of silicon representing insensitive layer. During a simulation of particle track, data collection algorithm stores the data from the strip with the biggest energy deposition. All materials used in construction of setup are used from GEANT4 internal database. Low energy electromagnetic physics was simulated employing PENELOPE model (penetration and energy loss of positrons and electrons) including processes of atomic relaxation and de-excitation (XRF, Auger, PIXE). Particles were created using GEANT4 General Particle Source. Energy calibration correction were performed by comparison of ^{226}Ra atoms decay with zero initial kinetic energy in point source and isotropic point source emitting alpha particles of defined energy, both covered by 100 nm layer of titanium dioxide. Geometrical efficiency was estimated by comparison of beam of 100% alpha decaying self-defined isotope in point source directing into various points on focal plane and also by isotropic point source of alpha particles lying 10 nm under surface of dead layer of well-shaped detector (only efficiency for first decay).

Adapted decay chains radon isotopes were used in simulation of radon isotopes decay instead of GEANT 4 radioactive decay database. Tabular values [8] were used and all isotopes were set to 100% single β^- emitters. Energy spectrum was acquired from implantation of 40 keV ions in beam of 5 mm radius standard deviation of beam position profile $\sigma_r = 0.5$ mm, pointing to geometrical centre of track detector volume. Information about energy spectra of all single isotopes in decay chains together with all particle track lengths and corresponding energies per track were stored. In the case of ^{231}Rn only 4 β^- decays were simulated, while coincidence detection probability for 5 β^- tracks is $3.5.10^{-4}$ and half-life of ^{231}Th is 25.5 hours. Searching for only 4 β^- tracks raised this probability by one order of magnitude. In the simulations with multi-strip detector 10^7 primary events were created and to understand processes β particles registration the statistics was 10^6 primary events.

6. Conclusion

Experiments with calibrated noble gas leaks and experiments with ^{40}Ar beam showed that MASHA separator has 7 ± 1.4 % transport efficiency and 1.8 ± 0.3 s separation time for mercury isotopes. Mass resolution is $M/\Delta M = 1300$ for natural xenon isotopes. MASHA is prepared for mass measurement of ^{283}Cn synthetized in reaction $^{238}U(^{48}Ca,3n)^{283}Cn$. Monte Carlo simulations in GEANT 4 outlined possibilities for measurement of neutron rich Rn isotopes decay by compact silicon tracking device. Characterization of registration processes inside tracking detector together with probabilities of detecting 4β event and geometrical properties of well-shaped silicon strip focal detector together with its energy calibration correction were evaluated.

7. Acknowledgement

This work was supported by the Russian Fond for Fundamental Research, grant № 11-02-12150 ofi m.

References

[1] A. Efremov et al.: Nucl. Instr. Meth. B **204** (2003) 368
[2] V.A. Shchepunov et al.: Nucl. Instr. Meth. B **204** (2003) 606
[3] http://www.canberra.com/products/detectors/pips-detectors-single-multiple.asp
[4] Rodin A.M. et al.: Preprint of the JINR Dubna, P15-2011-47 (2011)
[5] Yu.Ts.Oganessian et al.: Physical Review C **70**, 6, 064609-064609-14, 2004
Yu.Ts.Oganessian et al.: Physical Review C **74**, 4, 044602-044602-9, 2006
[6] M. Veselsky, G.A. Souliotis: Nucl. Phys. A **872** (2011) 1
[7] http://geant4.in2p3.fr/
[8] S. Goriely, F. Tondeur, J.M. Pearson: Atomic Data and Nuclear Data Tables **77** (2001) 311

Journal of Physics: Conference Series **533** (2014) 012049 doi:10.1088/1742-6596/533/1/012049

Electromagnetic transition strengths in ^{33}S

B. I. Dimitrov[1], N. Goutev[1], G. Tz. Gavrilov[1], D. Tonev[1], P. Petkov[1], G. de Angelis[2], F. Recchia[2], E. Farnea[3], C. A. Ur[3], S. Aydin[2,4], P. G. Bizzeti[5], A. M. Bizzeti-Sona[5], I. Deloncle[6], A. Gottardo[2], H. Laftchiev[1], S. Lunardi[3], D. Mengoni[2], C. Michelangoli[2], D. R. Napoli,[2],R. Orlandi[2], E. Sahin[2], E. A. Stefanova[1], J.J. Valente-Dobon[2], Tz. K. Marinov[1], M. S. Yavahchova[1]

[1]Institute for Nuclear Research and Nuclear Energy, BAS, BG–1784, Sofia, Bulgaria.
[2]INFN, Laboratori Nazionali di Legnaro, I–35020, Legnaro, Italy.
[3]Dipartimento di Fisica and INFN, Sezione di Padova, Padova, Italy.
[4]Aksaray University, Department of Physics, 68100 Aksaray, Turkey.
[5]Dipartimento di Fisica dell'Universitá and INFN Sezione di Firenze, Firenze, Italy.
[6]CSNSM, IN2P3/CNRS et Université Paris-Sud, Orsay, France.

Abstract. An experiment using the Doppler Shift Attenuation Method was performed for ^{33}S at Laboratori Nazionali di Legnaro using the multi-detector array GASP. Excited states were populated in the fusion-evaporation reaction ^{24}Mg(^{14}N,α p)^{33}S. The data were analyzed using the Differential Decay Curve Method with gates set on the shifted component of a directly feeding transition thus eliminating the problem of the unobserved feeding. Reliable and precise lifetimes were determined and the data derived for the reduced transition probabilities are going to be compared to the predictions of shell model calculations.

1. Introduction

The shell model calculations provide a good description of the level energies and transitions strengths in the nuclei around ^{33}S. For low spin states, sd-interactions are applicable but for higher spin states the fp shell has to be taken into account [2, 3, 4, 5]. As a part of the spectroscopy of high-spin states in ^{33}S, lifetimes were investigated using the Doppler-shift attenuation method (DSAM). These data provide absolute transition strengths which are crucial for testing the predictions of nuclear models. The data were analyzed using gates from above, set on shifted component of a directly feeding transition. In this way, the longstanding problem of the unknown (unobserved) feeding inherent to singles measurements is eliminated and much more reliable lifetimes are obtained [6]. The aim of the present work was to extract precise lifetime values of excited states in ^{33}S and to measure electromagnetic transition strengths in this nucleus.

2. Experiment

High-spin states of ^{33}S have been populated via the fusion-evaporation reaction ^{24}Mg(^{14}N,α p) at 40MeV bombarding energy. The beam, delivered by the LNL XTU-Tandem accelerator impinged on the target with an average beam current of 5 pnA. The target, 1mg/cm^2 thick, consisted of 99.7% isotopically enriched ^{24}Mg evaporated on a 8 mg/cm^2 gold foil. The γ rays emitted in the reaction were detected using the 4π-GASP array composed of 40 Compton-suppressed large-volume, high-purity Ge detectors arranged in seven rings at different angles with respect to the beam axis. Events were collected when at least two germanium detectors fired in coincidence. Energy and efficiency calibrations were performed with standard γ-ray sources of ^{56}Co and ^{152}Eu. After gain-matching and corrections for shifts, the data were sorted into 49 γ-γ matrices, corresponding to all possible combinations of pairs of detectors positioned at the seven rings at polar angles with respect to the beam axis of 34°, 60°, 72°, 90°, 120° and 146°.

3. Data analysis and results

In order to obtain lifetime information the data have been analyzed by applying a modern version of the Doppler-shift attenuation method.

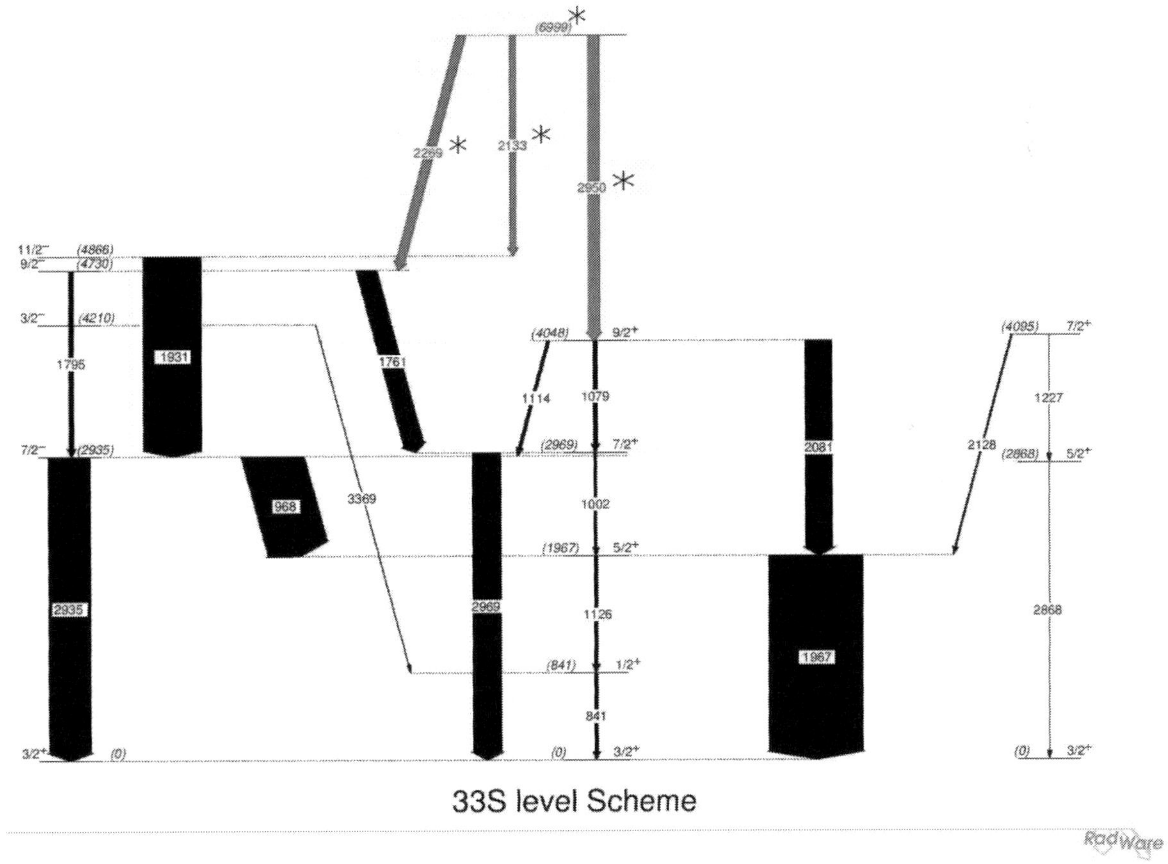

Figure 1. The analysis is based on the extended level scheme of ^{33}S. The new level is marked with a star as are the three new γ -ray transitions.

Figure 1 shows the extended level scheme of ^{33}S. More information can be found in [1] The gate was set only on the shifted component of a directly feeding transition. This component corresponds to emission during the slowing down of the recoiling nucleus in the target or the stopper and thus contains useful timing information. The resulting line-shape of the γ-transition depopulating the level of interest is then decomposed into shifted component associated with emission during the slowing down (SS), and unshifted component associated with emission at rest (U). The lifetime τ_a of the level of interest can be derived from the areas A_u and A_{ss} of these components through [8]:

$$\tau_a = \{B_{ss}, A_u\}/(d\{B_{ss}, A_{ss}\}/dt_s) \quad (1)$$

where t_s is the time at which the recoil comes to rest.

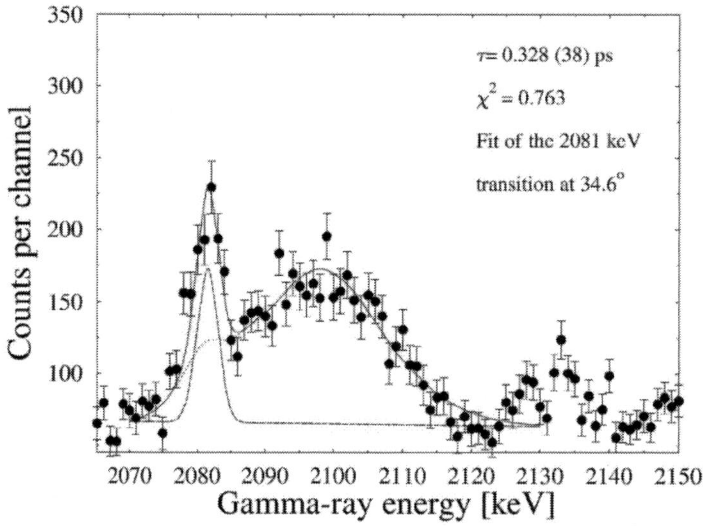

Figure 2. Line-shape of the 2081 keV transition at 34.6° generated by the sum of gates set on the shifted component of 2951 keV and determination of the lifetime.

The analysis was carried out within the framework of the Differential decay curve method (DDCM) [7] according to the procedure outlined in [8, 9] where details about the Monte-Carlo simulation of the slowing down process, determination of stopping powers and fitting of line shapes can be found. To increase the statistics, we used equivalent gates set at the four rings with appreciable Doppler-shifts and summed-up the resulting spectra. Thus, lifetimes were derived independently at these four rings. In this way, with two forward and two backward rigs, four values for the lifetime τ were determined. The final value of τ was derived by averaging the individual results.

In Figure 2 is shown an example of the line-shape analysis of the transition of 2081 keV at 34.6°. The derived value for τ for the $9/2^+$ level is shown too. In Figure 3 is shown again an example for the same transition at 145.4°. Our analysis confirms previously known lifetimes. More details on the lifetime analysis will be given in a forthcoming paper.

4. Summary and conclusions

For the investigation to the level-scheme of ^{33}S we have performed a DSAM experiment at the Laboratori Nazionali di Legnaro, Italy with the GASP array. For the data analysis, the Differential Decay Curve Method was used in coincidence, with a gate from above, set on the shifted component of a directly feeding transition. Two previously known lifetimes in ^{33}S were confirmed. Further work on the lifetime measurement is in progress.

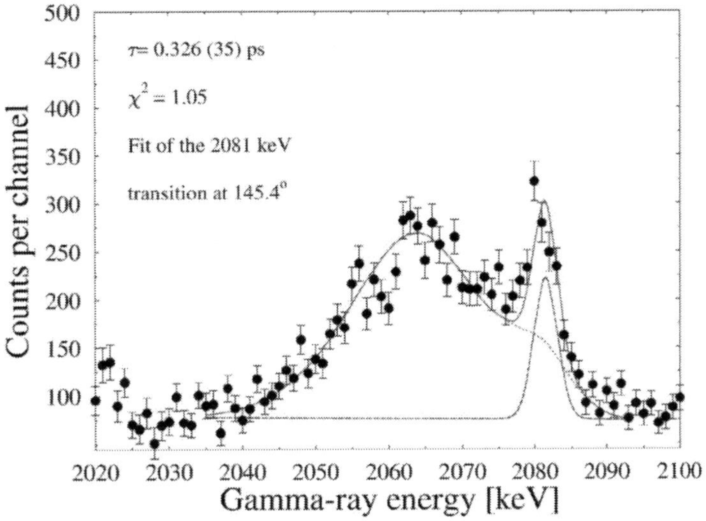

Figure 3. Example of the line-shape of the 2081 keV transition at 145.4° generated by the sum of gates set on the shifted component of 2951 keV and determination of the lifetime.

5. Acknowledgment

This research has been supported by Bulgarian Science Fund under contract DFNI-E 01/2 and by a NUPNET - NEDENSAA project funded by the Bulgarian Ministry of Education and Science.

References

[1] G Tz Gavrilov to be published
[2] M Ionescu-Bujor *et al.* 2009 *Phys. Rev.* C **80** 034314
[3] S Aydin *et al.* 2012 *Phys. Rev.* C **86** 024320
[4] S Aydin *et al.* 2014 *Phys. Rev.* C **89** 024320
[5] A Bisoi *et al.* 2014 *Phys. Rev.* C **89** 024303
[6] D Tonev *et al.* 2002 *Phys. Rev.* C **65** 034314
[7] G Böhm *et al.* 1993 *Nucl. Instr. Meth.* A **329** 248
[8] P Petkov *et al.* 1999 *Nucl. Instr. Meth.* A **A 437** 274
[9] P Petkov *et al.* 1998 *Nucl. Phys.* A **640** 293

Journal of Physics: Conference Series **533** (2014) 012050 doi:10.1088/1742-6596/533/1/012050

Level-scheme investigation of ^{33}S

G. Tz. Gavrilov[1], N. Goutev[1], B. I. Dimitrov[1], D. Tonev[1], P. Petkov[1], G. de Angelis[2], F. Recchia[2], E. Farnea[3], C. A. Ur[3], S. Aydin[2,4], P. G. Bizzeti[5], A. M. Bizzeti-Sona[5], I. Deloncle[6], A. Gottardo[2], H. Laftchiev[1], S. Lunardi[3], D. Mengoni[2], C. Michelangoli[2], D. R. Napoli,[2],R. Orlandi[2], E. Sahin[2], E. A. Stefanova[1], J.J. Valente-Dobon[2], Tz. K. Marinov[1], M. S. Yavahchova[1]

[1]Institute for Nuclear Research and Nuclear Energy, BAS, BG–1784, Sofia, Bulgaria.
[2]INFN, Laboratori Nazionali di Legnaro, I–35020, Legnaro, Italy.
[3]Dipartimento di Fisica and INFN, Sezione di Padova, Padova, Italy.
[4]Aksaray University, Department of Physics, 68100 Aksaray, Turkey.
[5]Dipartimento di Fisica dell'Universitá and INFN Sezione di Firenze, Firenze, Italy.
[6]CSNSM, IN2P3/CNRS et Université Paris-Sud, Orsay, France.

Abstract.
An angular correlation experiment was carried out for ^{33}S at Laboratori Nazionali di Legnaro with the gamma-ray detector array GASP. The reaction used was ^{24}Mg(^{14}N,α p)^{33}S at a beam energy of 40MeV. An analysis of DCO ratios and triple gamma coincidences was performed. So far, a new level depopulated by 3 γ -ray transitions has been found and its spin was determined. The work for further extension of the level scheme is in progress.

1. Introduction

The investigation of nuclei along the N=Z line is of considerable interest since it directly addresses the validity of the charge symmetry of the nuclear forces and the role of the Coulomb effects on nuclear structure [1, 2]. During the last years the interst to study nuclei in the mass region A $\sim$ 30 arose significantly. This fact is due to the efficient gamma-spectrometers and large variety of beams provided by accelerators. Level-schemes and lifetimes of low-spins states in this mass-region are known, but they need to be remeasured, especially lifetimes which are derived with limitted number of detectors, in some cases with one or two germanium detectors (see e.g. [3]). For low spin states, sd-interactions are applicable but for higher spin states the fp shell has to be taken into account [4, 5, 6, 7]. One may expect a number of interesting physical phenomena that can be studied in the hypothetical superdeformed configuration of ^{32}S, like the shape coexistence, proton neutron pairing and its deformation dependence, effects related to the time-odd components of nuclear mean fields, as well as nuclear - molecular and nuclear - cluster structures [8]. The nucleus ^{33}S is well studied and the relevant spectroscopic information is summarized in [9]. Nevertheless, the data on high-spin states is scarce. The aim of the present work was to provide such new data which allow a comparison with the predictions of shell model calculations at higher spins including fp-shell effects. Using the reaction ^{24}Mg(^{14}N,αp)^{33}S excited states in ^{33}S were populated and observed. The level scheme of ^{33}S was investigated in details.

2. Experiment

High-spin states in ^{33}S have been populated via a fusion-evaporation reaction at 40MeV bombarding energy. A ^{14}N beam, delivered by the LNL XTU-Tandem accelerator impinged on the target with an average beam current of 5 pnA. The 99.7% isotopically enriched ^{24}Mg target, 1mg/cm^2 thick, was evaporated on a 8 mg/cm^2 gold layer. The γ rays emitted in the reaction were detected by the 4π-GASP array composed of 40 Compton-suppressed large-volume, high-purity Ge detectors arranged in seven rings at different angles with respect to the beam axis. Events were collected when at least two germanium detectors fired in coincidence. Energy and efficiency calibrations were performed with standard γ-ray sources of ^{56}Co and ^{152}Eu. The data were sorted into a symmetric γ-γ-γ cube, a symmetric γ-γ matrix, and seven asymmetric matrices having the detectors at 34°, 60°, 72°, 90°, 120°, 146°, respectively, on the first axis and all detectors on the second axis.

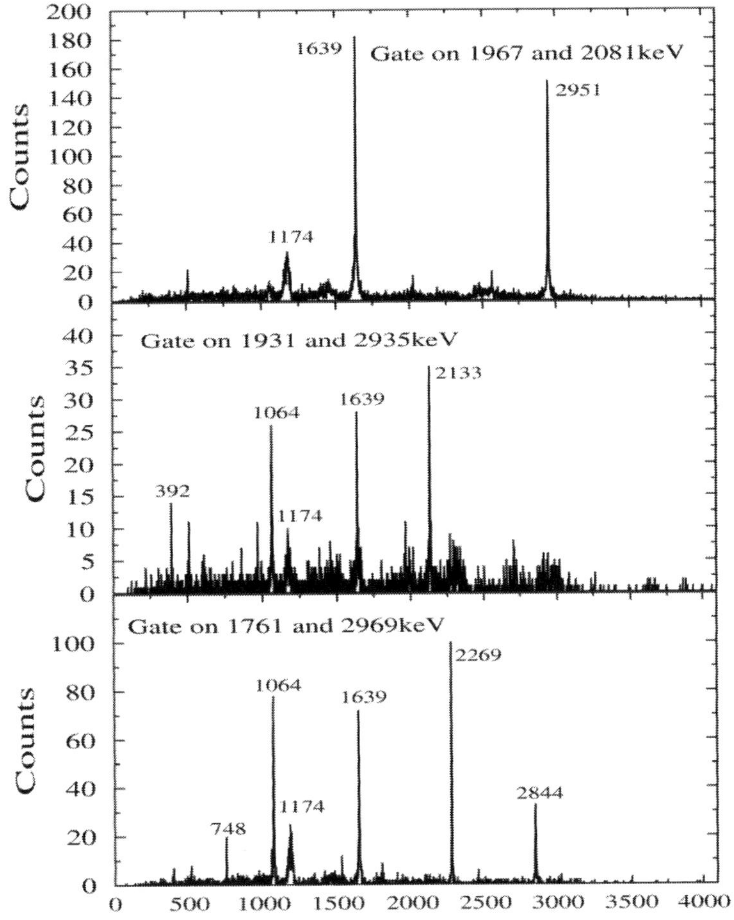

Figure 1. Examples of background-subtracted coincidence spectra obtained from the symmetric γ-γ-γ cube gated by (a) 1967- and 2081-keV γ -rays, (b) 2935- and 1931-keV γ -rays, and (c) 2969- and 1761-keV γ -rays (cf. Figure 2). The most intense peaks, taking into account the detector efficiency for the different energies, are marked and underlined (2951-, 2133- and 2269-keV). By adding their energies to the previously known levels, which are deexcited by the gated transitions, a brand new level with an excitation energy of 6999 keV was found.

The symmetric matrices have been used to study γ-γ coincidence relationships for the construction of the level scheme, while the asymmetric matrices were used to obtain information about the γ transition multipolarities and for lifetime determinations.

3. Data analysis and results

The level scheme of ^{33}S has been extended to higher spins and excitation energies with respect to previous studies. One new state and three γ-ray transitions have been added. The level scheme has been constructed based on double-gated γ-ray spectra. Typical examples of spectra obtained with double gate are shown in Figure 1.

In order to obtain information on the multipolarities and mixing ratios of the observed γ -ray transitions, an angular distribution analysis was performed by using efficiency-corrected γ -ray intensities, $I(\theta)$, obtained from the asymmetric matrices by gating on the axis where all detectors are included. The angular distributions were fitted with Legendre polynomials with free angular coefficients $A_{2,4}$. For transitions with lower intensity the multipolarity was determined by using the angular distribution ratio R_{ADO} defined by the formula in Ref. [5]:

$$R_{ADO} = (I(34.6°) + I(145.4°))/2I(90.0°) \tag{1}$$

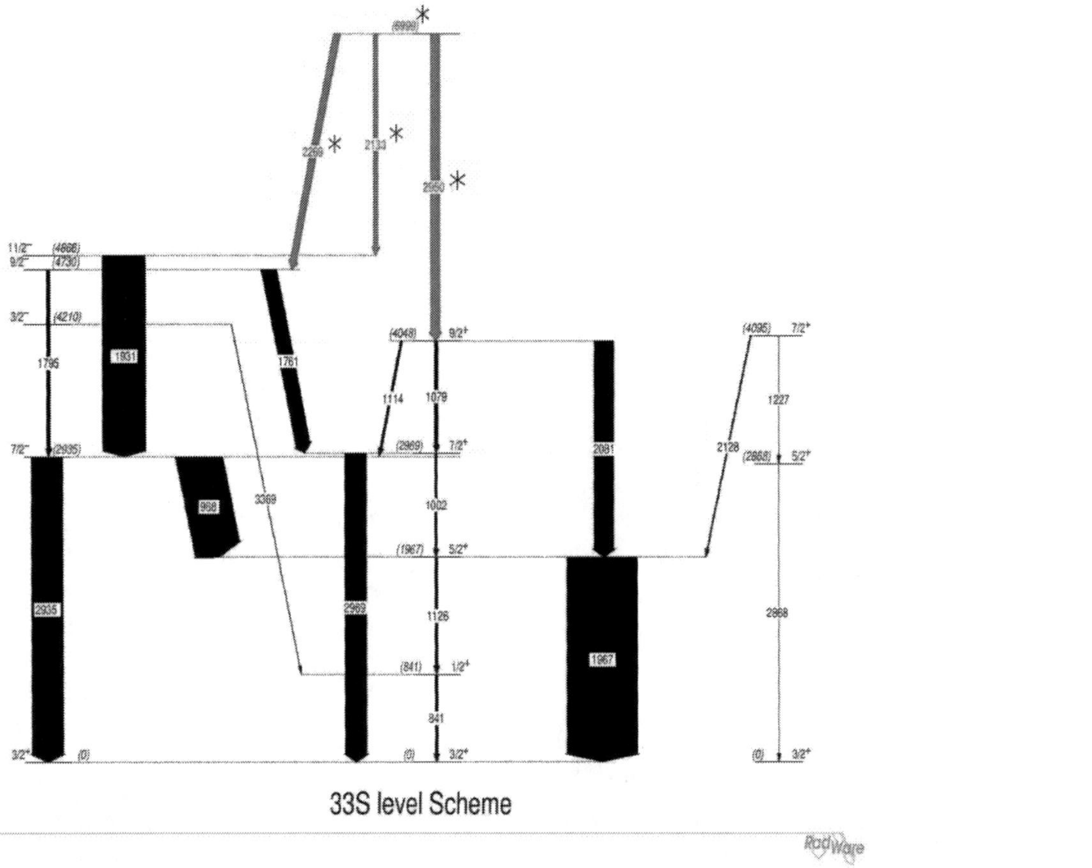

Figure 2. The extended level scheme. The new level is marked with a star as are the three new γ -ray transitions.

In the GASP geometry R_{ADO} values for stretched dipole and stretched quadrupole transitions are about 0.8 and 1.4, respectively. For transitions with mixed character the situation is not so well defined, because the angular distribution ratio depends on the value and sign of the mixing ratio δ. Using this method the multipolarities of previously known transitions have been redetermined and proved with full exactness. The multipolarities of the three new γ -ray transitions have also been determined, all leading to the conclusion that the newly found level at 6999 keV has a spin of $11/2^{+}\ \hbar$.

4. Summary and conclusions

A new high-spin level with an excitation energy of 6999 keV in ^{33}S was found along with three new gamma transitions which deexcite it. Further search for higher-lying levels is in progress. The results obtained so far are promising. Rich spectra with very good statistics guarantee that more results can be expected in the near future. Comparison with shell-model calculations is also going to be performed in order to support the proposed spins.

5. Acknowledgments

This research has been supported by Bulgarian Science Fund under contract DFNI-E 01/2 and by a NUPNET - NEDENSAA project funded by the Bulgarian Ministry of Education and Science.

References

[1] S M Lenzi *et al.* 2001 *Phys. Rev. Lett.* **87** 122501
[2] D Tonev *et al.* 2002 *Phys. Rev.* C **65** 034314
[3] J A J Hermans *et al.* 1974 *Phys. Lett.* **50B** 337
[4] M Ionescu-Bujor *et al.* 2009 *Phys. Rev.* C **80** 034314
[5] S Aydin *et al.* 2012 *Phys. Rev.* C **86** 024320
[6] S Aydin *et al.* 2014 *Phys. Rev.* C **89** 024320
[7] A Bisoi *et al.* 2014 *Phys. Rev.* C **89** 024303
[8] H Molique, J Dobaczewski and J. Dudek 2000 *Phys. Rev.* C **61** 044304
[9] J Chen and B Singh 2011 *Nuclear Data Sheets* **112** 1393

Journal of Physics: Conference Series **533** (2014) 012051 doi:10.1088/1742-6596/533/1/012051

Studies of neutron cross-sections important for spallation experiments using the activation method

J Vrzalová[1,2], P Chudoba[1], A Krása[1], M Majerle[1], M Suchopár[1,3], O Svoboda[1] and V Wagner[1,3]

[1] Nuclear Physics Institute of the ASCR, 250 68 Řež near Prague, Czech Republic
[2] Joint Institute for Nuclear Research, Joliot-Curie 6, 141 980 Dubna, Russia
[3] Faculty of Nuclear Sciences and Physical Engineering, Czech Technical University in Prague, Břehová 7, 115 19 Prague, Czech Republic

E-mail: vrzalova@ujf.cas.cz

Abstract. A series of experiments devoted to studies of neutron cross-sections by activation method was carried out. The cross-sections of various threshold reactions were studied by means of different quasi-monoenergetic neutron sources with energies from 14 MeV up to 100 MeV. Threshold reactions in various materials are among other used to measure fast neutron fields produced during accelerator driven system studies. For this reason our measurements of neutron cross-sections are crucial. At present, neither experimental nor evaluated data above 30 MeV are available for neutron threshold reactions in Au, I and In published in this proceedings. We studied materials in the form of thin foils and compared our data with the calculations preformed using the deterministic code TALYS 1.4.

1. Introduction

Suitable activation detectors are necessary for monitoring of neutron fields in different applications. One example is the study of the international collaboration "Energy and Transmutation of Radioactive Waste" at Joint Institute for Nuclear Research Dubna, Russia. The results of the experiments with activation detectors that were carried out with the spallation setup QUINTA [1] are presented also in these proceedings. This contribution is focused on the experiments carried out at The Svedberg Laboratory (TSL) in Uppsala, Sweden and Nuclear Physics Institute (NPI) in Řež, Czech Republic, especially on the determination of cross-section of threshold reactions, which were not published in [2-4]. We would like to mention that the values published in [2], denoted as $^{89}Y(n,3n)^{87}Y$ are actually the ground state ^{87g}Y production cross-section (not the total cross-sections).

2. Measurements and data analysis

Proton beam energies from 25 to 97 MeV were used at TSL and proton beam energies from 20 to 38 MeV at NPI. The activation materials Au, Bi, I, In and Ta were in the form of foils with dimensions from 1.5 x 1.5 cm^2 up to 3 x 3 cm^2 and thickness from 50 μm up to 1 mm. The iodine samples were in the form of KIO_3 powder pressed into pellets.

The activation method and γ-ray spectrometry were used for cross-section determination. Measured gamma spectra were analyzed by the DEIMOS code [5]. After identification of the isotope to a gamma peak the yield of activated material was calculated. The cross-sections were determined taking into account the number of atoms in a sample and neutron fluence [2]. Neutron spectra in NPI experiments were estimated by interpolating the data from an identical setup [6], neutron spectra in TSL experiments were calculated using the algorithm from [7] for the measured proton energies. The second approach was also applied to determining the neutron spectra in NPI with the help of MCNPX simulation [4]. The presented uncertainties of the neutron energies are actually FWHM of the mono-energetic peaks.

Table 1. Reactions on tantalum, bismuth, natural indium, iodine and gold.

	^{181}Ta	^{209}Bi	^{209}Bi	^{nat}In	^{nat}In	^{127}I	^{197}Au
Reaction	(n,2n)	(n,5n)	(n,10n)	(n,xn)	(n,xn)	(n,8n)	(n,7n)
Isotopes	**^{180g}Ta**	**^{205}Bi**	**^{200}Bi**	**^{113m}In**	**^{112m}In**	**^{120}I**	**^{191}Au**
Half-life	8.152(6) h	15.31(4) d	36.4(5) m	1.6582(6) h	20.56(6) m	81(6) m	3.18(8) h
E_γ [keV]	93.326(2)	1764.36(4)	1026.5(2)	391.690 (15)	156.56(10)	560.44(2)	586.45(3)
Intensity [%]	4.5	32.5(6)	100	64.2	13.2(3)	73	17
E_n [MeV]	**cross-section [barn]**						
17.5(8)	---	---	---	0.023(3)	0.043(7)	---	---
30.4(8)	---	---	---	---	0.159(22)	---	---
47.0(6)	0.139(25)*	0.81(13)*	---	---	0.28(7)	---	----
59.0(4)	0.059(10)	0.30 (5)*	---	0.019(3)*	---	---	0.094(19)
66.4(4)	0.040(7)	0.19 (3)*	---	0.016(4)*	---	---	0.24(4)
72.8(4)	0.042(10)	0.151(28)	---	0.0169(29)*	---	---	0.174(26)
89.3(7)	---	0.104(18)*	---	0.0158(26)*	---	---	0.078(12)
94.0(7)	---	0.152(25)*	0.040(6)	---	---	0.075(25)	---

The high energy neutron source based on proton reactions on a lithium target has the energy spectrum which contains besides the main neutron peak also lower-energy continuum. When determining the neutron fluence we subtracted the contribution of these "background" neutrons (all neutrons below the mono-energetic peak) using the TALYS code. It is estimated as the ratio between the folding of calculated cross-section and neutron spectrum in the peak and the same convolution in the whole spectrum (for more details, see [2]).

The procedure is insensitive to the absolute value of the TALYS [8] cross-section, but sensitive to the shape of cross-section dependency on neutron energy. If we use TALYS models with different nuclear level density, the shapes of cross-section dependency on energy will change. We can calculate with five preset options of level densities; with three phenomenological level density models and two options for microscopic level densities (see figure 2b). Uncertainty caused by the influence of different level density models does not exceed 10 % and was added to the total cross-section uncertainty.

3. Results

The numerical values of measured cross-sections are in table 1 and in figures 1-2 are their graphical representation. In table 1 are also published decay data [9] for the reactions used in this work. New data were obtained in the energy regions where no data existed up to now. The experimental results are compared with model calculations using the TALYS 1.4 code (basic setting). The cross-sections of the reactions ^{181}Ta(n,2n)^{180}Ta, ^{209}Bi(n,5n)^{205}Bi and ^{209}Bi(n,10n)^{200}Bi are also compared with the data from the EXFOR database (for Ta [10]; for Bi [11]). Our data show good agreement with TALYS in cases of In, I and reaction (n,5n) on Bi. In case of the ^{181}Ta(n,2n)^{180}Ta, ^{197}Au(n,7n)^{191}Au and ^{209}Bi(n,10n)^{200}Bi reactions the shapes agree with the TALYS calculation, however, the measured cross-sections are smaller (approximately by 40% for Ta, by 50% for Au and even by 80% for (n,10n)

* These data have been already published in [2].

reaction on Bi), but we can see in figure 3a that our data are in agreement with other experimental data [11].

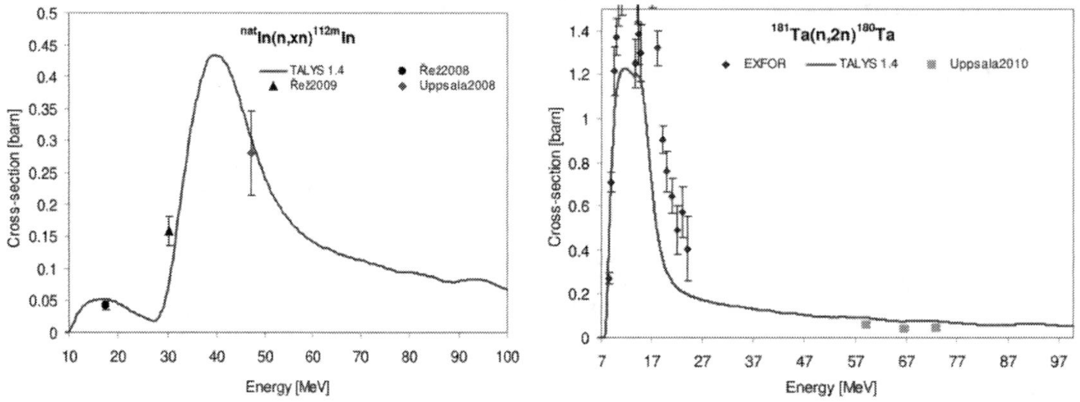

Figure 1. (a) Cross-section of ^{nat}In(n,xn)^{112m}In (b) Cross-section of ^{181}Ta(n,2n)^{180}Ta reaction.

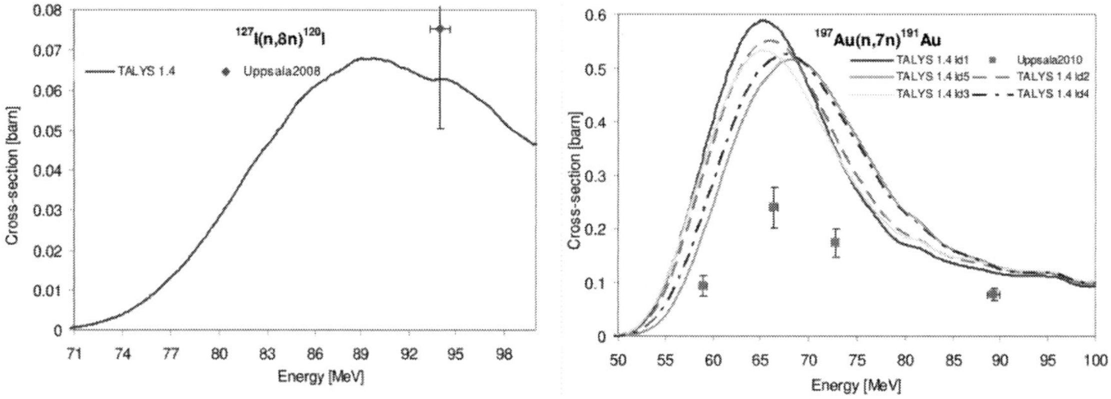

Figure 2. (a) Cross-section of ^{127}I(n,8n)^{120}I (b) Cross-section of ^{197}Au(n,7n)^{191}Au reaction (TALYS calculations with five available densities are shown as well).

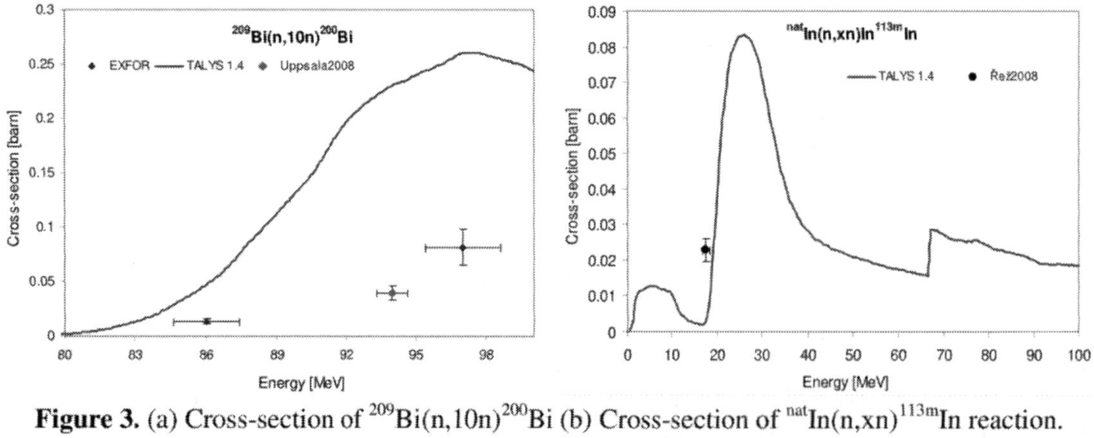

Figure 3. (a) Cross-section of ^{209}Bi(n,10n)^{200}Bi (b) Cross-section of ^{nat}In(n,xn)^{113m}In reaction.

The main contributions to the total uncertainty of cross-sections are uncertainties of Gauss-fit of peaks in DEIMOS (1-10%), detector efficiency (3%) beam intensity and neutron spectra (10% for both sources).

Journal of Physics: Conference Series **533** (2014) 012051 doi:10.1088/1742-6596/533/1/012051

4. Conclusion

We present measurements of neutron-induced reaction cross-sections of the reactions (n,xn) on natural indium, (n,2n) on tantalum, (n,5n) on bismuth, (n,10n) on bismuth, (n,8n) on iodine and (n,7n) on gold from 17 up to 94 MeV. Overall uncertainties in the best cases are near 15%. Our data were compared with data from the EXFOR database and with the TALYS 1.4 code. A good agreement between our data and TALYS was observed for In, I and (n,5n) on Bi, while TALYS predicts significantly higher values than the measured ones for Au, Ta and (n,10n) on Bi. The ^{209}Bi(n,5n)^{205}Bi and ^{209}Bi(n,10n)^{200}Bi are the only cases where earlier experimental data are available and they agree with each other. No evaluated data libraries contain cross-sections of the presented reactions on Au, In and I in the measured energy regions.

Using our measured cross-sections, it becomes possible to better analyze neutron fluxes in different places of experimental setups consisting of lead, natural uranium, and graphite irradiated by relativistic protons and deuterons to study transmutation of radioactive materials by spallation neutrons at JINR Dubna. Such studies are intended for providing important data serving for the construction of larger-scale Accelerator Driven Systems (ADS).

Acknowledgements

Authors are grateful to the operation crew of the TSL Uppsala, the personnel of fast neutron generators at the NPI Řež and the NPI cyclotron staff for irradiation and good beam parameters. This research was finnancially supported by the EFNUDAT program, grant number CTU0808214 and by the F4E program of the Nuclear Reaction Department of the Nuclear Physics Institute, F4E-2008-GRT-014.

References

[1] Suchopár M et al 2012 Monte Carlo Simulations of Natural Uranium Setups Irradiated With Relativistic Deuterons by Means of MCNPX Code *Proceedings of Science* (Baldin ISHEPP XXI **091**) http://pos.sissa.it/archive/conferences/173/091/Baldin%20ISHEPP%20XXI_091.pdf

[2] Vrzalová J et al 2013 Studies of (n,xn) cross-sections in Al, Au, Bi, Cu, Fe, I, In, Mg, Ni, Ta, Y and Zn by the activation method *Nuclear Instruments and Methods in Physics Research Section A: Accelerators, Spectrometers, Detectors and Associated Equipment* **726** 84-90 http://www.sciencedirect.com/science/article/pii/S0168900213007444

[3] Svoboda O et al 2010 Cross-section Measurements of (n,xn) threhold reactions in Au, Bi, I, In and Ta *Journal of the Korean Physical Society* **59** 1709-1712 http://www.nd2010.org/proceedings/viewfullpdf.php?no=91

[4] Chudoba P et al 2013 Measurement of cross-sections of yttrium (n,xn) threshold reactions by means of gamma spectroscopy *Physics Procedia*

[5] Frana J 2003 Program DEIMOS32 for gamma-ray spectra evaluation *Journal of Radioanalytical and Nuclear Chemistry* **257** 3 583-587

[6] Uwamino Y et al 1997 *Nuclear Instruments and Methods in Physics Research* A **389** 463

[7] Prokofiev A V et al 2002 Development and Validation of the ^{7}Li(p,n) Nuclear Data Library and Its Application in Monitoring of Intermediate Energy Neutrons *Journal of Nuclear Science and Technology* **1** 112-115

[8] Koning A J et al 2012 *Nuclear Data Sheets* **113** 2841

[9] LBNL Isotopes Project LUNDS Universitet http://ie.lbl.gov/toi/

[10] a) Frehaut J 1980 *U.S. report to the I.N.D.C* **84** 399
b) Vesser 1977 *Physical Review Part C Nuclear Physics* **16** 1792

[11] Kim E et al 1998 Measurements of neutron spallation cross-sections of ^{12}C and ^{209}Bi in the 20-150 MeV energy range *Nuclear Science and Engineering* **129** 209

Cross-section studies of important neutron and relativistic deuteron reactions

V Wagner[1,3], M Suchopár[1,3], J Vrzalová[1,2], P Chudoba[1], T Herman[1], O Svoboda[1], B Geier[1], A Krása[1], M Majerle[1], A Kugler[1], J. Adam[2], A Baldin[2], W Furman[2], M Kadykov[2], J Khushvaktov[2], A Solnyshkin[2], V Tsoupko-Sitnikov[2], S Tyutyunikov[2], L Zavorka[2], N Vladimirova[2], M Bielewicz[4], S Kilim[4], M Szuta[4], E Strugalska-Gola[4]

[1] Nuclear Physics Institute of the ASCR, 250 68 Řež near Prague, Czech Republic
[2] Joint Institute for Nuclear Research, Joliot-Curie 6, 141 980 Dubna, Russia
[3] Faculty of Nuclear Sciences and Physical Engineering, Czech Technical University in Prague, Břehová 7, 115 19 Prague, Czech Republic
[4] National Centre for Nuclear Research, 04-500 Otwock-Swierk, Poland

E-mail: wagner@ujf.cas.cz

Abstract. The cross-sections of relativistic deuteron reactions on natural copper were studied by the means of activation method. The deuteron beams produced by JINR Nuclotron (Russia) with energies from 1 GeV up to 8 GeV were used. Lack of such cross-sections prevents the usage of copper foils for beam integral monitoring. The copper monitors will help us to improve the beam integral determination during ADS studies. The yttrium samples are very suitable activation detectors for monitoring of neutron fields not only in the ADS studies. But experimental cross-section data for higher energy threshold neutron reactions are still missing. This situation is the reason why we have started to study neutron reactions on yttrium by the means of quasi mono-energetic neutron source based on NPI Řež cyclotron (Czech Republic).

1. Introduction

Suitable activation detectors are necessary for monitoring of proton and deuteron beams and also neutron fields in accelerator driven system (ADS) studies. One example is the study of the international collaboration “Energy and Transmutation of Radioactive Waste” (E&T RAW) at Joint Institute for Nuclear Research Dubna, Russia [1,2,3]. The results of the experiments with activation detectors that were carried out with the spallation setup QUINTA [2] are presented also in these proceedings. This contribution is focused on two subjects.

We use aluminium foils and ^{24}Na reaction for beam monitoring. Large distance from irradiated set-up is necessary in this case due to production of ^{24}Na by neutrons. On the other hand, the determination of the deuteron production of radionuclides on copper monitor is not influenced by neutron reactions and such monitor can be placed near set-up. Therefore we measured cross-sections of different radionuclides production by deuterons on copper.

We made also extensive studies of neutron reaction cross-sections by the means of quasi mono-energetic neutron sources at Nuclear Physics Institute (NPI) in Řež, Czech Republic and at The Svedberg Laboratory (TSL) in Uppsala, Sweden. These studies are described by J. Vrzalová in further contribution in these proceedings. The preliminary results of studies of neutron reactions on yttrium will be presented in this contribution.

2. Relativistic deuteron reactions on copper

Cross-sections of relativistic proton reactions on aluminium and copper are mostly known. Situation is very different for relativistic deuterons. Experimental cross-sections for ^{24}Na production by deuteron reaction on aluminium are very scarce for energies higher than 200 MeV and experimental cross-sections for deuteron reactions on copper are completely missing. For this reason, we have started a set of studies of these relativistic deuteron reactions.

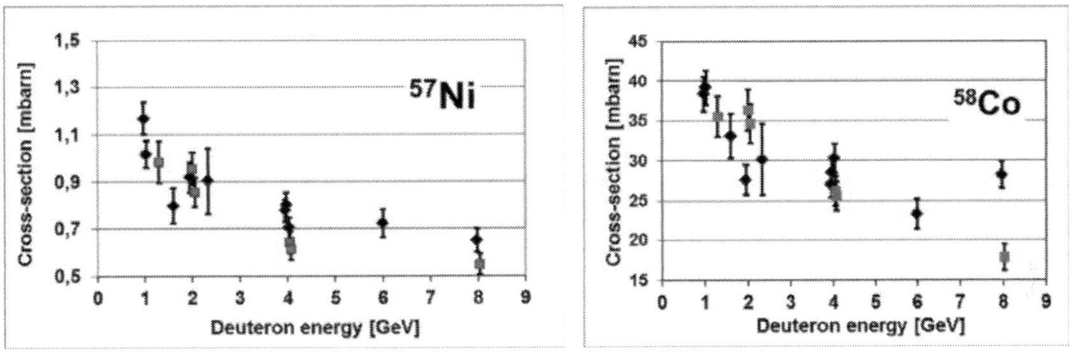

Figure 1. (a) Cross-section of ^{nat}Cu(d,x)^{57}Ni reaction (b) Cross-section of ^{nat}Cu(d,x)^{58}Co reaction.

The measurements were performed in the frame of E&T RAW collaboration during irradiations of QUINTA and GAMMA-3 set-ups with deuterons from Nuclotron accelerator at JINR Dubna. Sixteen irradiations were performed during five sets of experiments realized from 2011 up to 2013. The deuteron beam integrals were determined by production of ^{24}Na in aluminium foil. The aluminium and copper foils had the same sizes (10×10 cm) and thicknesses of copper and aluminium were 0.0128 cm and 0.0196 cm, respectively. Both foils were placed in the same position. The distance from set-up was sufficient to neglect the possible influence of neutrons and other particles produced to the direction of the beam monitors. The deuteron reaction cross-sections on copper were determined relatively to ^{27}Al(d,3p2n)^{24}Na in principle. The used copper foils have natural isotope composition (69.15 % of ^{63}Cu and 30.85 % of ^{65}Cu).

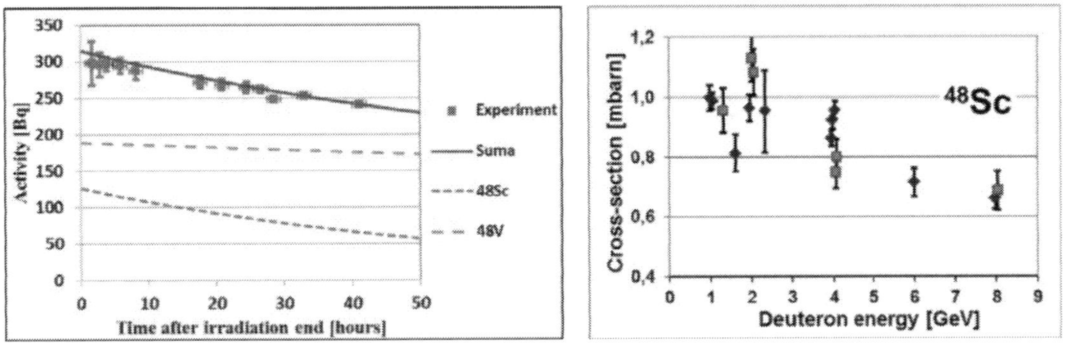

Figure 2. (a) Gamma decay curve of ^{48}Ti gamma lines (4 GeV deuteron irradiation in December 2012) (b) Cross-section of ^{nat}Cu(d,x)^{48}Sc reaction (results of last two set of irradiations are signed by red).

The activation method and gamma-ray spectrometry were used for the cross-section determination. Measured gamma spectra were analyzed by the DEIMOS code. After identification of the isotope by the means of gamma peaks the yield of activated material was calculated. All necessary spectroscopic corrections were taken into account. The detailed description is in [4].The foil was packed from original size to a smaller one with dimensions 2.5×2.5×0.3 cm^3 for the spectroscopy measurement. Activated foils were measured by two or more detectors and also in more different geometries which were in the range from 4 cm up to 10 cm far from detector. More measurements were done to detect and identify short lived and long lived radioisotopes. The cross-sections were determined taking into account the number of atoms in a sample and deuteron beam integral, see [5] for details.

Gamma lines of more than seventeen different radioisotopes (for example ^{24}Na, ^{42}K, ^{43}K, ^{43}Sc, ^{44}Sc+ ^{44m}Sc, ^{46}Sc, ^{47}Sc, ^{48}Sc, ^{48}V, ^{48}Cr, ^{52}Mn, ^{56}Mn, ^{55}Co, ^{56}Co, ^{57}Co, ^{58}Co + ^{58m}Co and ^{57}Ni) were identified in the obtained gamma spectra, for some examples see Fig. 1. Radioisotopes ^{48}Sc (43.7 hours) and ^{48}V (383.4 hours) decay both to the same daughter nucleus ^{48}Ti. The analysis of the A=48 isobars is complicated also by the sequence of the decays ^{48}Cr → ^{48}V → ^{48}Ti, see Fig. 2.

Similar case is decay of radionuclides ^{43}K (22.3 hours) and ^{43}Sc (3.9 hours) which decay both to the same daughter nucleus ^{43}Ca. The gamma line with energy 617.49 keV is produced only by ^{43}K decay and it is possible to use its intensity as norm to suppress contribution of ^{43}K to gamma line with energy 372.8 keV during first few hours after end of irradiation. We will obtain yield of ^{43}Sc decay using 372 keV line after subtraction of ^{43}K contribution, see Fig. 3.

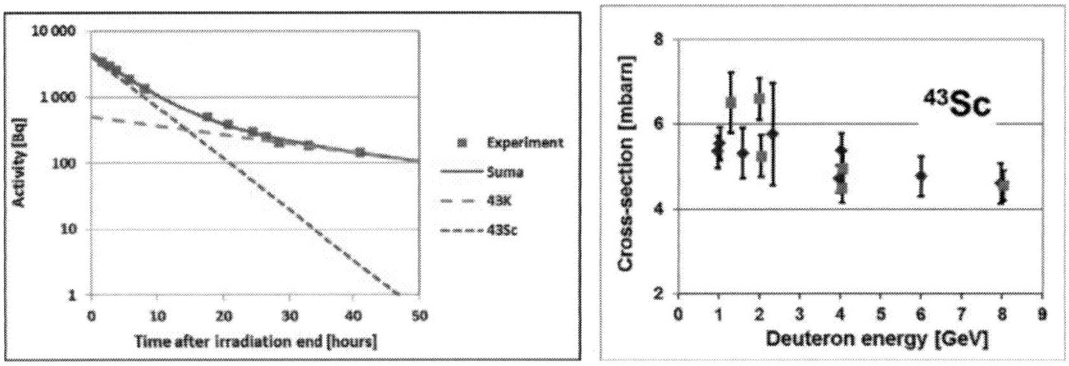

Figure 3. (a) Gamma decay curve of ^{43}Ca gamma line (b) Cross-section of ^{nat}Cu(d,x)^{43}Sc reaction.

The ground state ^{44g}Sc with half-life 3.9 hour and isomeric state ^{44m}Sc with energy 271.13 keV and halftime 58.6 hours are populated in the case of isotope ^{44}Sc, see Fig 4. The isomeric state decays by gamma transition to the ground state with probability 98.2 %. The beta decay of this state is within our accuracy negligible. Half-life of ground state beta decay is much shorter than half-life of isomeric state which is dominantly decaying by gamma decay to ground state of the same isotope. The number of ^{44g}Sc nuclei is very early done only by isomeric state decay.

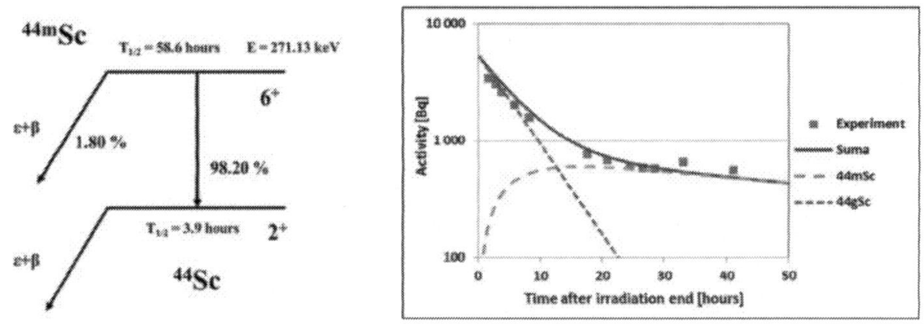

Figure 4. (a) Decay scheme of ^{44m}Sc and ^{44}Sc (the Lund/LBNL Nuclear Data Search web) (b) Activity of ^{44g}Sc produced by ^{44g}Sc states populated directly and through isomeric state ^{44m}Sc.

3. Neutron threshold reactions on yttrium

We have made extensive studies of (n,xn) reactions on materials used as activation detectors [4]. The two quasi mono-energetic neutron sources have been used. We are studying threshold (n,xn) reactions on yttrium just now. Preliminary results of first test experiment were shown during last Varna school [6]. We have six new irradiations with different energies on NPI neutron source for energies from 17.4 MeV up to 33.5 MeV. We analyse two accessible reactions $^{89}Y(n,2n)^{88}Y$ and $^{89}Y(n,3n)^{87}Y$. The second reaction is similar to mentioned reaction of ^{44}Sc production. The ground state has half-life 79.8 hours much longer than half-life (13.38 hours) of isomeric state with energy 380.79 keV. The beta decay of the isomeric state is again within our accuracy negligible, see more detailed description in [6].

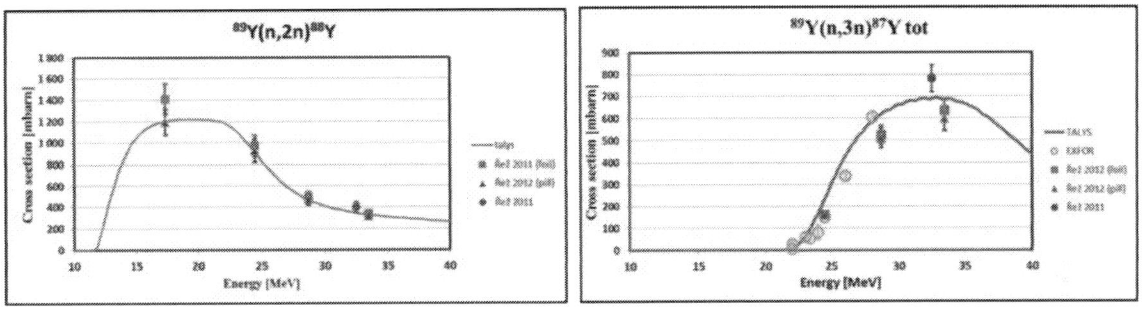

Figure 5. Excitation function of $^{89}Y(n,2n)^{88}Y$ and $^{89}Y(n,3n)^{87}Y$ (total production). Two types of samples (circular pill with diameter 10 mm and thickness 1.5 mm and square foil 25×25×0.64 mm) were used.

4. Conclusion

We present measurements of different radionuclide production on natural copper by relativistic deuterons. The excitation functions of more than seventeen different radionuclides production were obtained within energy range from 1 GeV up to 8 GeV. The excitation functions of reactions $^{89}Y(n,3n)^{87g}Y$ and $^{89}Y(n,3n)^{87m}Y$ were measured for the first time. Shape of excitation functions for all measured reactions is in good agreement with TALYS 1.4 code. Certain differences are for absolute magnitude of cross-sections.

Acknowledgements

Authors are grateful to the operation crew of the JINR Nuclotron and the staff of fast neutron generators and the cyclotron at the NPI Řež for irradiation and good beam parameters. This research was financially supported by the ERINDA program, the Czech JINR grants and by the F4E program of the Nuclear Reaction Department of the Nuclear Physics Institute, F4E-2008-GRT-014.

References

[1] Krása A et al 2010 *Nucl. Instr. and Meth. in Phys. Res.* A615 (2010) 70
[2] Furman W et al 2012 *Proceedings of Science* (Baldin ISHEPP XXI **086**) http://pos.sissa.it/archive/conferences/173/086/Baldin%20ISHEPP%20XXI_086.pdf
[3] Suchopár M et al 2012 *Proceedings of Science* (Baldin ISHEPP XXI **091**) http://pos.sissa.it/archive/conferences/173/091/Baldin%20ISHEPP%20XXI_091.pdf
[4] Vrzalová J et al 2013 *Nucl. Instr. and Meth. in Phys. Res.* **A726** 84-90
[5] Wagner V et al 2012 *Proceedings of Science* (Baldin ISHEPP XXI **090**) http://pos.sissa.it/archive/conferences/173/090/Baldin%20ISHEPP%20XXI_090.pdf
[6] Wagner V et al 2012 XIX International School on Nuclear Physics, Neutron Physics and Aplications (VARNA 2011) *Journal of Physics: Conference series 366* http://iopscience.iop.org/1742-6596/366/1/012047

Journal of Physics: Conference Series **533** (2014) 012053 doi:10.1088/1742-6596/533/1/012053

Calculations of ADS with deep subcritical uranium active cores – comparison with experiments and predictions

P. Zhivkov[1], **W. Furman**[2], **Ch. Stoyanov**[1]
[1] INRNE, BAS, Tsarigradsko shouse 72, 1874 Sofia, Bulgaria
[2] Joint institute for nuclear research, Joliot Curie 6, 141980 Dubna, Russia

E-mails: peter_jivkov@inrne.bas.bg

Abstract. The main characteristics of the neutron field formed within the massive (512 kg) natural uranium target assembly (TA) QUINTA irradiated by deuteron beam of JINR Nuclotron with energies 1,2,4, and 8 GeV as well as the spatial distributions and the integral numbers of (n,f), (n,γ) and (n,xn)- reactions were calculated and compared with experimental data [1]. The MCNPX 27e code with ISABEL/ABLA/FLUKA and INCL4/ABLA models of intra-nuclear cascade (INC) and experimental cross-sections of the corresponding reactions were used. Special attention was paid to the elucidation of the role of charged particles (protons and pions) in the fission of natural uranium of TA QUINTA. Extensive calculations have been done for quasi-infinite (with very small neutron leakage) depleted uranium TA BURAN having mass about 20 t which are intended to be used in experiments at Nuclotron in 2014-2016. As in the case of TA QUINTA which really models the central zone of TA BURAN the total numbers of fissions, produced ^{239}Pu nuclei and total neutron multiplicities are predicted to be proportional to proton or deuteron energy up to 12 GeV. But obtained values of beam power gain are practically constant in studied incident energy range and are approximately four. These values are in contradiction with the experimental result [2] obtained for the depleted uranium core weighting three tons at incident proton energy 0.66 GeV.

1. Introduction

The main characteristics of the neutron field formed within the massive (512 kg) natural uranium target assembly (TA) QUINTA surrounded by 10 cm lead blanket under irradiation by deuteron beam of JINR Nuclotron with energies 1, 2, 4, and 8 GeV as well as the spatial distributions and the integral numbers of (n,f), (n,γ) and (n,2n)- reactions were studied experimentally during last two years [1]. Here there are presented some results of simulations of these experiments by the code MCNPX 27.e [3] with INC model ISABEL, evaporation model ABLA and transport of the high energy particles by FLUKA (for detail see [4]). For the sake of precision the simulations were done also with combination INCL4/ABLA. In calculations all interactions of deuterons and produced by them neutrons, protons, pions and photons with whole TA QUINTA materials were considered.

2. The influence of charged particles on the fission of uranium in TA QUINTA

The deuteron beam generates in TA QUINTA not only neutrons, but high energy charged particles and photons too. In our simulations there were used the experimental cross sections for (n,γ), (n,f), (p,f),

(π,f) and (γ,f) reactions from TENDL 2009 up to 20 MeV for respective entrance channel energies and for higher energies they were calculated by TALYS code.The flux of the charged particles was generated by MCNPX code with options described above. Note that the real beam profile and its position within TA are strongly influence on results and should be taken into account for each experiment precisely.The calculations show that the reactions ^{238}U(p,f) and ^{238}U(π,f) have an essential significance in the central part of the uranium target close to the axis of the beam. The results are presented in figure 1. It is seen that just on the axis (R=0) and up to R=4cm about 25% of total fission rate is related to protons and pions but their contribution sharply decreases to R=8cm and plays a negligible role at R=12 cm. The influence of the reaction ^{238,235}U(γ,f) was insignificant.

3. The contribution of ^{235}U(n,f)-reaction in total fission rates for TA QUINTA

In natural uranium has the central part of TA QUINTA there is ~0.7% of ^{235}U. Due to presence in total energy spectrum of many slow neutrons it is necessary to account for ^{235}U(n,f)-reaction in spatial distribution of fission rate within TA. The calculations show that the ratios ^{235}U(n,f)/^{238}U(n,f) are from 0.03 till 0.14 depending on the positions within TA and on deuteron energy E_d. For E_d=4 GeV and R=0cm the ratios are from 0.03 up to 0.07depending on positions along beam axis, for radii R=4cm, 8cm and 12cm the ratios consist ~ 0.08, 0.1 to 0.12 and 0.12-0.14 respectively. For beam energy 8 GeV these ratios are rather similar. And the contribution of ^{235}U(n,f)-reaction into total fission number for TA QUINTA is in limits 2-4%.

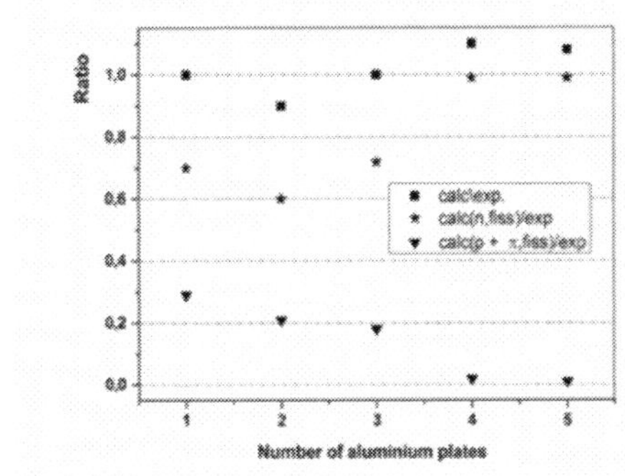

Figure 1. Influence of charge particle induced fission of ^{nat}U within TA QUINTA at E_d=8 GeV.

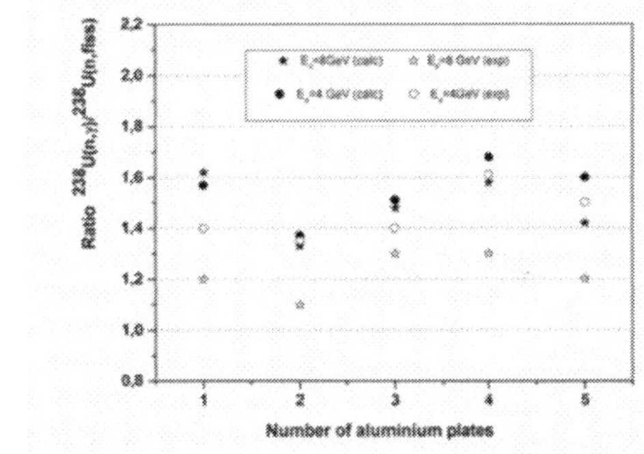

Figure 2. Spectral indices ^{nat}U(n,γ)/^{nat}U(n,f) are shown depending on the position along the incident beam for E_d=4 and 8 GeV for R=8cm.

4. The distributions of the neutron flux and its mean energy <E_n > over TA

The neutron flux and its mean energy <E_n> have been calculated for all detector positions within and on the surface of TA QUINTA. This is important for analysis of data obtained with the silicon detectors used to measure the spatial distribution of total neutron flux that are rather sensitive to the value of <E_n>. The calculated values <E_n> vary from 1 to 30 MeV and from 1 to 36 MeV for deuteron energies E_d = 4 GeV and E_d =8 GeV respectively. The neutron flux varies essentially over TA and in general does not follow R^2 dependence. It is maximal in the middle of TA and reduces to three-four times by the end of the target. The radial flux dependence becomes weaker (change up two times) for last uranium sections of TA.

5. Spectral indices ^{nat}U(n,f)/ ^{238}U (n,2n) and ^{238}U(n,γ)/^{238,235}U(n,f)

Natural uranium samples placed on the aluminum plates within TA QUINTA were used [1, 5] for simultaneous activation measurements of natural uranium fission rate, (n,2n) and (n,γ)-reactions. This provides a possibility to obtain so called spectral indices (SI) defined as the ratios of the respective reaction rates: ^{nat}U(n,γ)/^{nat}U(n,f) and ^{238}U(n,2n)/^{nat}U(n,f). By definition these values do not depend on the uncertainties of beam monitoring (error ~ 13%), on the intensity of neutron flux and on the exact knowledge of beam position on the target. So the spectral indices could be measured more precisely than absolute values of the particular reaction rates. In each TA position SI depend on the shape of neutron energy spectrum only and can be used to verify the correctness of the simulated spectrum. In figure 2 a comparison of experimental and calculated SI ^{nat}U(n,γ)/^{nat}U(n,f) is presented for deuteron energies E_d=4 and 8 GeV. One can see that calculations reproduce the whole character change the shape of the neutron spectra along the axis of the target. But the calculated values of SI and their dependence on incident energy do not agree with experiment. Namely experimental SI indicate some "hardening" of neutron spectra with growth of deuteron energy that does not reproduced by simulation.

6. Total numbers of fission and neutron radiative capture in TA QUINTA

By integration over the spatial distributions of measured (n,f) and (n,γ) reaction rates the total numbers of these reactions in TA QUINTA were deduced [1,5]. In figure 3 the calculated values of total numbers of fission N_f and the total amounts N_{Pu} of produced ^{239}Pu nuclei (normalized per one incident particle) are compared with experiment for deuteron energy range E_d = (1 – 8) GeV. As follows from the picture the respective experimental values grow with E_d remaining constant (within limit of their uncertainties) per unit of deuteron energy. If the calculated N_{Pu} values reasonably reproduce the measured ones so the calculated ratios N_f/E_d show a noticeable drop with growth of E_d. It means that simulated neutron spectra fail to reproduce high energy part of the experimental ones .

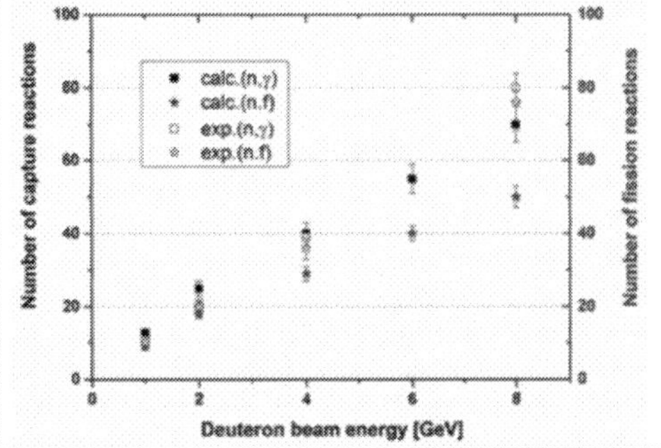

Figure 3. Total numbers of fissions and (n,γ)- reactions in TA QUINTA

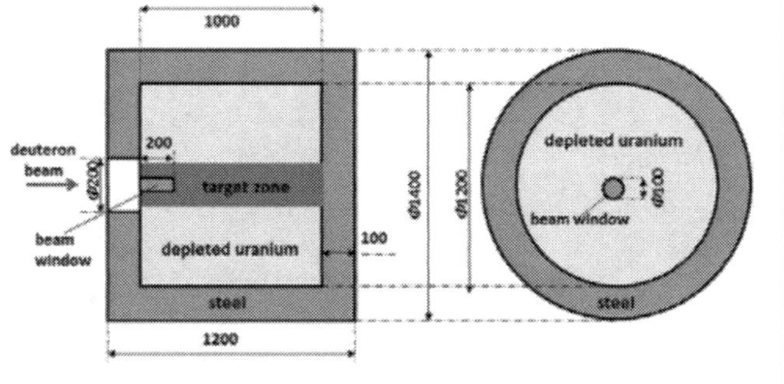

Figure 4. Schematic lay-out of TA BremaiURAN

7. Simulation of future experiments with quasi-infinite TA BURAN

In eighties of last century there was designed and constructed unique massive (about 21 t) TA made from metallic depleted (0.03% of ^{235}U) uranium – Big URANium (BURAN) target. It was intended for experiments on JINR Synchrophasotron beams. But these plans have not been realized. Now the project "Energy and Transmutation Spent Nuclear Fuel" endorsed at JINR for 2014-1026 is based on an upgrade and extensive use of TA BURAN. The schematic layout of it is shown in figure 4. Due to rather large dimensions of TA BURAN most part of neutrons and charged particles generated by incident high energy deuterons has to be absorbed in its volume. So this TA could be considered as deep-subcritical quasi-infinite active core proposed to use in new ADS (Accelerating Driving System) scheme [6]. For better planning of future experiment with TA BURAN the simulation of its main characteristics under irradiation by protons and deuterons has been done with the same variant of codes as described above. The results are given in Table 1. It is seen that all presented values excluding the neutron leakage stay near constant. It means that their absolute amounts per one incident deuteron grow linearly with increasing incident energy. The neutron leakage decreases essentially with growth of E_d that indicates on more effective "use" of neutrons inside of TA BURAN for higher incident energy. Beside present calculations confirm the assumption of a small neutron leakage from TA BURAN which is ~ 2% in comparison with ~ 60% for TA QUINTA.

Table 1. The integral characteristics of TA BURAN under irradiation by protons and deuterons (per one incident particle/per one GeV)

	Protons			**Deuterons**		
$E_{p(d)}$, (GeV)	**1**	**6**	**12**	**1**	**6**	**12**
Total neutron multiplicity	126	128	121	125	132	121
Total number of produced ^{239}Pu nuclei	70	73	69	70	75	70
Total number of fission	16	16.6	15	15	16.6	15.2
Number of leaked neutrons	3	2.3	1.9	3	2.3	1.8
BPG = $E_{release}/E_{p(d)}$	3.82	3.75	3.5	3.82	3.85	3.55

Note that if the total neutron multiplicity and total number of fission increased about two times in comparison with TA QUINTA so the total number of produced ^{239}Pu nuclei grew more than six times. This indicates an essential softening of the calculated neutron spectrum in TA BURAN.

Conclusion. The key elements of new ADS scheme [6] are the maximally hard neutron spectrum formed in the deep subcritical multiplying target from natural or depleted uranium and large enough the beam power gain (BPG $\geq$ 9). Both these parameters are necessary for effective utilization SNF with simultaneous energy production. But the calculated BPG $\approx$ 3.8 are lower even than the value of BPG $\approx$ 6 was obtained in [6] from the data [2] in which has been studied the massive (3t) metallic depleted uranium target irradiated by protons with energy up to 0.66 GeV. This energy is well below the threshold of fragmentation when the neutron multiplicity per unit of incident energy has not yet reached its asymptotic value. It seems probable that simulated neutron spectrum in TA BURAN is too soft underestimating its high energy part what in turn can lead to significant underestimation of BPW. So only future experiments with TA BURAN allow us to clear out real possibilities of new ADS scheme [6] for solving the problems of global energy.

Acknowledgments
This work has been partially supported by the Equipment Subsidies Program of the Alexander von Humboldt Foundation.

References

[1] Furman W., Adam J., Baldin A. et al., 2012, *Recent results of the study of ADS with 500 kg natural uranium target assembly QUINTA irradiated by deuterons with energies from 1 to 8 GeV at JINR NUCLOTRON.*, In Proceedings of XXI IHEP, Dubna 2012, PoS(MC2000)086
[2] Vassil'kov R., Gol'dansky V., Pimenov B. et al., 1978, *Multiplication of neutrons in uranium irradiated by protons with energy, 300-660 MeV.*, Atomic Energy, v.44, p. 329 - 334 (in Russian)
[3] https://mcnpx.lanl.gov/
[4] Filges D., Goldenbaum F., 2010, *Handbook of Spallation Reaction.*, WILEY – VCH, 2010
[5] Adam J., Artyushenko M., Baldin A. et al., 2013, *Investigation of spatial distributions of fission and radiative capture in massive uranium target irradiated by deuterons of energies 1 – 8 GeV.*, Preprint of JINR P1-2012-147, Dubna, (in Russian)
[6] Baldin A., Belov E., Galanin M. et al., 2011, *Nuclear relativistic technologies (NRT for energy production and utilization of spent nuclear fuel. Results of first experiments on physical background of NRT.*, 2011, Letters to PEPAN, v. 8(6), p.1007 - 1023

Journal of Physics: Conference Series **533** (2014) 012054 doi:10.1088/1742-6596/533/1/012054

Finite amplitude method in linear response TDDFT calculations

Takashi Nakatsukasa

RIKEN Nishina Center, Wako 351-0198, Japan

Center for Computational Sciences, University of Tsukuba, Tsukuba, 305-8571, Japan

E-mail: nakatsukasa@riken.jp

Abstract. The finite amplitude method is a feasible and efficient method for the linear response calculation based on the time-dependent density functional theory. It was originally proposed as a method to calculate the strength functions. Recently, new techniques have been developed for computation of normal modes (eigenmodes) of the quasiparticle-random-phase approximation. Recent advances associated with the finite amplitude method are reviewed.

1. Introduction

The nuclear energy density functional methods are extensively utilized in studies of nuclear structure and reaction [1]. For the linear response calculations, the quasiparticle-random-phase approximation (QRPA), based on the time-dependent density-functional theory [2, 3, 4, 5], is a standard theory [6]. However, for realistic energy functionals, it demands both complicated coding and large-scale computational resources. To resolve these issues, there have been several developments [7, 8, 9], including the finite amplitude method (FAM) [10].

The FAM allows us to avoid explicit evaluation of complex residual fields, which significantly reduces the necessary coding effort for realistic energy functionals. In addition, the use of iterative solution of the FAM equations also reduces the computational task and the memory requirement. The FAM was first adopted for the linear response calculations for the electric dipole mode, using the Skyrme energy functionals without the pairing correlations [11, 12]. The method was soon extended to superfluid systems [13]. Then, the FAM was adopted in the code HFBTHO to study monopole resonances in superfluid deformed nuclei with axially symmetry. Mario Stoitsov played a leading role in this project, which started during his visit to RIKEN in 2010 [14]. Later, there have been further developments in the FAM with relativistic and non-relativistic frameworks [15, 16, 17]

The FAM was originally proposed for a feasible method to calculate the response function of a given one-body operator. Recently, new methodologies have been developed, for calculating discrete eigenstates using the FAM. In this article, we review some of these recent developments.

2. The finite amplitude method

In this section, we briefly illustrate the essential idea of the FAM. Let us start from the Hartree-Fock-Bogoliubov (HFB) equations,

$$H\Phi_i = E_i\Phi_i, \qquad H\tilde{\Phi}_i = -E_i\tilde{\Phi}_i, \tag{1}$$

where H, $E_i > 0$, and $\Phi_i = \begin{pmatrix} U_i \\ V_i \end{pmatrix}$ are the HFB Hamiltonian including the particle-number cranking term $(-\mu N)$, the quasiparticle energies, and the quasiparticle states associated with the HFB ground state, respectively. $\tilde{\Phi}_i = \begin{pmatrix} V_i^* \\ U_i^* \end{pmatrix}$ are conjugate to Φ_i with negative quasiparticle energies [6]. They are orthonormalized as $\Phi_i^\dagger \Phi_j = \tilde{\Phi}_i^\dagger \tilde{\Phi}_j = \delta_{ij}$ and $\Phi_i^\dagger \tilde{\Phi}_j = \tilde{\Phi}_i^\dagger \Phi_j = 0$. The generalized density matrix R [6] at the ground state can be written in a simple form as

$$R \equiv \begin{pmatrix} \rho & \kappa \\ -\kappa^* & 1-\rho^* \end{pmatrix} = 1 - \sum_i \Phi_i \Phi_i^\dagger = \sum_i \tilde{\Phi}_i \tilde{\Phi}_i^\dagger. \tag{2}$$

In the small-amplitude (linear) approximation, the density fluctuation δR can be decomposed into normal modes $\delta R^{(n)}$,

$$\delta R(t) = \sum_n \delta R^{(n)} e^{-i\omega_n t} + \text{h.c.}. \tag{3}$$

In the linear order, the matrix elements of $\delta R^{(n)}$ between Φ_i and $\tilde{\Phi}_j$ are denoted by $X_{ij}^{(n)}$ and $Y_{ij}^{(n)}$ as

$$X_{ij}^{(n)} \equiv \Phi_i^\dagger \delta R^{(n)} \tilde{\Phi}_j = -\Phi_j^\dagger \delta R^{(n)} \tilde{\Phi}_i = -X_{ji}^{(n)}, \tag{4}$$

$$Y_{ij}^{(n)} \equiv \tilde{\Phi}_j^\dagger \delta R^{(n)} \Phi_i = -\tilde{\Phi}_i^\dagger \delta R^{(n)} \Phi_j = -Y_{ji}^{(n)}. \tag{5}$$

The other matrix elements of δR vanish in the quasiparticle basis:

$$\Phi_i^\dagger \delta R^{(n)} \Phi_j = \tilde{\Phi}_i^\dagger \delta R^{(n)} \tilde{\Phi}_j = 0. \tag{6}$$

Note that $\delta R^{(n)}$ are in general non-Hermitian, nevertheless $\delta R(t)$ is Hermitian.

The density fluctuation $\delta R^{(n)}$ induces an residual field in the HFB Hamiltonian, $H = H + \delta H^{(n)}$, where $\delta H^{(n)}$ linearly depends on $\delta R^{(n)}$. The essential idea of the FAM is that, instead of explicitly expanding $\delta H^{(n)}$ in the linear order in $\delta R^{(n)}$, we compute $\delta H^{(n)}$ by the finite difference using a small parameter η as

$$\delta H^{(n)} = \frac{1}{\eta} \left(H[R + \eta \delta R^{(n)}] - H \right). \tag{7}$$

The Hamiltonian $H[R+\eta\delta R^{(n)}]$ should be evaluated at the density, $R+\eta\delta R^{(n)} = 1 - \sum_i \Psi_i' \Psi_i^\dagger = \sum_i \tilde{\Psi}_i' \tilde{\Psi}_i^\dagger$, using the following quasiparticles

$$\Psi_i^\dagger \equiv \Phi_i^\dagger - \eta \sum_j X_{ij}^{(n)} \tilde{\Phi}_j^\dagger, \qquad \Psi_i' \equiv \Phi_i - \eta \sum_j Y_{ij}^{(n)} \tilde{\Phi}_j, \tag{8}$$

or

$$\tilde{\Psi}_i^\dagger \equiv \tilde{\Phi}_i^\dagger + \eta \sum_j \Phi_j^\dagger Y_{ji}^{(n)}, \qquad \tilde{\Psi}_i' \equiv \tilde{\Phi}_i + \eta \sum_j \Phi_j X_{ji}^{(n)}. \tag{9}$$

For a given set of forward and backward amplitudes, $\{X^{(n)}, Y^{(n)}\}$, the calculation of Eq. (7) is relatively easy because all we need to calculate are the one-body quantities with two-quasiparticle indices, such as $\delta H_{ij}^{(n)} = \Phi_i^\dagger \delta H \tilde{\Phi}_j$. In contrast, the QRPA matrices have four-quasiparticle indices [6], $A_{ij,kl}$ and $B_{ij,kl}$. The calculation of these matrix elements demands significant coding effort.

3. Calculation of the QRPA matrices

In this section, we recapitulate the matrix-FAM (m-FAM) proposed in Ref. [15]. This provides a simple numerical method to calculate the QRPA matrices using the principle of the FAM, Eqs. (7), (8), and (9).

The QRPA equation in the matrix form is given by $\mathcal{H}\vec{Z}_n = \omega_n \mathcal{N}\vec{Z}_n$, where

$$\mathcal{H} \equiv \begin{pmatrix} A & B \\ B^* & A^* \end{pmatrix}, \quad \mathcal{H} \equiv \begin{pmatrix} 1 & 0 \\ 0 & -1 \end{pmatrix}, \quad \vec{Z}_n \equiv \begin{pmatrix} X^{(n)} \\ Y^{(n)} \end{pmatrix}. \tag{10}$$

The most demanding part is the calculation of the matrix elements, $A_{ij,kl}$ and $B_{ij,kl}$, which are formally defined by the derivative of the HFB Hamiltonian with respect to the density. However, since the FAM allows us to evaluate $\mathcal{H}\vec{Z}$ for a given vector $\vec{Z}$, these matrix elements are provided by the FAM in the following way.

Let us define the "forward" unit vector $\hat{e}_{kl}$ as $X_{ij} = \delta_{ik}\delta_{jl}$ and $Y_{ij} = 0$, and the "backward" one $\tilde{e}_{kl}$ as $Y_{ij} = \delta_{ik}\delta_{jl}$ and $X_{ij} = 0$. Namely, these unit vectors are

$$\hat{e}_{kl} = \begin{pmatrix} X \\ Y \end{pmatrix} = \begin{pmatrix} 0 \\ \vdots \\ 1 \\ 0 \\ \vdots \end{pmatrix} \leftarrow X_{kl}, \qquad \tilde{e}_{kl} = \begin{pmatrix} X \\ Y \end{pmatrix} = \begin{pmatrix} 0 \\ \vdots \\ 1 \\ 0 \\ \vdots \end{pmatrix} \leftarrow Y_{kl}. \tag{11}$$

Then, it is trivial to see that the upper component of $(\mathcal{H}\vec{e}_n)_{ij}$ and $(\mathcal{H}\tilde{e}_n)_{ij}$ are identical to $A_{ij,kl}$ and $B_{ij,kl}$, respectively.

$$(\mathcal{H}\hat{e}_{kl})^{\rm up}_{ij} = A_{ij,kl}, \qquad (\mathcal{H}\tilde{e}_{kl})^{\rm up}_{ij} = B_{ij,kl} \tag{12}$$

On the other hand, using the FAM, $(\mathcal{H}\hat{e}_{kl})^{\rm up}$ reads

$$A_{ij,kl} = (\mathcal{H}\hat{e}_{kl})^{\rm up}_{ij} = (E_i + E_j)\delta_{ik}\delta_{jl} + \Phi_i^\dagger \delta H \tilde{\Phi}_j, \tag{13}$$

where $\Phi_i^\dagger \delta H \tilde{\Phi}_j$ in the right hand side can be computed according to Eq. (7). Here, $R + \eta\delta R$ is given by the ground-state quasiparticles, $\Psi_i^\dagger = \Phi_i^\dagger$ and $\Psi_i' = \Phi_i$, except for $i = k$ and l,

$$\Psi_k^\dagger = \Phi_k^\dagger - \eta\tilde{\Phi}_l^\dagger, \quad \Psi_l^\dagger = \Phi_l^\dagger + \eta\tilde{\Phi}_k^\dagger. \tag{14}$$

Following the same procedure with the "backward" $\tilde{e}_{kl}$, we can calculate $B_{ij,kl}$.

The numerical coding of the present method is extremely easy. All we need to calculate is the HFB Hamiltonian at the density $R + \eta\delta R$ which is defined by Eq. (14). After constructing the QRPA matrix, the QRPA normal modes of excitation are obtained by diagonalizing the QRPA matrix [6].

4. Iterative FAM with a contour integral in the complex frequency plane

In this section, we present a contour integral technique combined with the FAM developed in Ref. [17].

The iterative FAM (i-FAM) with a complex frequency ω provides a solution of the linear response equation $(\mathcal{H} - \omega\mathcal{N})\vec{Z}(\omega) = -\mathcal{F}$, where $\mathcal{F}$ is the one-body external field [10, 13].

$$\mathcal{F} \equiv \begin{pmatrix} F \\ F^* \end{pmatrix}, \quad \hat{F} = \sum_{i<j} F_{ij} a_i^\dagger a_j^\dagger + \text{h.c.} + \cdots \tag{15}$$

Here, we assume $\hat{F}$ is a Hermitian operator. The QRPA response function is calculated as [10]

$$S(F;\omega) = \sum_{i<j} \left\{ F_{ij}^* X_{ij}(\omega) + F_{ij} Y_{ij}(\omega) \right\}. \tag{16}$$

This can be expressed in terms of the QRPA normal modes as

$$S(F;\omega) = \sum_n |\langle n|\hat{F}|0\rangle|^2 \left(\frac{1}{\omega - \omega_n} - \frac{1}{\omega + \omega_n} \right), \tag{17}$$

This shows that the transition strength for the n-th normal mode, $|\langle n|\hat{F}|0\rangle|^2$, is a residue at $\omega = \omega_n$. Therefore, if we choose the contour C_n in the complex ω-plane that encloses $\omega = \omega_n > 0$, it reads

$$|\langle n|\hat{F}|0\rangle|^2 = \frac{1}{2\pi i} \oint_{C_n} S(F;\omega) d\omega. \tag{18}$$

The corresponding QRPA normal modes, $X^{(n)}$ and $Y^{(n)}$, are also given by the contour integral as [17]

$$X_{ij}^{(n)} = |\langle n|\hat{F}|0\rangle|^{-1} \frac{1}{2\pi i} \oint_{C_n} X_{ij}(\omega) d\omega, \tag{19}$$

$$Y_{ij}^{(n)} = |\langle n|\hat{F}|0\rangle|^{-1} \frac{1}{2\pi i} \oint_{C_n} Y_{ij}(\omega) d\omega. \tag{20}$$

The contour integral method with i-FAM is complementary to the m-FAM. In the present approach, the eigenmodes are obtained by solving the i-FAM equations for complex frequencies combined with the contour integral. In the m-FAM, we do not resort to an iterative algorithm to solve the FAM equations, but we need to diagonalize the QRPA matrix at the end. For a small model space, the m-FAM has a significant advantage, however, the computational task of the m-FAM strongly depends on the size of the QRPA matrix D, which typically scales as D^3.

Acknowledgments

The present work is supported by JSPS KAKENHI Grant numbers 24105006, 25287065, and 25287066. We thank all the collaborators on the present subject, including Dr Mario Stoitsov. We shall always remember him for his excellent contributions to the field of theoretical and computational nuclear physics.

References

[1] Bender M, Heenen P H and Reinhard P G 2003 *Rev. Mod. Phys.* **75** 121
[2] Nakatsukasa T and Yabana K 2001 *J. Chem. Phys.* **114** 2550
[3] Nakatsukasa T and Yabana K 2005 *Phys. Rev. C* **71** 024301
[4] Nakatsukasa T 2012 *Progress of Theoretical and Experimental Physics* **2012** 01A207
[5] Simenel C 2012 *The European Physical Journal A* **48** 1
[6] Ring P and Schuck P 1980 *The nuclear many-body problems*(New York: Springer-Verlag)
[7] Johnson C, Bertsch G and Hazelton W 1999 *Comp. Phys. Comm.* **120** 155
[8] Toivanen J, Carlsson B G, Dobaczewski J, Mizuyama K, Rodríguez-Guzmán R R, Toivanen P and Veselý P 2010 *Phys. Rev. C* **81**(3) 034312
[9] Carlsson B G, Toivanen J and Pastore A 2012 *Phys. Rev. C* **86**(1) 014307
[10] Nakatsukasa T, Inakura T and Yabana K 2007 *Phys. Rev. C* **76** 024318
[11] Inakura T, Nakatsukasa T and Yabana K 2009 *Phys. Rev. C* **80** 044301
[12] Inakura T, Nakatsukasa T and Yabana K 2011 *Phys. Rev. C* **84** 021302
[13] Avogadro P and Nakatsukasa T 2011 *Phys. Rev. C* **84** 014314
[14] Stoitsov M, Kortelainen M, Nakatsukasa T, Losa C and Nazarewicz W 2011 *Phys. Rev. C* **84** 041305
[15] Avogadro P and Nakatsukasa T 2013 *Phys. Rev. C* **87** 014331
[16] Liang H, Nakatsukasa T, Niu Z and Meng J 2013 *Phys. Rev. C* **87** 054310
[17] Hinohara N, Kortelainen M and Nazarewicz W 2013 *Phys. Rev. C* **87** 064309

Hartree-Fock-Bogoliubov calculations of deformed weakly bound nuclei using continuous bases

M. Stoitsov[1,2,3,4], **N. Michel**[1,5], **K. Matsuyanagi**[1,6]

[1]Department of Physics, Graduate School of Science, Kyoto U., Kyoto 606-8502, Japan
[2]Department of Physics and Astronomy, U. of Tennessee, Knoxville, Tennessee 37996, USA
[3]Physics Division, ORNL, P.O. Box 2008, Oak Ridge, Tennessee 37831, USA
[4]Institute of Nuclear Research and Nuclear Energy, Bulg. Acad. Sc., Sofia-1784, Bulgaria
[5]Grand Accélérateur National d'Ions Lourds, Caen, BP 55027-14076, France
[6]Theoretical Nuclear Physics Laboratory, RIKEN Nishina Center, Wako 351-0198, Japan

Abstract. The Hartree-Fock-Bogoliubov equations for deformed weakly bound nuclei are solved using a basis of Pöschl-Teller-Ginocchio and Bessel/Coulomb wave functions, which possess proper asymptotic behavior. Results compare well with standard Hartree-Fock-Bogoliubov calculations using box boundary condition or transformed harmonic oscillator basis.

1. Introduction

The precise description of weakly bound and unstable nuclei is a very challenging task in nuclear structure theory. The fundamental theoretical problem is to consistently describe both structure and reaction aspects of weakly bound nuclei, which requires a rigorous treatment of the many-body continuum. Many facilities have been or are built throughout the world in order to generate nuclei close to drip-lines, such as the projects FAIR in GSI, Germany and SPIRAL2 in GANIL, France, as well as the RIBF facility at RIKEN, Japan and the ISAC and ARIEL accelerators at TRIUMF, Canada.

In order to systematically study the whole nuclear chart, it is necessary to use density functional theory (DFT), used in conjunction with the Hartree-Fock-Bogoliubov (HFB) theory, as other methods, such as configuration interaction, become much too expensive numerically for nuclei bearing many nucleons. HFB equations can be solved on the one hand in coordinate space using box boundary condition (denoted in this paper as HFB/Box) [1, 2, 3] and on the other hand with diagonalization techniques by expanding HFB states in a harmonic oscillator (HO) or transformed harmonic oscillator (THO) basis [4]. These methods (denoted in this paper as HFB/HO or HFB/THO) are very fast and allow large-scale calculations [5, 6]. However, the HO and THO bases cannot reproduce simultaneously all asymptotic properties. Differences between calculations using the THO and the coordinate-space bases have been noticed for example in pairing properties of nuclei (see Sec. 3 and Ref. [7]).

The inclusion of continuum at basis level is definitely the most precise method in order to mitigate these problems. For this, the exactly solvable Pöschl-Teller-Ginocchio (PTG) potential [8] provides a numerically stable continuous basis. Phase shifts of real-energy scattering states vary indeed smoothly therein so that real-energy continua can be precisely discretized with Gauss-Legendre quadrature. This method to solve HFB equations with PTG basis is denoted as HFB/PTG in this paper.

Numerical calculations have been performed for a strongly deformed nucleus ^{110}Zr, and two HFB solutions for ^{40}Mg with different, prolate and oblate, deformations.

The paper is organized as it follows. The HFB/PTG algorithm is described in Sec. 2. Results of numerical calculation are presented in Sec. 3. Brief summary and conclusions are given in Sec. 4. All shown tables and figures were previously published in Ref. [9].

2. The HFB/PTG approach

Our aim is to develop an efficient method of solving the HFB equations

$$\begin{pmatrix} h-\lambda & \tilde{h} \\ \tilde{h} & -h+\lambda \end{pmatrix} \begin{pmatrix} U \\ V \end{pmatrix} = E \begin{pmatrix} U \\ V \end{pmatrix} \qquad (1)$$

for deformed weakly bound nuclei. h and $\tilde{h}$ denote the particle-hole and the particle-particle (hole-hole) components of the single-particle Hamiltonian, respectively, U and V the upper and the lower components of the single-quasiparticle wave function, of energy E, and λ is the chemical potential [3]. We now outline the calculational scheme of the HFB/PTG method:

(1) One starts with deformed HFB calculations in the HO basis (HFB/HO). This provides a good approximate solution for the HF potential and the effective mass. They are then averaged over angular and spin degrees of freedom.

(2) For each ℓj subspace, one fits the associated PTG potential to the averaged HF potential when it possesses bound or narrow resonant states in this ℓj subspace. If no such states appear in the HF ℓj spectrum, a set of Bessel/Coulomb wave functions [10] is selected for the ℓj partial wave basis. Bound and resonant states of the averaged HF potential are always bound in the PTG basis, and one takes into account only scattering states of real energy, so that the PTG basis does not make use of complex-valued states.

(3) One diagonalizes the HFB eigenvalue equations in the basis composed of the PTG and Bessel/Coulomb wave functions. This step continues until self-consistency is achieved.

3. Numerical examples

We have made a feasibility test of the HFB/PTG method for deformed neutron-rich nuclei ^{110}Zr and ^{40}Mg. All calculations were done using the SLy4 density functional. For the pairing interaction, we use the surface-type delta pairing with the strength $t_0' = -519.9$ MeV fm^3 for the the density-independent part and $t_3' = -37.5t_0'$ MeV fm^6 for the density-dependent part with a sharp energy cut-off at 60 MeV in the quasiparticle space. They have been fitted to reproduce the neutron pairing gap of ^{120}Sn. These values are consistent with those given in Ref. [11]; the slight difference is due to different cut-off procedures, sharp cut-off in our case and smooth cut-off in Ref. [11].

In the case of axially deformed nuclei, few HFB/Box calculations are available to check the HFB/PTG results. We consider the well-deformed nucleus ^{110}Zr (deformation $\beta \approx 0.4$), already studied in Ref. [7] and two states with different deformations for the drip line nucleus ^{40}Mg.

Table 1 compares the three approaches with respect to ground state properties of ^{110}Zr. In general they yield similar values. The differences seen in Table 1 are partially due to different structure of the model spaces adopted and the associated fitting of the pairing strength.

Proton and neutron densities for nuclei ^{110}Zr and ^{40}Mg are displayed in Fig. 1, with comparison with THO results (circles) for ^{110}Zr, in normal scale (left column panels) and logarithmic scale (right column panels).

While agreement between the PTG and THO densities for ^{110}Zr is good in normal scale, we can notice discrepancies in asymptotic properties, which are visible from the figure in logarithmic scale (see Fig. 1). It is obvious that all densities calculated with the THO basis eventually follow the common asymptote dictated by the scaling function, while they are well reproduced with

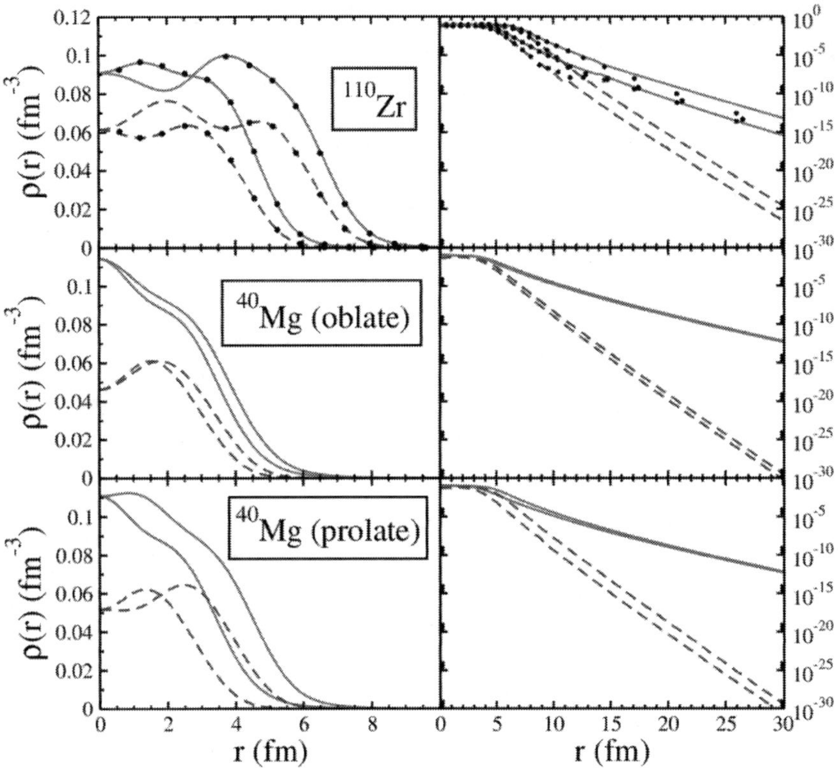

Figure 1. (color online) The neutron and proton densities of the prolately deformed nucleus ^{110}Zr ($\beta = 0.40$), respectively calculated by the HFB/PTG (respectively solid and dashed lines) and HFB/THO (circles) methods in normal (top left) and logarithmic (top right) scale. They are given along the long and short axes of deformation, easily identified from the figure. The neutron and proton densities of ^{40}Mg calculated by the HFB/PTG method for two states with different deformations (oblate $\beta = -0.09$ and prolate $\beta = 0.26$) in normal (middle and bottom left) and logarithmic (middle and bottom right) scale are also provided with the same line convention.

Table 1. Comparison of ground state properties of ^{110}Zr calculated with the HFB/Box, HFB/PTG and HFB/THO approaches. The rms radii are in fm, quadrupole moments are in barn, and all other quantities are in MeV.

	HFB/Box	HFB/PTG	HFB/THO
Q_{tot}	12.088	12.53	12.303
Δ_n	0.480	0.626	0.562
E_n^{pair}	-1.53	-3.015	-2.05
r_n	4.82	4.836	4.831
r_p	4.55	4.560	4.556
E_{tot}	-893.93	-893.952	-893.711

use of the PTG basis. This comparison also confirms the presence of deformation effects even in the far asymptotic region.

4. Conclusions

The HFB/PTG method is based on configuration-space diagonalization of the HFB Hamiltonian in the complete set of analytical PTG and Bessel/Coulomb wave functions, thus restoring the correct asymptotic properties of the HFB wave functions. The PTG potential is chosen to fit the nuclear HF potential and effective mass. The resulting PTG wave functions are close to the bound and continuum states of the related HF potential while the resonance states are substituted by the bound PTG states. Partial waves of high angular momentum are very well represented by Bessel/Coulomb wave functions. The examples shown, with a strongly deformed nucleus ^{110}Zr and ^{40}Mg with prolate and oblate deformations, compare well with coordinate space and THO results and thus have demonstrated the efficiency of the HFB/PTG method.

Acknowledgments

The authors acknowledge Japan Society for the Promotion of Science for awarding The Invitation Fellowship for Research in Japan (Long-term) to M. S. and The JSPS Postdoctoral Fellowship for Foreign Researchers to N. M., which make our collaboration possible. This work was supported by the JSPS Core-to-Core Program "International Research Network for Exotic Femto Systems." This work was carried out as a part of the U.S. Department of Energy under Contracts Nos. DE-FG02-96ER40963 (University of Tennessee), DE-AC05-00OR22725 with UT-Battelle, LLC (Oak Ridge National Laboratory), and DE-FG05-87ER40361 (Joint Institute for Heavy Ion Research), the UNEDF SciDAC Collaboration supported by the U.S. Department of Energy under grant No. DE-FC02-07ER41457.

References

[1] Bulgac A Preprint FT-194-1980 1980 Central Institute of Physics Bucharest ; nucl-th/9907088
[2] Dobaczewski J, Flocard H and Treiner J 1984 *Nucl. Phys.* **A422** 103
[3] Dobaczewski J, Nazarewicz W, Werner T R, Berger J F, Chinn C R and Dechargé J 1996 *Phys. Rev.* C **53** 2809
[4] Stoitsov M V, Dobaczewski J, Nazarewicz W, Pittel S and Dean D J 2003 *Phys. Rev.* C **68** 054312
[5] Goriely S, Samyn M, Heenen P H, Pearson J M and Tondeur F 2002 *Phys. Rev.* C **66** 024326
[6] Stoitsov M V, Dobaczewski J, Nazarewicz W and Ring P 2005 *Comput. Phys. Commun.* **167** 43
[7] Blazkiewicz A, Oberacker V E, Umar A S and Stoitsov M V 2005 *Phys. Rev.* C **71** 054321
[8] Ginocchio J 1985 *Ann. of Phys.* **159** 467
[9] Stoitsov M V, Michel N and Matsuyanagi K 2008 *Phys. Rev.* C **77** 054301
[10] Abramowitz M and Stegun I A 1970 *Handbook of Mathematical Functions* (Dover, New York)
[11] Bennaceur K and Dobaczewski J 2005 *Comput. Phys. Commun.* **168** 96

Isovector and isoscalar pairing in the nuclear shell model

S. Pittel[1], Y. Lei[2,1], N. Sandulescu[3], A. Poves[4], Y.M. Zhao[2] and B. Thakur[1]

[1] Bartol Research Institute and Department of Physics and Astronomy, University of Delaware, Newark, DE 19716 USA
[2] INPAC, Department of Physics and Shanghai Key Lab for Particle Physics and Cosmology, Shanghai Jiao Tong University, Shanghai, 200240, China
[3] Institute of Physics and Nuclear Engineering, 76900 Bucharest, Romania
[4] Departamento de Fisica Teorica and IFT UAM/CSIC, Universidad Autónoma de Madrid, 28049, Madrid, Spain

E-mail: pittel@bartol.udel.edu

Abstract. Shell-model calculations based on a schematic hamiltonian are reported for several $N = Z$ nuclei in the $2p1f$ shell to assess the relative importance of isoscalar and isovector pairing in the presence of nuclear deformation and a spin-orbit interaction.

1. Introduction

Proton-neutron (pn) pairing correlations in $N \approx Z$ nuclei are typically treated using the generalized Hartree Fock Bogoylubov (HFB) approximation, with the pn pairing field included in addition to the alike-nucleon nn and pp fields [1]. Unfortunately, this method is not able to appropriately describe the physics of these competing modes without full restoration of symmetries [2].

The nuclear shell model, in contrast, can treat all pairing modes on an equal footing, without violation of symmetries. Furthermore, it can treat them in the presence of nuclear deformation, as is often present in nuclei with $N \approx Z$. Thus, in this paper we report a study of the various modes of pairing in $N = Z$ nuclei, using the nuclear shell model. We focus on nuclei in the $2p1f$ shell and consider a schematic hamiltonian that not only includes the various pairing modes, but also includes nuclear deformation and single-particle-splitting effects.

2. Our model

The model we use involves an inert ^{40}Ca core and an equal number of valence neutrons and protons interacting via a schematic hamiltonian

$$H = \chi \left(Q \cdot Q + aP^{\dagger} \cdot P + bS^{\dagger} \cdot S + \alpha \sum_i \vec{l}_i \cdot \vec{s}_i \right) . \tag{1}$$

The first term in (1) is a quadrupole-quadrupole interaction (Q is the mass quadrupole operator) which by itself produces deformation. The second and third terms involve

XX International School on Nuclear Physics, Neutron Physics and Applications (Varna2013) IOP Publishing
Journal of Physics: Conference Series **533** (2014) 012056 doi:10.1088/1742-6596/533/1/012056

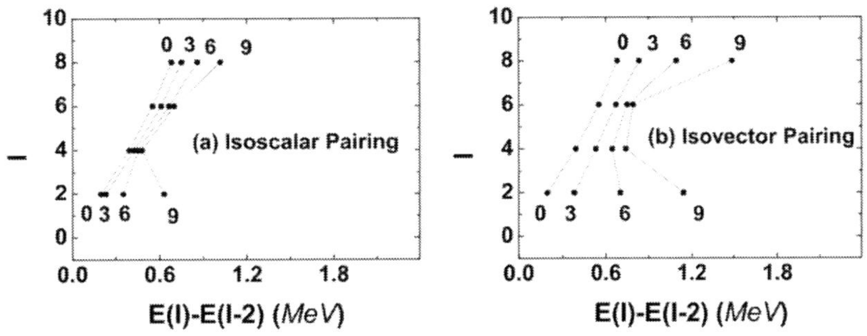

Figure 1. Spectra of the ground band of ^{44}Ti as a function of the strength of (a) isoscalar pairing interaction and (b) the isovector pairing interaction, with the optimal spin-orbit term present.

isoscalar and isovector pairing, respectively. The final term is the one-body part of a spin-orbit force. We carry out calculations as a function of the strength parameters a, b and α for a series of $N = Z$ nuclei, to assess how the various terms affect the rotational properties derived from the pure quadrupole-quadrupole force.

For an appropriate set of interaction parameters, the above hamiltonian in fact provides a fairly realistic description of the properties of nuclei in the lower part of the $2p1f$ shell. With the parameters $\chi = -0.05\ MeV$, $a = b = 12$ and $\alpha = 20$, we obtain a qualitative description of all nuclei up to ^{48}Cr, including the known backbend in ^{48}Cr. We refer to this as our *optimal Hamiltonian*. Note that in the optimal Hamiltonian, the strength of the isoscalar and isovector pairing interactions are the same.

3. Results and discussion

We first consider the nucleus ^{44}Ti, which has two active neutrons and two active protons. In the absence of a spin-orbit interaction, our results confirm the well-known result that isoscalar and isovector pairing produce precisely the same effect on the rotational properties of the system. More specifically, when we separately ramp up the strengths of the isoscalar pairing interaction and the isovector pairing interaction, with the optimal quadrupole strength but with $\alpha = 0$, we find exactly the same effect on the properties of the ground YRAST band from the two pairing interactions.

When we include the realistic spin-orbit interaction ($\alpha = 20$), however, this ceases to be the case. As can be seen from Figure 1, in the presence of the realistic spin-orbit splitting, isovector pairing dominates even for $N = Z$ nuclei. This conclusion, though already known from other studies, is borne out nicely in our model.

The next nucleus we consider is ^{46}V, an odd-odd nucleus with one additional neutron and one additional proton. We focus on the splitting between the lowest two states of the system, a $J^{\pi} = 0^{+}$, $T = 1$ state and a $J^{\pi} = 1^{+}$, $T = 0$ state. We consider equal isoscalar and isovector strengths ($a = b$), as in the optimal Hamiltonian, and study the effect of ramping up these two strengths together, first without and then with the spin-orbit term. When there is no spin-orbit term, $\alpha = 0$, these two states are degenerate in energy for any value of the common pairing strength $a = b$. In contrast,

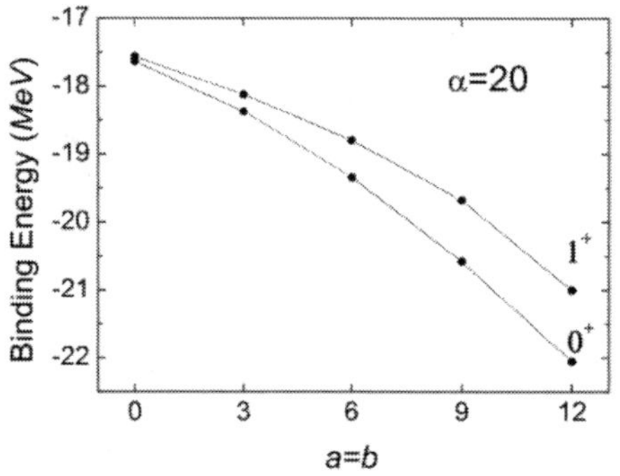

Figure 2. Calculated energies of the lowest $J^\pi = 0^+$ and $J^\pi = 1^+$ states of ^{46}V as a function of the equal strengths of isoscalar and isovector pairing, with the optimal spin-orbit term present.

in the presence of the optimal spin-orbit force ($\alpha = 20$), the degeneracy is removed. As seen in Figure 2, when the pairing strength is turned on, the 0^+ $T = 1$ state becomes the ground state, with its energy gain with respect to the 1^+ $T = 0$ state growing as the strength is increased. For the optimal pairing strength $a = b = 12$, the 0^+ state is at 1.05 MeV, fairly near the experimental value of 1.23 MeV. We conclude, therefore, that the fact that ^{46}V has a 0^+ ground state derives primarily from the spin-orbit interaction and its suppression of isoscalar pairing.

Lastly, we turn to ^{48}Cr, with four active neutrons and four active protons. Here we focus on a scenario with the optimal spin-orbit term ($\alpha = 20$) included. When we consider separately the effects of isoscalar pairing and isovector pairing on the YRAST band, we find that isoscalar pairing has essentially no effect whereas isovector pairing has a dramatic effect, producing a backbend at $J = 12$, as in experiment.

The backbend in ^{48}Cr was discussed earlier in the context of a a realistic shell-model study [3], where it was shown to derive from isovector pairing. Our results support that conclusion. This can be seen in Figure 3, where we show the number of collective isoscalar and isovector pairs in the YRAST band for the optimal hamiltonian. For low angular momenta, the states are dominated by isovector pairs, with only few isoscalar pairs present. As we crank up the angular momenta, the number of isovector pairs gradually decreases, reaching a value comparable to the number of isoscalar pairs just prior to the backbend. In the backbend region, both the number of isoscalar and isovector pairs suddenly increase. After the backbend region, both then decrease to zero as rotational alignment is achieved.

We also studied the $K^\pi = 2^+$ YRARE band and found no evidence that it crosses the YRAST band in the vicinity of the backbend. We thus conclude that the backbend is not driven by band crossing. Indeed, the only unique feature we find in the region of the backbend is an anomalous growth in the number of collective pairs (both isovector and isoscalar), suggesting that perhaps this is driving the backbend. Further study of this result is called for.

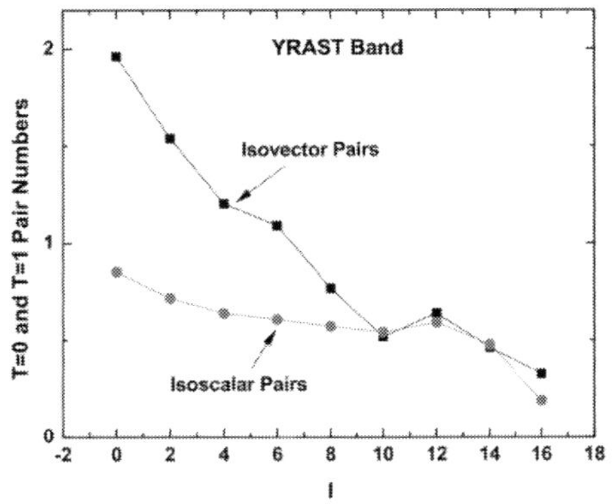

Figure 3. Calculated numbers of isovector pairs and isoscalar pairs in the ground band of ^{48}Cr for the optimal values of the hamiltonian parameters.

4. Closing remarks

We have reported a shell-model study of pn pairing in the $2p1f$ shell using a schematic hamiltonian that includes deformation, spin-orbit effects and both isoscalar and isovector pairing [4]. Working in a shell-model framework, we were able to assess the role of the various modes of pn pairing in the presence of nuclear deformation without violating symmetries.

Some of the key conclusions that emerged are: (*i*) the symmetry between isoscalar and isovector pairing disappears even at $N = Z$ in the presence of a spin-orbit force and isovector pairing dominates, (*ii*) that ^{46}V has a 0^+ ground state derives from the spin-orbit interaction and its relative effect on isoscalar and isovector pairing, (*iii*) isovector pairing dominates in ^{48}Cr and produces its backbend, and (*iiii*) isoscalar and isovector pairing exhibit an unusual behavior in the vicinity of the ^{48}Cr backbend.

Acknowledgments

This is based on a talk presented by S.P. at the XXth International School on Nuclear Physics, Neutron Physics and Applications and is dedicated to the memory of his very dear friend and valued collaborator, Mario Stoitsov. The work of S.P., B.T. and Y.L. was supported by US National Science Foundation grant # PHY-0854873, that of N.S. by the Romanian Ministry of Education and Research through CNCSIS grant Idei nr 57, that of A.P. by projects FPA2009-13377 MICINN(Spain) and HEPHACOS S2009/ESP-1473 Comunidad de Madrid(Spain), and that of Y.M.Z. and Y.L. by the National Science Foundation of China under grant # 10975096 and by Chinese Major State Basic Research Developing Program under grant # 2007CB815000.

References

[1] Goodman A L 1979 *Adv. Nucl. Phys.* **11** (1979) 263
[2] Dobeš J and Pittel S 1998 *Phys. Rev.* C **57** 688
[3] Poves A and Martinez-Pinedo G 1998 *Phys. Lett.* B **430** 203
[4] Lei Y, Pittel S, Sandulescu N, Poves A, Thakur B. and Zhao YM 2011 *Phys. Rev.* C **84** 044318

Journal of Physics: Conference Series **533** (2014) 012057 doi:10.1088/1742-6596/533/1/012057

Exactly solvable pairing models in nuclear and mesoscopic physics

J. Dukelsky

Instituto de Estructura de la Materia, C.S.I.C., Serrano 123, 28006 Madrid, Spain

E-mail: j.dukelsky@csic.es

Abstract. The exact solution of the BCS pairing Hamiltonian dates back to the work of Richardson in 1963. Little attention was paid to this exactly solvable model for almost 40 years. However, at the beginning of the 21th century, there was a burst of work focusing on its applications in different areas of quantum physics. In this contribution we introduce the generalized integrable Richardson-Gaudin models, and discuss a recent application of the hyperbolic model to heavy nuclei.

1. Introduction

After a long struggle to explain from a microscopic description the phenomenon of superconductivity, Bardeen, Cooper and Schrieffer published in 1957 [1] a successful theory for treating variationally a pairing Hamiltonian. The BCS theory was able to explain quantitatively most of the superconducting properties from the associated BCS wave function, leading to complete microscopic explanation of superconductivity.

The success of the BCS theory quickly spread to other quantum many-body systems, including the atomic nucleus. Soon thereafter Bohr, Mottelson and Pines published a paper [2] suggesting that the gaps observed in even-even nuclei could be due to superconducting correlations. They noted, however, that these effects should be strongly influenced by the finite size of the nucleus. Since then, and up to the present, number projection and in general symmetry restoration in the BCS and Hartree-Fock-Bogoliubov approximations have been important issues in nuclear structure.

At the beginning of the sixties, while several groups were developing numerical techniques for number-projected BCS calculations [3, 4], Richardson provided an exact solution for the reduced BCS Hamiltonian [5, 6]. In spite of the importance of his exact solution, this work did not have much impact in nuclear physics with just a few exceptions. Later on, his exact solution was rediscovered in the framework of ultrasmall superconducting grains [7] where BCS and number-projected BCS were unable to describe appropriately the crossover from superconductivity to a normal metal as a function of the grain size. Since then, there has been a flurry of work extending the Richardson exact solution to families of exactly-solvable models, now called the Richardson-Gaudin (RG) models [8, 9], and applying these models to different areas of quantum many-body physics including mesoscopic systems, condensed matter, quantum optics, cold atomic gases, quantum dots and nuclear structure [10]. In this contibution, we introduce the generalized RG integrable models and a recent application to nuclear physics.

2. Richardson-Gaudin integrable models

The fully integrable and exactly solvable RG models are based on the $SU(2)$ algebra. We first introduce the generators of $SU(2)$, using a basis familiar to nuclear structure,

$$K_j^0 = \frac{1}{2}\left(\sum_m a_{jm}^\dagger a_{jm} - \Omega_j\right), \; K_j^+ = \sum_m a_{jm}^\dagger a_{j\overline{m}}^\dagger, \; K_j^- = (K_j^+)^\dagger. \tag{1}$$

Here $a_{jm}^\dagger$ creates a fermion in single-particle state jm, $j\overline{m}$ denotes the time reverse of jm, and $\Omega_j = j + \frac{1}{2}$ is the pair degeneracy of orbit j. These operators fulfill the $SU(2)$ algebra $[K_j^+, K_{j'}^-] = 2\delta_{jj'}K_j^0$, $[K_j^0, K_{j'}^\pm] = \pm\delta_{jj'}K_j^\pm$.

We now consider a general set of L Hermitian and number-conserving operators that can be built up from the generators of $SU(2)$ with linear and quadratic terms,

$$R_i = K_i^0 + 2g\sum_{j(\neq i)}\left[\frac{X_{ij}}{2}\left(K_i^+K_j^- + K_i^-K_j^+\right) + Y_{ij}K_i^0K_j^0\right] . \tag{2}$$

Following Gaudin[11], we then look for the conditions that the matrices X and Y must satisfy in order that the R operators commute with one another. It turns out that there are essentially two families of solutions, referred to as the rational and hyperbolic families, respectively.

i. The rational family

$$X_{ij} = Y_{ij} = \frac{1}{\eta_i - \eta_j} \tag{3}$$

ii. The hyberbolic family

$$X_{ij} = 2\frac{\sqrt{\eta_i\eta_j}}{\eta_i - \eta_j} \; , \; Y_{ij} = \frac{\eta_i + \eta_j}{\eta_i - \eta_j} \tag{4}$$

Here the set of L parameters η_i are free real numbers.

The reduced BCS Hamiltonian with a constant pairing interaction is an example of the rational family. It can be obtained as a linear combination of the integrals of motion, $H_P = \sum_j \varepsilon_j R_j(\eta_j)$, with $\eta_j = \varepsilon_j$.

The complete set of eigenstates of the rational and hiperbolic integrals of motion can be found in Ref. [8, 9, 10].

The key point of the RG models is that any Hamiltonian that can be expressed as a linear combination of the R operators can be treated exactly using this method. The most general exactly solvable Hamiltonian has $2L + 1$ free parameters: a set of L internal parameters η_i, a set of L free coefficients ε_i defining the linear combination of integrals of motion, and the paring strength g. As a result, the RG Hamiltonians have an enormous flexibility to model physical situations.

3. The hyperbolic model

The rational model was extensively used to model different mesoscopic systems like superconductiong grains [12], cold atoms [13], quantum dots [14], nuclei [15], etc. The hyperbolic family of models did not find a physical realization until very recently when it was shown that they could model a p-wave pairing Hamiltonian in a 2-dimensional lattice [16], such that it was possible to study with the exact solution an exotic phase diagram having a non-trivial topological phase and a third-order quantum phase transition [17]. A slightly modified version of the p-wave pairing Hamiltonian gives rise to a separable pairing Hamiltonian with 2 extra free parameters that can be adjusted to reproduce the properties of heavy nuclei as described by a Gogny HFB treatment [18]. Both applications are based on a simple linear combination of hyperbolic

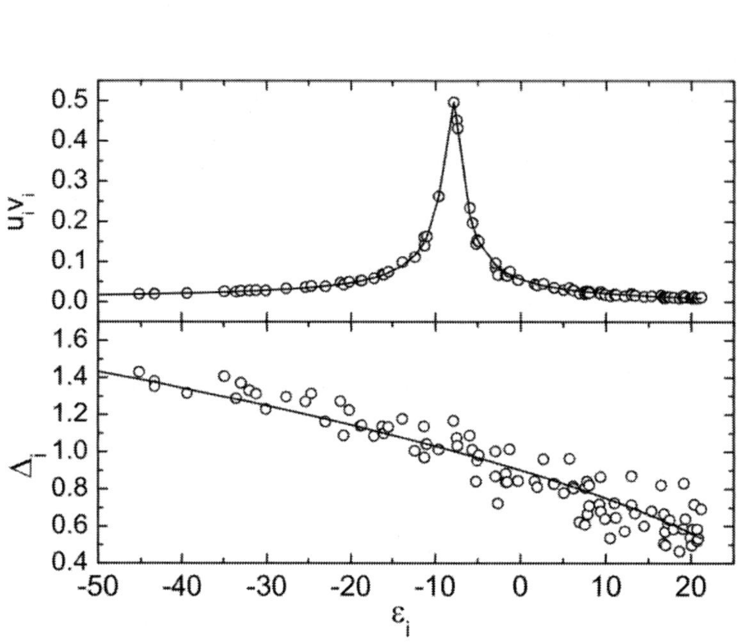

Figure 1. Gaps Δ_i and pairing tensor u_iv_i for protons in ^{154}Sm. Open circles are Gogny HFB results in MeV. Solid lines are BCS results of the hyperbolic Hamiltonian in MeV.

integrals of motion. In the nuclear application the exactly solvable pairing Hamiltonian reduces to

$$H = \sum_i \varepsilon_i \left(c_i^\dagger c_i + c_{\bar{i}}^\dagger c_{\bar{i}}\right) - 2G \sum_{ii'} \sqrt{(\alpha - \varepsilon_i)(\alpha - \varepsilon_{i'})}\; c_i^\dagger c_{\bar{i}}^\dagger c_{\bar{i}'} c_{i'}, \tag{5}$$

where the free parameter α plays the role of an energy cutoff and ε_i is the single-particle energy of level i. The complete set of eigenvalues are $E = 2\alpha M + \sum_i \varepsilon_i \nu_i + \sum_\beta E_\beta$. Where the pair energies E_β correspond to a particular solution of the set of non-linear Richardson equations

$$\frac{1}{2}\sum_i \frac{1}{\eta_i - E_\beta} - \sum_{\beta'(\neq\beta)} \frac{1}{E_{\beta'} - E_\beta} = \frac{Q}{E_\beta}, \tag{6}$$

with $Q = \frac{1}{2G} - \frac{L}{2} + M - 1$. Each particular solution of Eq. (6) defines a unique eigenstate.

Due to the separable character of the hyperbolic Hamiltonian, in BCS approximation the gaps $\Delta_i = 2G\sqrt{\alpha - \varepsilon_i}\sum_{i'}\sqrt{\alpha - \varepsilon_{i'}}u_{i'}v_{i'} = \Delta\sqrt{\alpha - \varepsilon_i}$ and the pairing tensor $u_iv_i = \frac{\Delta\sqrt{\alpha-\varepsilon_i}}{2\sqrt{(\varepsilon_i-\mu)^2+(\alpha-\varepsilon_i)\Delta^2}}$ have a very restricted form. In order to test the validity of the exactly solvable Hamiltonian (5) we take the single-particle energies ε_i from the HF energies of a Gogny HFB calculation and we fit the parameters α and G to the gaps and pairing tensor in the HF basis. Figure 1 shows the comparison for protons in ^{154}Sm between the Gogny HFB results in the HF basis and the BCS approximation of the hyperbolic model. As it can be seen, there is a remarkable agreement between the Gogny force and the hyperbolic Hamiltonian for the pairing tensor. The Gogny gaps exhibit large fluctuations due to the details of the two-body Gogny force. However, the general trend of the gaps is very well described by the square root $\sqrt{\alpha - \varepsilon_i}$ of the hyperbolic model. From these results we extracted the values $\alpha = 32.7\ MeV$ and $G = 2.24 \times 10^{-3} MeV$. The valence space determined by the cutoff α corresponds to $L = 91$

Journal of Physics: Conference Series **533** (2014) 012057 doi:10.1088/1742-6596/533/1/012057

levels with $M = 31$ proton pairs. The size of the Hamiltonian matrix in this space is 1.98×10^{24}, well beyond the limits of exact diagonalization. However, the integrability of the hyperbolic model provides an exact solution by solving a set of 31 non-linear coupled equations. Moreover, the exact solution shows a gain in correlation of $\sim 2\ MeV$ suggesting the importance of taking into account correlations beyond mean-field.

4. Conclusions

The key feature of the RG integrable models, is that they transform the diagonalization of the hamiltonian matrix, whose dimension grows exponentially with the size of the system, to the solution of a set of M coupled non-linear equations where M is the number of pairs. This makes it possible to treat problems that could otherwise not be treated and in doing so to obtain information that is otherwise inaccessible. The exactly solvable RG Hamiltonians also provide excellent benchmarks for testing approximations beyond HFB in realistic situations. Moreover, the RG integrable models have been extended to larger rank algebra making possible the exact solution of a proton-neutron pairing Hamiltonian with isospin $T = 1$ [19] or $T = 0$ and $T = 1$ [20].

Acknowledgments

This work was supported by the Spanish MINECO under Grant FIS2012-34479.

References

[1] J. Bardeen, L. N. Cooper, and J. R. Schrieffer, Phys. Rev. **108**, 1175 (1957).
[2] A. Bohr, B. R. Mottelson, and D. Pines, Phys. Rev. **110**, 936 (1958).
[3] A. K. Kerman, R. D. Lawson, and M. H. Macfarlane, Phys. Rev. **124**, 162 (1961).
[4] K. Dietrich, H. J. Mang, and J. H. Pradal, Phys. Rev. **135**, 22 (1964).
[5] R. W. Richardson, Phys. Lett. **3**, 277 (1963).
[6] R. W. Richardson, Nucl. Phys. **52**, 221 (1964).
[7] G. Sierra, J. Dukelsky, G. G. Dussel, J. von Delft, and F. Braun, Phys. Rev. B **61**, 11890 (2000).
[8] J. Dukelsky, C. Esebbag, and P. Schuck, Phys. Rev. Lett **87**, 066403 1 (2001).
[9] J. Dukelsky, S. Pittel, and G. Sierra, Rev. Mod. Phys. **76**, 643 (2004).
[10] G. Ortiz, R. Somma, J. Dukelsky, and S. Rombouts, Nucl. Phys. B **707**, 421 (2005).
[11] M. Gaudin, J. Phys. (Paris) **37**, 1087 (1976).
[12] G. Sierra, J. Dukelsky, G. G. Dussel, J. von Delft, and F. Braun, Phys. Rev. B **61**, R11890 (2000).
[13] G. Ortiz and J. Dukelsky, Phys. Rev. A **72**, 043611 (2005).
[14] A. Faribault, and D. Schuricht, Phys. Rev. Lett. **110**, 040405 (2013).
[15] G. G. Dussel, S. Pittel, J. Dukelsky, and P. Sarriguren, Phys. Rev. C **76**, 011302 (2007).
[16] M. Ibañez, J. Links, G. Sierra, and S-Y Zhao, Phys. Rev. B **79**, 180501 (2009).
[17] S. M. A. Rombouts, J. Dukelsky, and G. Ortiz, Phys. Rev. B **82**, 224510 (2010).
[18] J. Dukelsky, S. Lerma H., L. M. Robledo, R. Rodriguez-Guzman, and S. M. A. Rombouts, Phys. Rev. C **84**, 061301 (2011).
[19] J. Dukelsky, V. G. Gueorguiev, P. Van Isacker, S. Dimitrova, B. Errea, and S. Lerma H., Phys. Rev. Lett. **76**, 072503 (2006).
[20] S. Lerma H., B. Errea, J. Dukelsky, and W. Satula, Phys. Rev. Lett. **79**, 032501 (2007).

Journal of Physics: Conference Series **533** (2014) 012058 doi:10.1088/1742-6596/533/1/012058

Probing pairing correlations in Sn isotopes using Richardson-Gaudin integrability

S De Baerdemacker[1,2], V Hellemans[3], R van den Berg[4], J-S Caux[4], K Heyde[2], M Van Raemdonck[1,2], D Van Neck[1,2], P A Johnson[5]

[1] Ghent University, Center for Molecular Modeling, Technologiepark 903, 9052 Ghent, Belgium
[2] Ghent University, Department of Physics and Astronomy, Proeftuinstraat 86, 9000 Ghent, Belgium
[3] Université Libre de Bruxelles, PNTPM, CP229, 1050 Brussels, Belgium
[4] Institute for Theoretical Physics, University of Amsterdam, Science Park 904, Postbus 94485, 1090 GL Amsterdam, The Netherlands
[5]Department of Chemistry and Chemical Biology, McMaster University, Hamilton, Ontario, Canada

E-mail: `stijn.debaerdemacker@ugent.be`

Abstract. Pairing correlations in the even-even $A = 102 - 130$ Sn isotopes are discussed, based on the Richardson-Gaudin variables in an exact Woods-Saxon plus reduced BCS pairing framework. The integrability of the model sheds light on the pairing correlations, in particular on the previously reported sub-shell structure.

1. Introduction

Pairing is an important component of the correlations in atomic nuclei at low-excitation energy [1, 2, 3]. The Sn isotopes provide a unique laboratory to probe the neutron-neutron pairing correlations, because the large proton shell gap at $Z = 50$ ensures that the low-lying nuclear structure is largely unaffected by proton particle-hole excitations across the shell gap. Moreover, experimental data of the Sn isotopes in three major shells have become available in recent years thanks to intensive experimental activity with radio-active beam facilities. There exist several theoretical approaches to investigate pairing correlations in atomic nuclei, ranging from fundamental ab initio calculations to studies based on a more phenomenological footing [2]. In the present contribution, we will employ a Woods-Saxon [4] plus level-independent Bardeen-Cooper-Schrieffer (BCS) pairing Hamiltonian [5, 6] as a global probe for pairing correlations in the ground state of Sn. The level-independent, or reduced, BCS Hamiltonian has a complete basis of Bethe Ansatz eigenstates [7, 8], and belongs to the class of Richardson-Gaudin (RG) integrable models [9, 10]. Integrability offers unique opportunities to investigate pairing correlations. On the one hand, the RG variables in the pair-product structure allow for a transparent graphical representation, as well as a clear-cut connection with bosonization approximations [11] via a pseudo-deformation of the quasi spin algebra [12]. On the other hand, physical observables related to particle removal and addition properties [13] can be obtained conveniently using Slavnov's theorem for the RG model [14].

2. Richardson-Gaudin integrability for Sn isotopes

The reduced BCS Hamiltonian is given by [1]

$$\hat{H} = \sum_{i=1}^{m} \varepsilon_i \hat{n}_i + g \sum_{i,k=1}^{m} \hat{S}_i^\dagger \hat{S}_k, \qquad (1)$$

with $\hat{S}_i^\dagger = \sum_{m_i>0} (-)^{j_i - m_i} a^\dagger_{j_i m_i} a^\dagger_{j_i - mi}$ the nucleon-pair creation operator in a single-particle level ε_i with (spherical) quantum numbers $(i \equiv n_i, l_i, j_i)$ and of degeneracy $\Omega_i = 2j_i + 1$. This Hamiltonian supports a complete set of Bethe Ansatz eigenstates parametrised by the set of RG variables $\{x\}$ that are a solution of the RG equations [7, 8]. The associated eigenstate energy is then given as $E = \sum_{\alpha=1}^{N_p} x_\alpha + \sum_{i=1}^{m} \varepsilon_i v_i$, with v_i the seniority [15], and N_p the number of pairs.

The single-particle levels are provided by a Woods-Saxon potential [4], for which we used a recent global parametrisation [16], and the single-particle energy spectrum for ^{100}Sn is given in Table 1. We followed a global prescription $g = g_0/\sqrt{A}$ for the pairing interaction, in order to reproduce the 3-point pairing gaps $\Delta^{(3)}(A) = (-)^A[BE(A) - 2BE(A-1) + BE(A-2)]$ [4], presented in Figure 1b. The two-neutron separation energies $S_{2n} = [BE(A) - BE(A-2)]$ [4] are

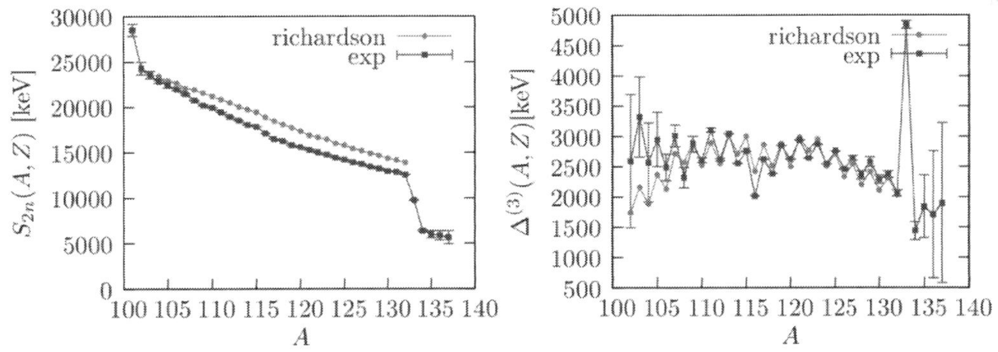

Figure 1. Experimental (squares) and theoretical (circles) two-neutron separation energies S_{2n} (a) and three-point pairing gaps $\Delta^{(3)}$ (b). Experimental data taken from [17].

given in Figure 1a, following a general linear trend, with the exception of a small kink around mid shell, signaling a sub-shell closure. The calculated curve is smoother than the experimental values at this point, consistent with the overestimated pairing gaps $\Delta^{(3)}$ around mid shell. Recent measurements showed a decrease in the $B(E2: 0_1^+ \to 2_1^+)$ strength around mid shell [18], which was qualitatively attributed [19] to this sub-shell effect in the seniority scheme [15]. Figure 2 depicts the RG variables for the ground state of the even-even $^{102\text{-}130}$Sn isotopes, and sheds more light on the sub-shell structure. Weakly correlated pair states give rise to a clustering of RG variables around the single-particle poles in the complex plane, whereas collective pairing states organise the RG variables along a broad arc in the complex plane [10, 12]. The pairing interaction in the lighter isotopes is strong enough to distribute the RG variables along an arc in the complex plane, however the arc only extends over the $d_{5/2}$ and $g_{7/2}$ sub-shell single particle poles. For the heavier nuclei, the pairs separate into two distinct sets, with seven RG variables clustering around the $d_{5/2}$ and $g_{7/2}$ sub-shell poles and the remaining forming a collective arc around the other poles. For medium-heavy nuclei, there is a gradual transition between both situations. This structure can be quantified using the pseudo-deformation scheme, where all RG variables can be labeled according to their collective behaviour in the Tamm-Dancoff Approximation (TDA) (see Table 1) [12]. From the table, it can be seen that the TDA structure is consistent

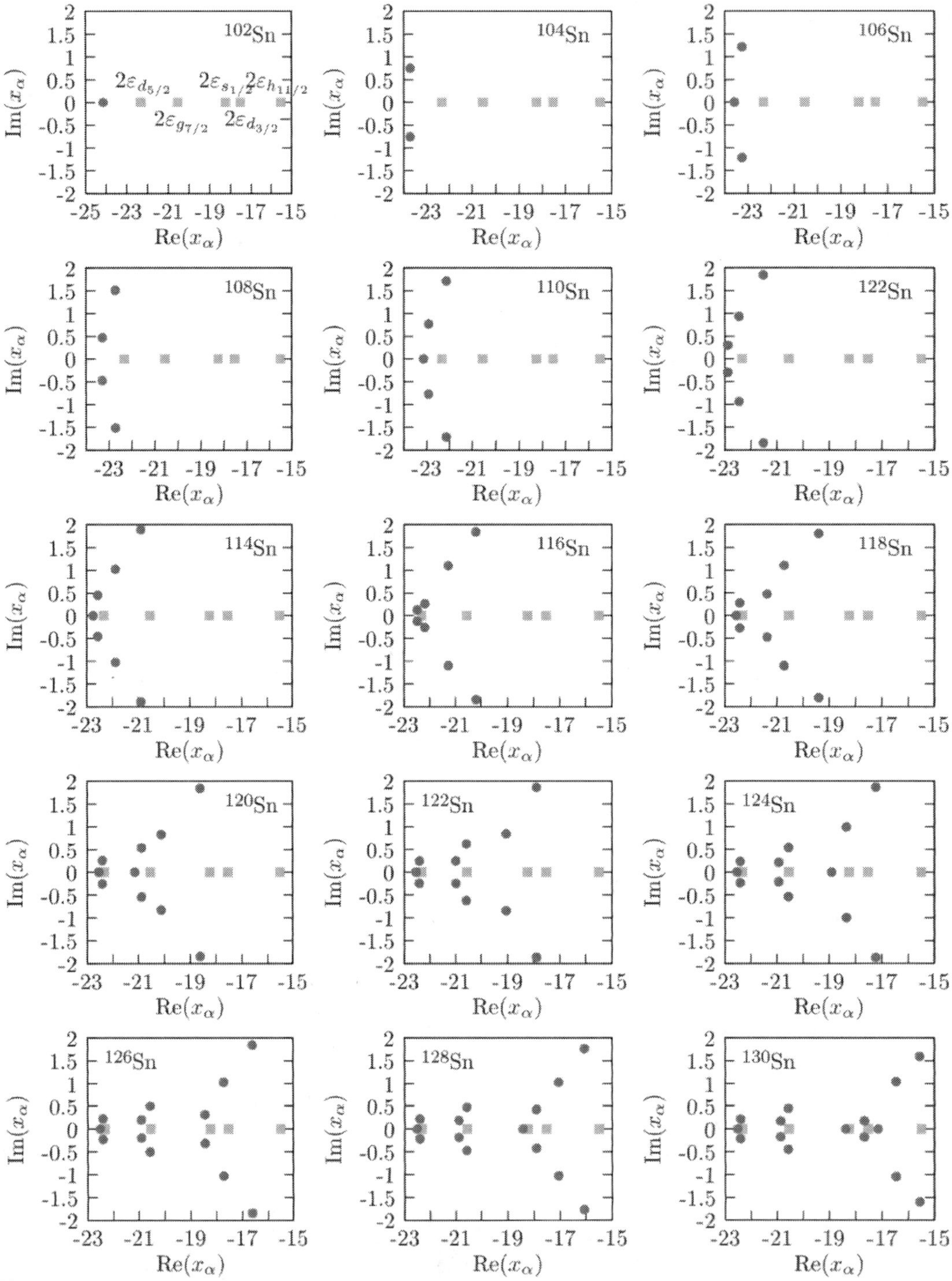

Figure 2. The RG variables (circles) and single-particle poles (squares) of the even-even $^{102\text{-}130}Sn$ isotopes.

with the discussed sub-shell structure. The lightest isotopes are consistent with a collective TDA condensation in the $d_{5/2}$ and $g_{7/2}$ sub shell, whereas the TDA structure of the heavier isotopes points towards a normal filling of the $d_{5/2}$ and $g_{7/2}$ sub shell, with the additional pairs collectively distributed over the $s_{1/2}$, $d_{3/2}$ and $h_{11/2}$ sub shell.

level	ε_i [MeV]	1	2	3	4	5	6	7	8	9	10	11	12	13	14	15
$d_{5/2}$	-11.164	1	2	3	4	5	6	7	8	3	3	3	3	3	3	3
$g_{7/2}$	-10.275	0	0	0	0	0	0	0	0	6	7	4	4	4	4	4
$s_{1/2}$	-9.124	0	0	0	0	0	0	0	0	0	0	4	5	6	1	1
$d_{3/2}$	-8.766	0	0	0	0	0	0	0	0	0	0	0	0	0	6	2
$h_{11/2}$	-7.754	0	0	0	0	0	0	0	0	0	0	0	0	0	0	5

Table 1. The single-particle energies ε_i obtained from a Woods-Saxon potential [16], and the TDA eigenmode decomposition of the 0^+ground state for the even-even isotopes $^{102\text{-}130}$Sn. The number of active pairs N_p in the isotope ^{A}Sn is given in the upper row ($N_p = (A-50)/2$).

3. Conclusions

We have investigated pairing correlations in the Sn isotopes by inspecting the location of the RG variables with respect to the single-particle poles in the complex plane, generated by a schematic Woods-Saxon plus reduced BCS Hamiltonian. The results point towards a sub-shell structure, consistent with previous studies. We expect this structure to also be reflected in the relevant transition rates; this will be investigated in future publications

Acknowledgments

SDB is an FWO-Vlaanderen post-doctoral fellow and acknowledges an FWO travel grant for a "long stay abroad" at the University of Amsterdam (The Netherlands). VH acknowledges financial support from the FRS-FNRS Belgium. This project is also supported by Belspo IAP Grant no P7/12 (SDB, VH, and KH).

References

[1] Heyde K 1994 *The nuclear shell model* (Berlin: Springer-Verlag)
[2] Dean D J and Hjorth-Jensen M 2003 *Rev. Mod. Phys.* **75** 607
[3] Brink D and Broglia R 2005 *Nuclear superfluidity, pairing in finite systems* (Cambridge: University Press)
[4] Bohr A and Mottelson B 1998 *Nuclear Structure, Vol.2* (Singapore: World Scientific)
[5] Bardeen J, Cooper L N and Schrieffer J 1957 *Phys. Rev.* **108** 1175
[6] Bohr A, Mottelson B R and Pines D 1958 *Phys. Rev.* **110** 936
[7] Richardson R W 1963 *Phys. Lett.* **3** 277
[8] Richardson R W and Sherman N 1964 *Nucl. Phys.* **52** 221
[9] Gaudin M 1976 *J. Phys. (Paris)* **37** 1087
[10] Dukelsky J, Pittel S and Sierra G 2004 *Rev. Mod. Phys.* **76** 643
[11] Ring P and Schuck P 2004 *The Nuclear Many-Body Problem* 3rd ed (Berlin: Springer)
[12] De Baerdemacker S 2012 *Phys. Rev. C* **86** 044332
[13] Grasso M, Lacroix D and Vitturi A 2012 *Phys. Rev. C* **85** 034317
[14] Faribault A, Calabrese P and Caux J S 2008 *Phys. Rev. B* **77** 064503
[15] Talmi I 1993 *Simple models of complex nuclei* (Chur: Harwood academic publishers)
[16] Schwierz N, Wiedenhöver I and Volya A 2007 (*Preprint* `nucl-th/0709.3525`)
[17] Audi G, Wapstra A H and Thibault C 2003 *Nucl. Phys. A* **729** 337
[18] Jungclaus *et. al.* 2011 *Phys. Lett. B* **608** 110
[19] Morales I O, Van Isacker P and Talmi I 2011 *Phys. Lett. B* **703** 606

Journal of Physics: Conference Series **533** (2014) 012059 doi:10.1088/1742-6596/533/1/012059

Relation between pairing gaps and transition probabilities in ^{132,136}Te

Jun Terasaki
Division of Physics and Center for Computational Sciences, University of Tsukuba, Tsukuba 305-8577, Japan
E-mail: terasaki@ccs.tsukuba.ac.jp

Abstract. Nuclear masses were measured very precisely in the region around ^{132}Sn recently. One of the new insights obtained from the measurements is the sudden decrease in the neutron pairing gap as the neutron number N increases across 82. There was an indication of that decrease based on the experimental data of the anomalous first 2^+ states of ^{132,136}Te and the numerical results of the quasiparticle random-phase approximation (QRPA). We show a simple analytical explanation of the relation between the anomaly and the neutron pairing gap referring to the most recent experimental gaps and equations of the QRPA.

1. Introduction

One of the frontiers of nuclear physics is study of nuclear structures of very neutron-rich heavy-mass (e.g. mass number larger than 100) region. Information at this frontier is indispensable for nucleosynthesis, in particular the r-process [1]. Recently, significant progress has been made in the measurements of the nuclear masses around ^{132}Sn [2, 3]. The neutron pairing gaps Δ_n's were deduced [2] from the measurements, using the three-point formula, and two remarkable properties were reported; one is the different Z-dependences of Δ_n between $N = 81$ and 83, and another is the siginificant decrease in Δ_n in $N > 82$ compared to $N < 82$.

The explanation of the former property remains to be an open problem, while the latter has been discussed about 10 years ago from an abnormal relation of the first 2^+-state energies $E_{2_1^+}$ and the reduced transition probabilities $B(E2; 0^+ \to 2_1^+)$ (hereafter denoted by $B(E2)\uparrow$) of ^{132,136}Te by the calculation of the quasiparticle random-phase approximation (QRPA) [4] and independently by a shell-model study [5] but without the recent very precise experimental pairing gaps. The purpose of this study is to re-examine the relation between the anomaly and the decrease in Δ_n by considering a simplified analytic explanation, not shown in Ref. [4], with the most recent Δ_n's in Ref. [2].

2. Analytical explanation

The experimental values of $E_{2_1^+}$ and $B(E2)\uparrow$ of ^{132,136}Te are shown in Tab. 1. Both $E_{2_1^+}$ and $B(E2)\uparrow$ decrease by adding 4 neutrons to ^{132}Te; this relation dose not obey an empirical rule [8, 9] that $E_{2_1^+}$ and $B(E2)\uparrow$ are inversely proportional in many experimental data. We explain this anomalous behavior of the 2_1^+ states of those two nuclei based on the QRPA with several approximations for simplicity.

	$E_{2_1^+}$ (keV)	$B(E2)\uparrow$
^{132}Te	974	0.172(17)
^{136}Te	606	0.103(35)

Table 1. Experimental $E_{2_1^+}$ and $B(E2)\uparrow$ of ^{132,136}Te [6, 7]. The $B(E2)\uparrow$ of ^{136}Te may be larger by 20% according to more recent data.

First, we assume that those states consist of only a two-neutron quasiparticle excitation and a two-proton quasiparticle excitation approximately. This noncollective assumption comes from a fact that nuclei with the proton and the neutron numbers close to the magic numbers do not have many low-energy particle-hole (p-h) excitations compared to the mid-shell nuclei. Indeed, the $B(E2)\uparrow$ of ^{132,136}Te are much smaller than those of the mid-shell nuclei [6]. Second, we introduce the separable approximation, e.g. [10], of the interaction used in the QRPA; the interaction is assumed to be $Q^\dagger Q$ with the quadrupole moment operator Q. With these approximations the forward neutron ($X^n_{\mu\nu}$) and proton ($X^p_{\mu'\nu'}$) amplitudes of the QRPA solution of the 2_1^+ state of ^{132}Te can be written

$$X^n_{\mu\nu} = \frac{1}{\mathcal{N}}\frac{\langle\mu_n||Q||\nu_n\rangle}{\mathcal{E}^n_\mu + \mathcal{E}^n_\nu - E_{2_1^+}}, \quad X^p_{\mu'\nu'} = \frac{1}{\mathcal{N}}\frac{\langle\mu'_p||Q||\nu'_p\rangle}{\mathcal{E}^p_{\mu'} + \mathcal{E}^p_{\nu'} - E_{2_1^+}}, \tag{1}$$

where $\mathcal{E}^n_\mu$ denotes the quasiparticle energy of the neutron state μ relevant to the 2_1^+ state, and $\langle\mu_n||Q||\nu_n\rangle$ is the reduced matrix element of Q. $\mathcal{N}$ stands for the normalization factor. The backward amplitudes are ignored here.

We introduce 4 approximate equations:

$$\langle\mu'_p||Q||\nu'_p\rangle(^{132}\mathrm{Te}) \simeq \langle\mu'_p||Q||\nu'_p\rangle(^{136}\mathrm{Te}), \quad \langle\mu_n||Q||\nu_n\rangle(^{132}\mathrm{Te}) \simeq \langle\mu''_n||Q||\nu''_n\rangle(^{136}\mathrm{Te}), \tag{2}$$

$$\mathcal{E}^p_{\mu'} + \mathcal{E}^p_{\nu'} \simeq 2\Delta_p, \quad \mathcal{E}^n_\mu + \mathcal{E}^n_\nu \simeq 2\Delta_n, \tag{3}$$

The first equation of the approximation (2) is justified, because the proton configuration does not change. The same approximation for the neutrons is also used, because the average angular momenta of a few neutron levels above and below the $N = 82$ shell gap are close; the levels above are $1h_{9/2}$, $3p_{3/2}$ and $2f_{7/2}$, and those below are $2d_{3/2}$, $1h_{11/2}$ and $3s_{1/2}$. Thus, the average node quantum numbers are also close, and it is assumed that the average matrix element of Q with the relevant neutron states for ^{132}Te and ^{136}Te are close to each other. The third and fourth approximate equations (3) are based on an assumption that the single-particle levels corresponding to the μ'_p, ν'_p, μ_n and ν_n are close to the Fermi level with Δ_p being the proton pairing gap. By inserting Eq. (3) to Eq. (1) for ^{132}Te and Eqs. (2) and (3) to the corresponding equations for ^{136}Te, the ratio of the proton and neutron amplitudes of the two nuclei are obtained

$$(X^p_{\mu'\nu'}/X^n_{\mu\nu})(^{132}\mathrm{Te}) = R\frac{(2\Delta_n - E_{2_1^+})_{^{132}\mathrm{Te}}}{(2\Delta_p - E_{2_1^+})_{^{132}\mathrm{Te}}}, \tag{4}$$

$$(X^p_{\mu'\nu'}/X^n_{\mu''\nu''})(^{136}\mathrm{Te}) = R\frac{(2\Delta_n - E_{2_1^+})_{^{132}\mathrm{Te}} + 2\delta\Delta_n + \delta E_{2_1^+}}{(2\Delta_p - E_{2_1^+})_{^{132}\mathrm{Te}} + \delta E_{2_1^+}}, \tag{5}$$

$$\delta E_{2_1^+} \equiv E_{2_1^+}(^{132}\mathrm{Te}) - E_{2_1^+}(^{136}\mathrm{Te}), \quad \delta\Delta_n \equiv \Delta_n(^{136}\mathrm{Te}) - \Delta_n(^{132}\mathrm{Te}), \tag{6}$$

$$R \equiv \langle \mu_p' || Q || \nu_p' \rangle / \langle \mu_n || Q || \nu_n \rangle. \tag{7}$$

R is a constant according to one of the approximations mentioned above. The experimental data [2, 6, 7] yield the values

$$\delta E_{2_1^+} = 0.37\text{MeV}, \quad \delta \Delta_n = -0.21\text{MeV}. \tag{8}$$

From these values and Eqs. (4) and (5), we see

$$\left(X^p_{\mu'\nu'} / X^n_{\mu\nu} \right)^2 (^{132}\text{Te}) > \left(X^p_{\mu'\nu'} / X^n_{\mu''\nu''} \right)^2 (^{136}\text{Te}). \tag{9}$$

With this relation and the normalization condition

$$\left(X^p_{\mu'\nu'} \right)^2 + \left(X^n_{\mu\nu} \right)^2 = 1, \tag{10}$$

it is found that

$$\left(X^p_{\mu'\nu'} \right)^2 (^{132}\text{Te}) > \left(X^p_{\mu'\nu'} \right)^2 (^{136}\text{Te}), \tag{11}$$

and this equation explains the decrease in $B(E2)\uparrow$ with the increase in N across 82. It is emphasized, as is seen from Eqs. (4) and (5), that the negative $\delta\Delta_n$ plays a key role in decreasing the $B(E2)\uparrow$ in ^{136}Te. Therefore, the anomaly of the 2_1^+ states of ^{132,136}Te reflects on the change in Δ_n and is consistent with the experimental Δ_n's obtained by the recent very precise mass measurements.

3. Summary

We have shown the analytical explanation of the anomalous relation of the 2_1^+ states of ^{132,136}Te in terms of Δ_n based on the QRPA and the several approximations for simplicity. The only input data for this explanation are $E_{2_1^+}$'s and Δ_n's, and how the Δ_n affects the $B(E2)\uparrow$ is clear; this explanation is still consistent with the most recent experimental paring gaps. This study was inspired by the recent very precise mass measurements, which confirmed the N-dependence of the Δ_n indicated by the discussion on the 2_1^+ states. The anomaly of the 2_1^+ states is an interesting example that information on the ground states can be obtained through properties of the excited states. Finally, it is mentioned that the mixed-symmetry state is also a topic of the excited states around ^{132}Sn [7] recently.

References

[1] Iliadis C 2007 *Nuclear Physics of Stars* (Weinheim: Wiley-VCH)
[2] Hakala J *et al* 2012 *Phys. Rev. Lett.* **109** 032501
[3] Van Schelt J *et al* 2012 *Phy. Rev. C* **85** 045805
[4] Terasaki J *et al* 2002 *Phys. Rev. C* **66** 054313
[5] Shimizu N *et al* 2004 *Phys. Rev. C* **70** 054313
[6] Radford D C *et al* 2002 *Phys. Rev. Lett.* **88** 222501
[7] Danchev M *et al* 2012 *Phys. Rev. C* **84** 061306(R)
[8] Grodzins L 1962 *Phys. Lett.* **2** 88
[9] Raman S *et al* 2001 *At. Data Nucl. Data Tables* **78** 1
[10] Bes D R and Sorensen R A 1969 *Advances in Nuclear Physics vol. 2* (New York: Plenum Press)

AUTHOR INDEX

AUTHOR INDEX

AUTHOR INDEX

AUTHOR INDEX